INTERNATIONAL REVIEW OF COMPARATIVE PUBLIC POLICY

Volume 9 • 1997

ISLAM AND PUBLIC POLICY

EDITORIAL REVIEW BOARD

INTERNATIONAL REVIEW OF COMPARATIVE PUBLIC POLICY

ISLAM AND PUBLIC POLICY

Series Editors: NICHOLAS MERCURO
Michigan State University

Volume Editors: SOHRAB BEHDAD
Department of Economics
Denison University

FARHAD NOMANI
Department of Economics
American University of Paris

VOLUME 9 • 1997

JAI PRESS INC.

Greenwich, Connecticut *London, England*

55 Old Post Road No. 2
Greenwich, Connecticut 06836

JAI PRESS LTD.
38 Tavistock Street
Covent Garden
London, WC2E 7PB
England

ISBN: 0-7623-0268-2

ISSN: 1051-4694

Transferred to Digital Printing 2006

CONTENTS

LIST OF CONTRIBUTORS

Nahla Abdo	Department of Sociology and Anthropology Carlton University
William Ballantyne	Barrister and Department of Arab Law School of Oriental and African Studies University of London
Sohrab Behdad	Department of Economics Denison University
Timur Kuran	Department of Economics and King Faisal Professor of Islamic Thought and Culture University of Southern California
Ann Elizabeth Mayer	Department of Legal Studies The Wharton School University of Pennsylvania
Haideh Moghissi	Department of Sociology Atkinson College York University
S.V.R. Nasr	Department of Political Science University of San Diego
Volker Nienhaus	Department of Economics Ruhr-Universität Bochum

Farhad Nomani — Department of International Economics, American University of Paris

Karen Pfeifer — Department of Economics, Smith College

Marsha Pripstein Posusney — Department of Political Science, Bryant College

Ali Rahnema — Department of International Economics, American University of Paris

STATEMENT OF SCOPE

The *International Review of Comparative Public Policy* is interdisciplinary and dedictated to publishing original scholarly contributions that systematically analyze international public policy issues. It is a nationally refereed, annual, sponsored by the Institute for Public Policy and Social Research at Michigan State University. Each yearly edition is devoted to a different public policy issue that can benefit from comparative political, legal, economic, and social scientific examination. The topic, together with the volume editors(s), will be selected through a competitive prospectus-selection process.

An individual who is interested in editing a future volume of the *International Review of Comparative Public Policy* should submit three copies of a 5-7 page, doublespaced, typed prospectus to the series editor. Each prospectus must include: (1) identification of the issue/topic and a prospective title of the volume; (2) a detailed description of the international dimensions of the particular policy issue; (3) a statement as to why and how the chosen issue can benefit from a comparative political, legal, economic, and social scientific examination; and, most important; (4) a list of potential contributors along with tentative titles of their contributions.

Each edition is expected to contain contributions and scholars from both the United States and abroad. Please note that the Series Editor does not accept individual manuscripts for publication consideration.

Send your volume proposal to:

Nicholas Mercuro, Series Editor
Institute for Public Policy
and Social Research
Michigan State University
321 Berkey Hall
East Lansing, MI 48824
phone: (517) 353-6469

FOREWORD

Public policy determines the scope, limits, and forms of social action. It is the government's means for encouraging, discouraging or prohibiting some acts or behaviors. It is also what governments do or do not do. Laws, decrees, regulations, orders, pronouncement, public persuasion, and propaganda, as well as direct actions of governments, are means of public policy. The sociopolitical structure determines the process of formulation and implementation, as well as the degree of legitimacy of public policy. In secular democracies, expressions of public consensus, such as elections, polls, and public debates in legislative process and in other forums of political negotiation lead to the formulation of public policy; and, in theory, the majority wins.

In Islamic states the process of definition and the source of legitimacy of public policy is more complex and potentially more controversial. The complexity arises from applying the interpretations of the divine law, the Shari'a, to contemporary social issues. Islam is concerned not only with the profound theological issues but just as intensively with the mundane aspects of everyday life. In an Islamic state, the Shari'a must be implemented in its entirety, as a means for the realization of the will of God on earth. The Shari'a is, therefore, the code of social conduct and the means of regulating social discourse as well as social change. Since definition of the Shari'a is itself a matter of interpretation, two fundamental issues arise.

First, what is the legitimate source and process of interpretation of the precepts of Islamic law, the Shari'a? In other words, who and in what process can the "true" precepts of Islam be determined? Second, how are the specific dictums of the Shari'a affected by the fundamental changes in the social and technological circumstances of life? Public consensus and adaptability of the

Shari‘a to change have been major sources of debate in Islamic public policy. The rise of Islamic revivalism in the past decades has added new dimensions to this politico-theological issue. The problem has evolved into the more fundamental questions of "What is Islam?" and "Whose Islam?" frequently asked by Muslim revivalists groups themselves, who claim that theirs is the "true Islam."

Unavoidably, comparisons are made, approvingly or disapprovingly, by the proponents or opponents of an Islamic state between elements of secular and Islamic public systems of policy. Of course, such comparisons could be made between the ideal, actual, or caricature of each of two forms of state and their system of public policy. Consequently, the number of possibilities that would emerge would be quite large.

The papers in this collection were commissioned specifically for this volume. We approached the recognized authorities in each arena of public policy with our proposal. We asked that each paper concern itself with the theoretical and practical applications of the issue, and with the controversies that may exist on the interpretation of Islamic precepts on the topic. In this respect, this volume is planned to be a cohesive and integrated collection. Obviously, we could have included other topics as well, except for the limitation of space, which has forced us to make some hard choices. Meanwhile, we do not pretend to have a uniformity of methodological approaches and viewpoints in this volume. We made no attempt in that direction. In fact, we believe that the resulting diversity of approaches and views is a valuable feature of this volume.

Sohrab Behdad
Farhad Nomani
Volume Editors

ISLAM, REVIVALISM AND PUBLIC POLICY

Sohrab Behdad

...the detailed structure of social life which ensues from the [Islamic] doctrine is inherently opposed to all others existing in the world....[A]ll problems of human life, whether big or small, are viewed and treated by Islam in its own way.

—Sayyid Abul Ala Maududi (1972/1940)[1]

...everywhere the best reformers and profound thinkers believe that a glorious renaissance and reconstruction is achievable on the basis of a pure Islamic ideology which planned to organize a society of free men, where exploitation of man by man would not be possible, where monarchies would give place to democratic republics, where feudalism and blood-sucking landlordism would be wiped out, where the State would become a welfare-state and not a police-state and where people would not be tyrannized for thinking freely.

—Khalifa Abdul Hakim (1953/1951)[2]

INTRODUCTION

Islamic revivalism marches on, and its tempo rises with its pledge to equity, social justice, freedom, and human dignity. Islamic revivalism condemns

International Review of Comparative Public Policy, Volume 9, pages 1-41.

ISBN: 0-7623-0268-2

"exploitation of man by man" and attacks in more recent decades not only "blood-sucking landlordism," which has diminished for one reason or another in most countries, but "blood-sucking," "plundering" capitalists and capitalism, imperialists and imperialism. It demands means of subsistence for all and denounces monopolies and accumulated fortunes. It praises simplicity and discretion in wants and condemns conspicuous consumption. In short, Islamic revivalism is a march of radicalism, demanding a dramatic change in social structure and the organization of production, with a heavy handed state in control of the economy. Some revivalists have asked for "Islamic socialism" or a "classless Islamic society."

Yet Islam is also the religion of flourishing bazaars and prosperous merchants. It is well-known for promoting trade and entrepreneurship, sanctioning private property rights, accumulation of wealth and profit, and frowning upon the state's intrusions in the market. It is the voice of the status quo and conservatism in the economic arena. To many, it is laissez faire capitalism. As Jama'at-i Islami in Pakistan and Taliban in Afghanistan have shown, some Muslim revivalists see Islamization of the economy as promoting free market capitalism.

These extreme representations of Islam closely reflect the universal expression of the range of social contestations in the political arena. It may be argued that Islam, as an encompassing ideology, like most other religions, lends itself to a wide range of interpretations, reflecting the formation of social forces in the theater of politics in society. There is some truth to this claim since no ideological doctrine can be immune to variations in interpretations arising from differences in social perspectives, or changes in social-historical processes and circumstances. But this assertion, which has gained wide popularity among the post-"Orientalist" Islamicists and Middle Eastern scholars, by way of rejecting the monolithic presentation of Islam put forth by the "Orientalists," reduces any analysis of Islamic ideology to the truism that Islam is what Muslims are. So, then, Buddhism is what Buddhists are and socialism is what socialists are, and there have been wide variations among Buddhists and among socialists. Then, one may conclude that "there are as many Islams as there are situations that sustain it" (Al-Azmeh 1993, p. 1). But, then, can one distinguish Islam from Buddhism and capitalism from socialism, as distinct ideological or social systems, in spite of historical and circumstantial variations? Or, to put it differently, is it possible to characterize an ideological system such as Islam in terms of its social ethos and public policy?[3] This is of special interest in an age when Islamic revivalism has become the banner of social change and beacon of a new social order in many Muslim countries.

The contention of this paper is that the contrasting interpretations of Islam represent two distinct visions of Islamic ideology corresponding to two specific historical circumstances of Islam. The first vision is that of Islam the rebellious, the idol smasher. It is Islam of the *mustaz'afin* (oppressed) and the banner of the mobilized masses. The second is the vision of order in an Islamic state, a class society. It is the Pax Islamicus that lasted for several centuries in a vast empire. While the populism and egalitarianism of Islam of the mustaz'afin generate the revolutionary thunder of Islamic revivalism, Islamic order, reflected in the laws and the historical traditions that ruled over the rigidly stratified Muslim societies of the past, brings together a diversified alliance of the propertied and privileged classes, Islamic or not. In this way, variability and instability of the newly established Islamic states generally reflect the predicaments of transition from Islamic populism to Pax Islamicus (or Islamic order). In this context, formulation of public policy in a newly established Islamic state becomes an arena of confrontation between these two visions of Islam. Whether a newly established state survives, collapses or degenerates, and how each case may exactly evolve, will obviously depend on the specific historical circumstances. Yet, the historical legacy of these two visions of Islam defines the terms and format of this social confrontation. The battle is fought in the arena of Islamic jurisprudence, where, by definition, the conservative vision of Islamic order is dominant.

Let us first consider the two contradictory representations of Islam and their historical contexts.

FROM ISLAM THE REBELLIOUS TO PAX ISLAMICUS

Islamic revivalism attempts to establish a social order according to the precepts of Islam, which are stated in the holy book, the Quran, and expressed by the words and actions of the Prophet Muhammad, Tradition (*Sunna*). Moreover, Islam, as a religion, has been a social movement that began when Muhammad preached a new social order while negating the spiritual legitimacy of the idols of the Meccans. In their place, he asked the Meccans to praise no god but Allah. This was a challenge against the dominant clans and families of Mecca and the power and social order that they enjoyed. Mecca was a flourishing commercial center, where merchants were sophisticated financiers and commercial activities were well-developed (Donner 1981, p. 52). If Muhammad's call to Islam is not to be regarded as merely an expression of a class conflict between the underprivileged and the ruling oligarchy of Mecca, it certainly was true that, as Bernard Lewis suggests, Muhammad drew his support "mainly from the poorer classes" and that his "opposition of the Meccans was largely economic in origin" (Lewis 1960, pp. 39-40). Many chapters (*suras*) of the Quran revealed in Mecca reflect Muhammad's antagonistic confrontation with the wealthy and powerful Meccan clans and families (Rodinson 1971, pp. 69-147). Muhammad brings God's "Woe" for the idolater of Mecca "Who hath gathered wealth" and who "thinketh that his wealth will render him immortal" (Sura CIV, 2-3). He threatens his own uncle,

Abu Lahab, who had joined the opposition boycotting Muhammad's clan, that when "He will be plunged in flaming fire," and "His wealth and gains will not exempt him" (CXI:1-3).[4] And in *Al-Fajr*, a very early Meccan sura, the Meccans are reminded of the fate of Pharaoh and his ilk:

> Dost thou not consider how thy Lord dealt with (the tribe of) A'ad,
> With many-columned Iram,...
> And with (the tribe of) Thamud,...
> And with Pharaoh,...
> Who (all) were rebellious (to Allah) in these lands,...? (LXXXIX:6-11).

They are told that they will receive the "disaster of His punishment," for as the Quran says,

> Nay, but ye...honour not the orphan, And urge not on the feeding of the poor, And ye devour heritage with devouring greed, And love wealth with abounding love (LXXXIX:17-20).

Muhammad's inability to make progress against the opposition led him to migrate (*hijra*) with his followers to Medina in 622 A.D. Many citizens of Medina came to the support of Muhammad and his followers, who formed a community of Muslims in that city. Thus, Muhammad, an idol smasher, an "anti-establishment" leader of an opposition movement preaching Islam in Mecca, became the chief magistrate of a community, practicing Islam in Medina (Lewis 1960, p. 41). In fact, it is in Medina that the first Islamic state is formed, and it is the ten years (622-632) of Muhammad's rule over this community that are viewed by Islamic revivalists as the Golden Age of Islam, which they wish either to reconstruct or to view as a model for contemporary social reform.

The revelations of the Quran to Muhammad in this period, too, reflect the change in the position of the Prophet, preaching a new religion, but also the leader of the community, the arbiter among groups and individuals, and the defender of the new community against its enemies. It is in Medina that some very important suras are revealed, such as, *Al-Anfal* (Spoils of War—Sura VIII), dealing with the distribution of the booty of the battle of Badr; *An-Nisa* (Women—Sura IV), dealing mainly with the treatment of women (widows and orphans of the battle of Uhud) and the internal dissidents (*khawarij*—hypocrites); and *Al Ma'ida* (The Table Spread—Sura V), addressing the matter of eating as well as the relations between the Muslims and Christians. The Tradition of Muhammad in this period also reflects the attempt of the Prophet to establish civil order, while trying to maintain social and military preparedness to fend off attacks of the enemy and to score victories against them.

The rule of Muhammad, as well as those of Abu Bakr (632-634) and Umar (634-644), the first two of the Four Rightly Guided Caliphs, was a period of state building and expansion. It was a society in transition and on the make. While the internal order was being established in the face of conflicts and challenges posed against the authority of Muhammad and the caliphs, the realm of Islam was expanding. Two giant imperialist powers of Persia and Byzantium collapsed under the blow of Islam, which, especially under the rule of Umar, had become a revolutionary thunder (Donner 1981). The booty of the wars certainly helped to empower the Islamic movement and the infant and fragile Islamic state. Soon after his arrival in Medina, Muhammad had begun his predatory raids against the Meccan cravens, both to blockade Mecca and to improve the economic well-being of the community (*umma*) in Medina (Lewis 1960, p. 44). But taking booty was easily accepted as a custom of the period. What the Quran needed to set straight was that the state receives a share, too.

> And know that whatever ye take as spoils of war, lo! a fifth thereof is for Allah, and for the messenger and for the kinsman (who hath need) and orphans and the needy and the wayfarer (VIII:41).

More importantly, as it was the tradition of the Arab tribes, the Quran and Sunna reveal repeatedly that life and property may be taken away in the name of Islam. Many wealthy merchants were killed or taken as prisoners in the raids led by Muhammad. It was made clear that it is legitimate to loot (confiscate) their merchandise (property) and to take them as prisoners and to release them in return for high ransoms (Rodinson 1971, pp. 162-170). The newly instituted state of Islam assumed the right of appropriation of property, from whoever threatens the security of the Islamic state.

When caliph Umar was assassinated in 644, Uthman was chosen as the new leader of the Islamic state. The rule of Uthman, the third Rightly Guided Caliph, which lasted until 656, was a period of intense internal conflict in the Islamic state, as it was transformed from a rebellious Islam to Pax Islamicus of the Umayyads and Abbasids dynasties. The election of Uthman, a member of the powerful Umayyad family of Mecca, as the caliph was a victory for the Meccan oligarchy, who benefited from his nepotism in taking many of the high posts of the empire and dominating its commercial relations (Lewis 1960, p. 59). By the end of Umar's rule, however, the Islamic state was entering a fiscal crisis. The Arab tradition was that the spoils of war would be distributed among the warriors to share with the tribe. This tradition was followed in Muhammad's time and during the rule of Abu Bakr and Umar. In addition to the loot that was carried away from the battlefield, in the Islamic conquests, there were other worthy prizes, namely land, which could be possessed, or taxed, and people, who could also be taxed. Until the rule of Umar, the second caliph, these revenues were also distributed equally among the warriors. But by Umar's time it was becoming obvious that the revenue from taxation of the conquered lands was not sufficient to keep the Arab recruits "on the pay roll" (Cameron 1973, p. 50). Thus, the office of treasurer of the state was established by Umar for regularizing the collection and distribution of the revenues of the state among those who had a right to pension. This imposed the power of the purse on the army, which believed the conquered land, and the tributes from it, belonged to them. Thus, what was viewed by the army as property of the Muslims (*mal al-Muslimin*) became the property of the state (*mal Allah*), or the "public treasury." The large band of warriors of Islam detested this transformation because they now had to depend on the state for their pension (Wellhausen 1902, p. 27ff).[5]

Uthman's nepotism and the corruption in the administration of the conquered territory only accentuated the friction that was forming between the state and the opposition groups, particularly those in the army. Uthman put his cousin Marwan in charge of the public treasury and gave him one-fifth of the loot of the North African conquest and offered one brother ("who was proscribed by Muhammad at the capture of Mecca" in 630) the governorship of Egypt, and to the other ("who had spat on Muhammad's face"), the governorship of Kufa (Cameron 1973, p. 52). His cousin Mu'awiya, the son of Abu Sufyan and Hind, was already appointed to the Governor of Syria by Umar.[6] In 656, in response to the economic crisis, and unsatisfied with policies of Uthman, rebels from Iraq and Egypt moved toward Medina. When the Egyptians arrived a crowd broke into Uthman's house and the 80-year old caliph was murdered as he was reading the Quran, one that he had commissioned to be edited and which is considered "the most important and fruitful achievement of [his] reign" (Spuler 1960, p. 31).

A strong representation of the conflict in the transition of Islam toward Pax Islamicus of Umayyads (and later that of the Abbasids) is the protests of Abu Dharr (also pronounced Abu Zarr) and Ali to what they viewed as the betrayal of Muhammad's Islam by Mu'awiya and the house of Umayyads. Abu Dharr was a companion of Muhammad. He objected to the luxurious life of Mu'awiya, and to his misuse of the public treasury. Abu Dharr pointed out to Mu'awiya that the Quran has condemned those who "hoard up gold and silver and spend it not in the way of Allah," and it warned that "unto them give tiding (O Muhammad) of a painful doom" (IX:34). Abu Dharr contended that the accumulated wealth at the house of Mu'awiya must be distributed among the poor at once. He claimed that Muhammad told him once that "should this mountain of Uhud be turned into gold for me, I would not wish that the nightfall comes before I have disposed of it all in charity except two carats" (Abdul-Rauf 1979, pp. 14-15). Abu Dharr's trouble making was unbearable to Mu'awiya, who asked his uncle, Uthman, to recall him to Medina. Thus, Uthman summoned Abu Dharr to Medina and from there sent him to al-Rabadha to exile, where he died in 652 or 653 (Gibb et al. 1960, pp. 114-115).

The struggle to establish the pure and unadulterated precepts of the ideology by confronting the "usurpers" and "impostors" who hold political power is not unique to Islam. It is not altogether surprising, therefore, to see that, twenty-four years after the death of Muhammad, Uthman, the third Rightly Guided Caliph of the Muslims is murdered by a group of disenchanted Muslims. The confrontation of Ali, the fourth Rightly Guided Caliph, against Mu'awiya and the house of the Umayyads, and the battle of Husayn (Ali's son) with Yazid (Mu'awiya's son), particularly to the Shi'is, is the continuation of the struggle between Abu Dharr and Uthman, between good and evil, liberation and oppression.

After the assassination of Uthman, Ali became the fourth caliph of the Muslims. Ali was not successful in leading the Islamic state effectively, as Mu'awiya and others, who blamed Ali for Uthman's death, challenged his leadership. In the heat of the tension between Ali and his opponents, Ali was murdered in 661 by one of his former followers. As there was no one to lead Ali's followers, they dispersed. Some historians of Islam suggest that Ali's assassination brought the "eminent danger of a civil war" among Muslims to an end (Spuler 1960, p. 34). Soon after Ali's death, Hasan, Ali's older son, signed a treaty with Mu'awiya, abdicating any claim over the caliphate (Momen 1985, pp. 26-28).[7] Thus, Mu'awiya, the son of the most powerful enemy of Muhammad and a member of the Meccan oligarchy, established himself as the caliph of the Muslims and continued to extend the Pax Islamicus, that he had established in Syria, to the rest of the Islamic Empire.

The last effort in reestablishing Muhammad's Islam of the populace in these early days of Islamic history was the struggle of Husayn, Ali's second son, and Muhammad's grandson. Mu'awiya, who died in April 680, had arranged for his son Yazid to succeed him. Among the followers of Ali, Yazid was known as a sacrilegious man. Husayn challenged the rule of the Umayyads, regarding both Mu'awiya and Yazid as the impious usurper of the caliphate. Yazid confronted Husayn in Karbela, where Husayn and seventy-two of his followers were brutally slain by Yazid's forces on October 10, 680 (Muharram 10, 61 A.H.) (Momen 1985, pp. 28-33; Spuler 1960, p. 39). This proved to be a crucial historical event for the Shi'is, who view Karbela as the theater of confrontation between good and evil and regard Husayn and his followers in that battle among the great martyrs of all ages, symbolizing the selfless, heroic struggle for Islamic justice in a battle with the satanic power of the wealthy, arrogant, and corrupt.

Abu Dharr's confrontations with Mu'awiya and Uthman, followed by Ali's and Husayn's battles against Mu'awiya and Yazid, were the manifestations of the tension in the newly established state of Islam. This tension was between populism, idealism, and egalitarianism that was preached by Muhammad, or at least understood as such by many converts to Islam, and the emerging social order in Medina and in the conquered lands. When the society of Muslims was set up in Medina, and as it expanded into the conquered territories, the imperatives of establishing a viable social order with a legal framework for public policy became clear to the political leaders, the caliphs. The issue became more clearly apparent with the fiscal crisis facing Umar's newly established Islamic empire. The populism and egalitarianism of early Islam, quite compatible with the collectivism of the tribal social order in Arabia, was obviously irreconcilable with the oligarchic order that was being established (or in some cases re-established) in the new imperial conglomerate of the highly complex agrarian-commercial societies of Mesopotamia, Persia, and Egypt.

The Umayyads succeeded in completing the transformation from Islamic populism to Pax Islamicus, a process which had been begun by Umar and made Uthman, Abu Dharr, Ali, and Husayn its main casualties. This made the Umayyads seem the principal revisionists and violators of the tenets of Islam to the followers of egalitarian Islamic movements as well as to the orthodox Muslims who believed that the Umayyads (661-750) denigrated the faith while promoting their own power. Obviously the Shi'is, not only from the above perspectives, but also for their partisan position on the intense and brutal confrontation of Mu'awiya and Yazid with Ali and his son, Husayn, view the Umayyads as the arch enemy of the "true Islam." Even the Abbasids, who ruled over the Islamic Empire for more than five centuries (750-1258) in a manner not much different than the Persian kings and who abandoned enforcing many precepts of Islam, such as the ban against wine drinking (Spuler 1960, pp. 49-52), viewed the Umayyads as adversaries of Islam (Schacht 1982, p. 23).

But both for the Umayyads and the Abbasids the main preoccupation of the state was political administration and public policy, and not religious purity. It was in this rather long period, from the seventh to the thirteenth century, that Islamic law (Shari'a) was fully formulated. The formulation of Islamic law was accompanied with the centralization and rationalization of the system of administration of the state and reforms in public finance, the monetary system, laws of contracts and commercial obligations, inheritance and family laws and the penal code. Many treatises were written by scholars on these matters, such as that written by Abu Yusuf (1969) at the request of the Abbasid caliph, Harun al-Rashid (786-809).[8] This trend also led to the centralization of the judicial system by the early Abbasid period, when judges (*qadis*) were appointed by the central government, and the position of the Chief Justice (*qadi al-qudat*) was created. The caliph, who ruled over the state as the sovereign, was expected to possess the same attributes as those of the qadis and to be constrained in the same way by the Shari'a. But the caliph also had the power to legislate in the matters of public policy to promote the welfare of the state. By the early Abbasid period this came to be known as *siyasa* (Schacht 1982, pp. 50-54).

Siyasa: Public Policy

Development of the concept of siyasa was the manifestation of a fundamental transformation in the institution of societal leadership, from prophethood to kingship. Caliphs were gradually evolving into kings or sultans, if not in name at first, but in practice. Dynasties were established. First the Umayyads and Abbasids, and then Fatemids and Ayyubids, before the Seljuks and the sons of Changiz took over the caliphate. One may regard this transformation as "monarchization of the caliphdom." This transformation was accelerated especially by the influence of the theories of administration espoused by Iranians, who reflected upon the administrative practices of the courts of the Persian kings. Yassin Essid (1995) examines the significance of the administrative literature, known as "the Mirror for Princes" written by scholars, wazirs or kings in the first centuries of Islamic rule. Among these are *Adab al-Kabir* (circa 750)[9] of Roozbeh (Abd Allah) Ibn al-Muqaffa', *Kitab al-Sultan* (889) of Ibn Qutayba, *Siyasat Nama* (1091) of Nizam al-Mulk, and *Nasihat al-Muluk* (1105-11) of Imam Muhammad Ghazali. In about 1121 al-Turtushi (also known as Ibn Randaqa) wrote *Siraj al-Muluk*, in an attempt to present an Arabized mirror (Essid 1995). These treatises are discourses on a wide array of subjects dealing with the conduct of the affairs of the state. Their range is from the protocol of the court of the king, the manners of conducting the administrative duties of the sovereign, and the intricacies of power plays in the court, to justification for the power of the sovereign, to discussion of the social stratification and occupational composition of society, and means of achieving prosperity and social tranquillity.

If these treatises reflect the discourse on the principles and wisdom of governance of the sovereign, the institution of *hisba* and the position of the *muhtasib* provided the means for implementation of public policy in the public space, namely the market (*suq*). The institution of hisba was to maintain law and order, to ensure collective welfare of the community (umma). It was responsible for regulating market activities, by guarding against fraud, dishonesty, and extortion. It was to guard the orderly flow of commerce and maintenance of market stability and, most importantly, the availability of food staples and stability of their prices. In addition to commercial activities, the office of hisba was the guardian of public morality.[10] In other words, the office of hisba was to promote good—*ma'ruf*—and to suppress evil—*munker* (Khan 1982, p. 135) Thus, the office of hisba was to provide both temporal tranquillity and spiritual purity in the Muslim city, as a corollary of the unitary conception of Islam (Essid 1995, p. 138).

The rules of conduct of the market were presented as the "laws of the market" (*ahkam al-suq*). The first known treatise of this kind is written by Ibn Umar (1975) in the third century of Islam (Essid 1995, p. 123). The responsibilities of the office of hisba were carried out by the muhtasib. He was expected to be a man "with a high degree of integrity, insight, reverence and social status," to oversee the organization of religious rituals, market activities, and the administration of affairs of the municipality (Khan 1982, p. 137). Thus, the muhtasib was to see that commerce is flowing smoothly, without any dishonesty, that no immoral acts or other disruptive behaviors were demonstrated, that building codes and zoning restrictions were observed, that public passages were not obstructed, and that roads were clear and streets were safe.

These duties of the muhtasib reflected the high degree of sensitivity of government to urban tranquillity and its fear of "enemies of the interior", and acts of protest and violence that the population of the suq had a tendency to exhibit, especially in response to acute shortage of provisions and the sharp increase of their prices (Essid 1995, pp. 129-130). The muhtasib was also responsible for seeing that riba was not practiced, weights were not short and scales were accurate, no transaction was carried out with children and mental incompetents, and hoarding was not exercised. One of the most significant aspects of the duties of the muhtasib was maintaining price stability. This was done although price regulation is a highly controversial issue in Islam.

Market and Prices

In spite of much controversy about price regulations, it is accurate to say that Islamic jurisprudence and scholarly discourse express a clear preference for a smoothly functioning and stable free market, where the main provisions of life are provided at a "reasonable price," and where profitability for the merchant and adequate livelihood for the populace are guaranteed. All of this is expected to be the general norm of a free market with no intervention by the government. The cornerstone of the opposition to price regulation in Islam is a hadith (statement) by Muhammad. It is reported that on the occasion of rising prices some asked him to fix prices. He replied:

> God is the Taker, the Disposer, the Succourer and the Controller of prices. I very much hope that when I meet God no-one will claim against me for an injury I have caused in blood and property (Ibn Taymiya 1982, p. 35).

There is some doubt about the validity of this hadith which is not reported in *Muwatta'* of Imam Malik or *Sahih* of al-Bukhari (Essid 1995, p. 153). Bernard Lewis asserts that "it is obvious that the Prophet said no such thing, for such a question could hardly have arisen in western Arabia during his lifetime" (Lewis 1974, p. xxi). Lewis adds that this "theological statement of laissez-faire economics...has been the predominant view among the professional men of religion, since virtually all the traditions dealing with prices take this line" (Lewis 1974, p. xxii). It could very well be that this is a fabricated hadith, but the reliability of its source and the chain of transmitters aside, one cannot reject it for its historical context and meaning.[11] Medina was a city much dependent on imported goods, and it is well-known that any economy reliant on imports is susceptible to exogenously induced price fluctuations. This could be the statement of a ruler of a city relying heavily on imported provisions. (With a high dependence on agricultural output, the statement could also be true in reference to the unpredictability of the climate.) If this is in fact Muhammad's statement, it cannot be construed simply as fatalistic submission to God's will. For a man who spent much of his life in the world of commerce, he was well aware of the workings of the market. Thus, this statement may mean that under normal circumstances the market price is "fair" and "just" (Essid 1995, p. 153). It may be also viewed as a way of discouraging the public from expecting the government to fix prices any time that they rise because there are many cases in the market where price control is ineffective and even possibly counterproductive.

Al-Maqrizi (d. 1442) points out that price fixing would be counterproductive because the merchants would view it as arbitrary and abusive, and will tend to react by overstocking their merchandise (Essid 1995, p. 153). It is on this basis that the Hanbali and Shafi'i schools of Islamic jurisprudence reject price fixing. Ibn Qudamah al-Maqdisi, a Hanbali scholar, states that government has no business fixing prices because it is an injustice inflicted upon the owners of the merchandise who have the right to sell their property at any price that they agree with the buyer. Furthermore, he argues that price fixing will in itself cause prices to increase. That is so, he states, because when government fixes prices in a market, merchants will not bring their goods into that market and those who have goods in that market will keep them in stock. But as the buyers cannot satisfy their demand, they bid up the price. Therefore, price increases and the buyers and sellers will both suffer. For this reason Ibn Qudamah maintains that price fixing is prohibited in Islam (Islahi 1988, p. 95).

Taqi al-Din Ahmad Ibn Taymiya (d. 1328), a prominent Hanbali theologian, however, underscores the views of some scholars who approve some price regulations.[12] He maintains that "if people refuse to sell what they are under obligation to sell," or if they refuse to sell at a "fair price," they may be ordered to do so and punished for noncompliance (Ibn Taymiya 1982, p. 50). In this way Ibn Taymiya maintains that rejecting price regulation by relying on Muhammad's statement above is a mistaken interpretation. He does not refute the authenticity of the statement attributed to Muhammad, but argues that it is a "particular judgment, not a general pronouncement," and is mainly related to the situation of Medina where there was little that the Prophet could do about the effects of fluctuations in import prices (Ibn Taymiya 1982, p. 51). Ibn Taymiya seems well aware of the price mechanism and is not disposed to price regulation when a price increase is the "natural" outcome of the dynamics of the market. He states:

> Rise and fall in prices is not always due to an injustice (*zulm*) by certain individuals. Sometimes the reason for it is deficiency in production or decline in import of the good in demand. Thus if desire for the good increases while its availability decreases, its price rises. On the other hand, if availability of the good increases and the desire for it decreases, the price comes down. This scarcity or abundance may not be caused by the action of any individuals; it may be due to a cause not involving any injustice, or sometimes, it may have a cause that does involve injustice. It is the Almighty God who creates desires in the hearts of the people (Islahi 1988, pp. 88-89).[13]

This and other similar passages indicate that Ibn Taymiya understood the dynamics of price determination and appreciated that prices may go up and down, depending on availability (supply) and desire (demand). He, too, attributed this to Almighty God who creates desires in our hearts. Thus, even the restrictive interpretation of Ibn Taymiya calls for the market to work freely.

Whether or not one accepts the validity of the hadith by Muhammad on free fluctuation of prices, Muslim scholars are generally proponents of a free market. Ibn Taymiya and all the other proponents of price regulation call for the intervention of the government only to rectify the effects of market irregularities (imperfections), such as monopoly and hoarding, on the price of the main provisions demanded by the public. Thus, the muhtasib is entitled, as Ibn Taymiya sees it, to compel the hoarder or the monopolists to sell their merchandise at their "fair value" (Ibn Taymiya 1982, pp. 31-33). When price stability is disturbed for any exogenous reasons, such as bad harvest, high price of imports, Ibn Habib suggests that the ruler should bring together all the merchants of the market and ask them how they buy and sell. Then the ruler may bring their price down to where it is acceptable to the merchants and the public (Ibn Taymiya 1982, p. 50). This is a form of mediation between the conflicting interests of the merchants and the public in order to evade a crisis. This voluntary price fixing among the merchants, under the sanction and control of the ruler, brings into focus the reconciliation of individual freedom of enterprise and the collective well-being of the society.

The proposed conditions for government control over prices is the most intrusive aspect of an Islamic ruler's market intervention in Ibn Taymiya's controversial views, which have gained him recognition as a maverick reformer in the long succession of great Muslim scholars (Islahi 1988; K. Ahmad 1982). The proposed criteria of Ibn Taymiya for market intervention are, however, not only insignificant in comparison to the "just price" doctrine of Medieval Christianity and the pre-Industrial Revolution mercantilist policies in Europe, but they are quite mild even in contrast to the restraints imposed on enterprises in their pricing and other business practices in the market economies of the post-Industrial Revolution and especially in modern capitalism. Predatory pricing, cornering the market, fraud and extortion are business practices that hamper the efficiency of the market and, more importantly, subvert its legitimacy. These practices are unequivocally condemned in Islam as they are viewed illegitimate in modern capitalism, too.

On the determination of wages and the rights of workers, Islam sanctions the dynamics of the labor market as defined in classical capitalism. As long as the laborers are paid the wages that they agreed upon with their employer, and are aware of the going wage rate in the market in reaching their agreement, there is little reason for the government to intervene. Therefore, an Islamic state is not expected to intervene in setting wages or determining the conditions of employment in any way such as determining the hours of work, minimum wage rate, or restricting child labor. Nor is the state expected to create any of the restrictions that the state of many modern capitalist economies (developed or otherwise) impose (Behdad 1989, 1992). The mechanism of the market is expected to maintain the well-being of the Muslim state except when special circumstances arise, for which special policies may be enacted.

Public Interest, Urgency, and Public Policy

When the society is functioning within its "normal" course and individuals live as devout Muslims, with honesty, decency, benevolence and in accordance with the rules of the Shari'a, minimal intervention of the state in the market would be necessary. The state would collect taxes for making public expenditures and would supplement the charity and alms giving of the individuals to help the poor and destitute. But should the need arise, an Islamic state may implement new or exceptional policies in order to deal with new or extraordinary circumstances. Theoretically, *ijtihad* (the process of deducing new laws from the basic sources by a qualified Muslim scholar, *mujtahid*) should be relied upon to deal with new circumstances. However, in the opinion of the major schools of Sunni jurisprudence, the "gates of ijtihad" have been closed since the fourth century of Islam (Schacht 1982, p. 69f). That is, by the time this view gained dominance, all essential questions dealing with the Shari'a had been dealt with by the great mujtahids of the past.

Therefore, according to this view, there is no need for independent reasoning in the Shari'a and the jurists are limited to the interpretation of "the doctrine as it had been laid down once and for all" (Schacht 1982, p. 71).[14] But since societies would confront new and extraordinary circumstances, the provisions of *istihsan* (juristic preference), *maslaha mursala* (unrestricted public interest), and *istislah* (consideration of public interest) have been devised by the major Sunni schools of jurisprudence (Paret 1978, pp. 255-259; Khadduri 1989, pp. 738-740).[15] These juristic practices involve setting aside an existing law, when its enforcement may harm the society of Muslims, in favor of an alternative ruling which would better serve public interest (Kamali 1991, p. 247). Imam Muhammad Ghazali (d. 1111) articulated the juristic concept of maslaha by suggesting that the "ultimate purpose of the Shri'a...[is] the maintenance of religion, life, offspring, reason and property. 'Anything which furthers these aims'...is maslaha" (Khadduri 1989, p. 739) Thus, in extraordinary circumstances, most importantly in the conditions of "urgency" (*daruriyyat*),[16] the objectives of the Shari'a may be relied upon in order to pursue public interest. Public interest here means safeguarding religion, life, future of the society, sanity, or property against an imminent danger in any of these regards for society. Ghazali points out that anything which furthers these aims is maslaha and anything which runs contrary to them is "mafsada" and cites the example of unbelievers shielding themselves among Muslims. He notes that the killing of innocent Muslims in this case is permissible in order to preserve Islam (Khadduri 1989, p. 739). Kamali, however, emphasizes that maslaha is invalid if it has been "nullified either explicitly or by an indication that could be found" in the Shari'a (Kamali 1991, p. 273). Therefore, for example, the argument that in modern times usury (riba) could be justified based on maslaha, because it is necessary for Muslims to engage in usury to participate in the world market, would not be acceptable because usury is explicitly prohibited in the text (*nass*) of the Quran (Kamali 1991, pp. 274-275).

Thus, as long as a ruling is not contrary to the text of the Quran and the objectives of the Shari'a, it could be upheld for the reason of expediency if it is viewed by Islamic jurists, or the Islamic state, that it is in the public interest (Kamali 1991, p. 280). With the wider scope of ijtihad in Shi'ism (Behdad 1989), a Shi'i state would have less need to rely on istihsan, maslaha, or istislah, which Shi'i scholars have rejected altogether (Kamali 1991, p. 246). But even a Shi'i state, when it needs to violate a primary rule of the Shari'a, finds a way of legitimizing its action in the name of public interest and preservation of the Islamic state. In 1981, in a context similar to that expressed in Imam Ghazali's example, Ayatollah Khomeini let the Consultative Assembly (*Majlis*) of the Islamic Republic of Iran invoke "secondary rules" (*ahkam thanaviyya*) in a condition that was declared "urgent," to redistribute land and take other actions, in violation of "primary ruling" (*ahkam avvaliyya*) which guarantees property rights (Behdad 1995a). This is similar to the example of Imam Ghazali, which is viewed as a troublesome case of maslaha by some Muslim scholars (Kamali 1991, p. 275). In each case a specific injunction of the Shari'a is violated; in Ghazali's case the sanctity of life of innocent Muslims, and in Khomeini's case the sanctity of property. In 1988, when Khomeini encountered the opposition of the jurists in the Council of Guardians to the passage of laws by the Majlis, even on grounds of urgency and only for the temporary duration of the urgency, he constituted the Council of Expediency (*Majma'-e Tashkhis-e Maslahat*) for reconsideration of the rulings of the Council of Guardians.

The issue points to the conclusion that in urgent circumstances an Islamic state can implement any policy measures to preserve itself, because the state views itself as the guardian of religion, life and even sanity and property, and any state claims that its collapse will bring chaos and destruction. This divinely sanctioned means of political expediency is especially convenient for the states that manage to incorporate the hierarchy of Islamic jurisprudence within their formal structure of the state.

REFORMISM AND REVIVALISM

From the tenth century (fourth century A.H.) to the fall of the Ottoman Empire after World War I, Islamic public policy was practiced within various local or national state entities based on what may be called traditional jurisprudence. This is viewed by many as the period of imitation (*taqlid*) and decline in Islamic jurisprudence, when original ijtihad was discouraged and scholars were preoccupied with writing commentaries on the works of the great theologians of the past (Kamali 1996, p. 67). From the late nineteenth century a movement began to reform and rejuvenate (*islah*) Islam. This movement was mainly in response to the colonial domination of Europe over the Muslim societies of the Middle East, and a reaction to the advances of Europe in science, technology and, most importantly, in acquiring an enviable economic affluence, in contrast to the poverty and stagnation of the Middle Eastern economies. Sayyid Jamal al-Din Afghani (d. 1897) is the pioneer in the path of Islamic reformism and anti-colonialism. As a political activist he traveled throughout the Middle East, and to India and England as well. He struggled to take away the control of the traditional ulema over the faith and to make Islam a means of political mobilization and a "source of solidarity" against Western imperialism (Keddie 1994, p. 23). The reformist trend, especially in promoting rationalism and a scientific outlook, was pursued by such influential thinkers as Muhammad Abduh (d. 1905) in Egypt (Kerr 1966), and Shariat-Sangalaji (d. 1944) and Ahmad Kasravi (d. 1946) in Iran (Richard 1988).[17]

In the early decades of the twentieth century a strong current in Islamic reformism evolved into a movement for the revival or renewal (*tajdid*) of Islamic social order.[18] That is, the Islamic movement, instead of attempting to "rationalize" Islam so that it would become compatible with the modernity and industrialism of the West, came to rely on the precepts of Islam to invigorate the umma toward construction of an ideal society, in accordance with the blueprints drawn from the Golden Age of Islam. As John Voll (1986, p. 168) observes, "revivalism is inherent in the logic and experience of Muslims in history," for Muhammad's Tradition (Sunna) is a pillar of the Shari'a and Islamic belief. The society that Muhammad organized in Medina, and the rule that he and the Rightly Guided Caliphs extended to the new territories of Islam, have been viewed by Muslims as examples of equitable and just societies to be followed. Moreover, the prosperous and powerful Islamic state of the Umayyads, the Abbasids and other Islamic dynasties that followed them on the vast territories spreading from India to Spain, and the great scientific contributions of Muslim scholars in medieval ages, when Europeans were struggling with the fanatical rigidities of the Christian church, have given many Muslims a sense of chauvinistic religious pride. This is in the face of the indignities imposed upon them by colonialism, foreign domination, and, not the least, by their own poverty and technological backwardness.

Thus, a utopianism, a wish and even a struggle to return to the morally upright, politically successful and economically prosperous golden ages of Islamic rule seems only a "natural" response to these circumstances. The reaction to the top-down, and at times coercive, secularization policies of the state in Turkey, Iran and Egypt, pursued particularly from the late 1920s, and the economic and social dislocations resulting from the industrialization and the subsequent urbanization of these and many other Middle Eastern societies, gave the urban and rural underclass and the traditional petty-producers and merchants ample reason to congregate under the unifying green flag of Islam. This congregation longed for two principal objectives. Foremost, it expressed a strong yearning for the values and morality of the "old-time religion." It opposed the "immorality and social decadence" that had been brought about by the secularization policies of the state and the imposition of Western culture on the Muslim society. Bureaucratic corruption, use of intoxicants, and sexual immorality were the main areas of concern, for all of which the state was held responsible. None of these maladies was new, except possibly the last. The objection was raised to the assertive and even coercive effort of the state, particularly in Turkey and Iran, in "bringing" women into the public sphere by de-veiling them and making available to them some educational and occupational opportunities.

No less significant, particularly to the urban and rural underclass, was the Islamic pledge to social fairness and economic justice. The peasants yearning for access to land, generally held by large landlords; the discontent of the recently urbanized laborers, working for meager wages with the constant fear, and often the reality, of chronic unemployment; and the struggle of the artisans and shopkeepers to survive the destructive waves of domestic or foreign competition and market instabilities, made any promise of social justice attractive.

Islamic Revivalism and Socialism

The Islamic revivalist movement was deeply influenced by the spread of socialism, as an ideological and a revolutionary movement, and by the cold war between the capitalist and socialist camps in the post-World War II decades. This influence was particularly accentuated by the anti-colonialist and anti-imperialist movements that many Muslim societies were engaged in for much of the twentieth century. As early as 1912, Mushir Hosain Kidwai, an Indian Muslim, an opponent of British rule over India and the Honorary

Secretary of the Pan-Islamic Society of London, wrote *Islam & Socialism.*[19] This may be the first pamphlet published on Islam and socialism. Kidwai advances a thesis that later became a prevalent position among many radical Islamic revivalists. He claims that the "idea of Socialism in Islam is not less than thirteen centuries old and cannot be attributed to European influence" (Kidwai 1912, p. ii). He states on the cover of his pamphlet that "[t]o us (Musalmans) Socialism means an organized, continuous and harmonious co-operation of individuals in all the affairs of life." He refers to the many verses of the Quran, Tradition of Muhammad, and the examples of the Rightly Guided Caliphs to demonstrate what he views as the many socialistic aspects of Islam. In Kidwai's view (1912, p. 33), Islam would have had a brilliant history if Ali had not been assassinated and would have ruled "on truly Socialistic lines." But he adds that "Muhammad had so profusely imbued Musalmans with Socialism that its spirit lingered tenaciously," even after the "destruction of democratic Muslim Socialism" by "the despotism of [Mu'awiya], the first Muslim 'king'" (Kidwai 1912, pp. 33, 66).

Ansari (1986) has studied the impact of socialism on Indian Muslims engaged in the struggle against the British rule in India in the early decades of the twentieth century. In the years following the end of the World War I, anti-British, pan-Islamist Muslims who left India to organize opposition to the British rule in India became aware of the October 1917 Revolution in Russia and the communist ideology. Many of these Muhajirin (expatriates) became converts to socialism. Some became the founders of the Communist Party of India. Many of these Muhajirin, communist or still remaining a Muslim, believed that there was much in common between Islam and communism. An important example is Maulana Barkat-Allah. He was a pan-Islamist Muslim, with tendencies very similar to Jamal al-Din Afghani (with whom he had met), and with a belief, as early as 1905, that the British rule over India is the cause of poverty of the Indian people (Ansari 1986, pp. 515, 519). Maulana Barkat-Allah, who remained a believer in Islam all his life, wrote strong words in support of communism in his *Bolshevism and the Islamic Body Politick* (Ansari 1986, p. 519). For example, "Oh, Muhammedans listen to this divine cry. Respond to this call of liberty, equality and brotherhood which Comrade Lenin and the Soviet Government of Russia are offering you" (Ansari 1986, p. 519). Or, Shaukat Usmani, one of the founders of the Communist Party of India, stated that "Islam preaches equality, so does Communism. That is why I am a Communist" (Ansari 1986, p. 530).

Muhammad Iqbal (d. 1938), the great poet and the pan-Islamic intellectual leader of Pakistan/India's independence movement, wrote in admiration of Karl Marx, Lenin, and the Russian Revolution (M. Iqbal 1955). He wrote in the *Song of the Worker*, "Let us overturn the foundation stones of old taverns" (Gordon-Polonskaya 1971, p. 119) and *To the Panjab Peasant*:

Break all the idols of tribe and caste,
Break the old customs that fetter men fast!
Here is true victory, here is faith's crown—
One creed and one world, division thrown down! (Iqbal 1955, p. 56)

He calls for a social revolution (Hakim 1961, p. 138) and summons the dispossessed to

Rise and arouse the poverty-stricken,
And shake the foundations of the palaces of the rich.
Warm up the blood of the slaves with the fire of faith
And make the sparrow dare to fight the eagle.
The time of sovereignty of the masses has arrived;
Wipes away all traces of ancient law and customs.

One wonders if Iqbal wanted to echo the International: "Arise ye prisoners of starvation/Arise ye wretched of the earth/For justice thunders condemnation/A better world's in birth" (IWW, p. 1976). It is well-known, however, that to him, socialism was quite compatible with Islam. He wrote to Muhammad Ali Jinnah, the political leader of Pakistan's independence, "[f]or Islam the acceptance of social democracy in some suitable form is not a revolution but a return to the original purity of Islam" (Hassan 1971, p. 154).

The enthusiastic celebration of socialism by Indian Muslims soon evolved into a general negation of communism and skepticism toward socialist ideals. This was an understandable evolution in the face of the declared atheism of the communist parties, the spreading of the news of the brutality of Stalinism in the Soviet Union, and the cold war that was waged in the post-World War II decades. Thus, Islamic revivalist movements, in India, Pakistan and elsewhere, evolved toward a third worldist, "Neither East, Nor West" ideology, with a clear attempt to draw the lines of demarcation of the Islamic movement with capitalism, on the one side, and with socialism, on the other. On the "left," Islamic reformists such as Khalifa Abdul Hakim (1953), Ghulam Ahmed Parwez (1968/1957), and Nasir Ahmad Sheikh (1961) have taken a position similar to Kidwai that socialism, in its best form, is the innovation of Islam. They all, however, concentrate on the issue of land ownership and the conditions of peasants and pay little attention to industrial production and wage workers (Malik 1971b, pp. 70-74).

On the "right," however, some important conservative Islamic revivalist organizations emerged as anti-communist forces confronting the spread of socialist ideology in the Muslim world. Notable among these were Ikwan al-Muslimin (Islamic Brotherhood) of Hasan al-Banna (d. 1949) in Egypt (Ramadan 1993), Fada'ian-e Eslam (Warriors of Islam) of Mojtaba Navvab-Safavi (d. 1956) in Iran (Behdad 1997), and Jama'at-i Islami of Abul Ala Maududi (d. 1979) in Pakistan (Nasr 1994). To these revivalists, communism and its negation of God, religion, and private property rights are manifestations of the decadence of the West.

Ali Shariati in Iran and Abul Ala Maududi in Pakistan have been the most eloquent proponents of these two tendencies in Islamic revivalism. The clear line of distinction between these two visions of an ideal Islamic society is the provision of social justice and the organization of the relations of production. Shariati recognizes Islam as the religion of liberation, and Islamic revivalism as a move toward formation of a classless society. On the other hand, to Maududi the foundations of an Islamic society are the principles of capitalism, where people's behavior has been modified by the Islamic rules of conduct. Here I will examine these two visions of Islam.

Vision I. Monotheistic Classless Society

Ali Shariati began his political activism in the late 1960s by giving a series of lectures at Mashhad University and the Hoseiniyeh Ershard (Tehran). He died in June 1977 in London, as the early signs of the revolutionary movement in Iran were becoming apparent. Shariati, who had received a doctorate in sociology from the University of Paris, like his precursors in Iran, Kasravi and Shariati-Sangalaji, rejected religious fanaticism and condemned political conservatism of the clergy (Behdad 1994). To Shariati, Islam is a rational and progressive world view, and a "a philosophy of liberation" (Shariati 1980a, p. 73, 1978a).

The most fundamental and controversial aspect of Shariati's thought is his conception of history, which may be regarded as a Quranic expression of Marx's historical materialism. Shariati (1979, p. 98) sees the war between Cain and Abel as "the war between two opposing fronts that have existed throughout the history, in the form of a historical dialectic." To him, the killing of Abel by Cain represents the end of the stage of communal ownership of a pastoral society (primitive communism) and the advent of private ownership of an agrarian society (a class society). He adds that this was the beginning of a permanent war between the owners and the dispossessed, the oppressors and the oppressed, the party of Cain and that of Abel. Then, he carries on a distinctly Marxian analysis of class formation and accepts Marx's formulation of the successive stages in the development of social formations, from primitive communes to advanced capitalism, and finally to "the triumph of the proletariat" (Shariati 1979, pp. 98-114). However, he contends that Marx is "confused" in his formulation of these stages since, according to Shariati, societies can have only one of two possible social structures. They can be a classless society, as in primitive and advanced communism. That would be "the structure of Abel." Or they can be a class society, as in slavery, feudalism, and capitalism. This is "the structure of Cain" (Shariati 1979, p. 112, 1980b, p. 37).

Shariati, then, concludes that the weapon of both Cain and Abel has been religion. Polytheism (*shirk*), the religion of a class society, is the religion of

Cain. Monotheism (divine unity, *towhid*), the religion of a classless s is the religion of Abel (Shariati 1979, pp. 106-109) In Shi'ism this opposi according to Shariati, is manifested in the struggle between Alavi Shi'ism (attributed to Imam Ali), the religion of struggle and liberation, and Safavi Shi'ism (attributed to the Safavid kings[20]), the religion of alienation and oppression. Shariati maintains that the existing (pre-revolutionary) Shi'i clergy represents Safavi Shi'ism, and expounds a conservative interpretation of Islam. It promotes religious fanaticism and trivializes the teaching of the Quran and Muhammad. This, he claims, is done to preserve the existing structure of social oppression. Shariati, thus, calls upon the Iranian intellectuals to understand Shi'ism as a religion of liberation and not as a stale and degenerated dogma (Shariati 1981a). He calls for an Islamic Protestantism, an Islamic reformation, toward the revival of Alavi Shi'ism with its revolutionary ideals and rejection of the establishment of the clergy (Shariati 1981b, p. 56).

Shariati is an Abu Dharrian revivalist. He praises Abu Dharr for "fighting exploitation and capitalism" (Shariati 1980b, p. 29), and in a voice not unlike William Godwin or Pierre Joseph Proudhon, negates private property rights as the basis and the manifestation of polytheism. He contends that Islam is in opposition to "capitalism, private ownership and class exploitation" (Shariati 1978b, p. 147). Thus, according to Shariati (1981c, p. 296), Islamic liberation is associated with rejection of private property and the struggle toward establishing a monotheistic classless society.

Many young followers of Shariati found themselves ideological and political sympathizers of the Organization of Mojahedin Khalq Iran (hereafter, the Mojahedin, a plural noun). A small number of these young activists joined this underground, Islamic revivalist, guerrilla organization, many of whose members were either killed in armed confrontations or spent a long time in the Shah's prisons, if they were not summarily executed (Abrahamian 1989). The Mojahedin were remarkably blunt in their regard for Marxist epistemology and methodology (Mojahedin 1972, 1980a). They accepted Marxist social thought but rejected atheism and emphatically objected to the charge by their Muslim opponents of being eclectic Islamic-Marxists (Mojahedin 1980b). In any case, the Mojahedin believed that in a monotheistic society "class contradictions" would have been resolved, "commodity relations" eradicated, the economy demonetized, and a "classless society" would have been established (Mojahedin 1980a, pp. 14, 22-23).

Shariati and the Mojahedin, although they recognized the progressiveness of capitalism as a stage in the process of historical development of the European societies, rejected the notion that the Iranian society must go through a similar historical path. According to Shariati, the Iranian society, with a weak "national bourgeoisie," and economic and cultural dependence on Europe and North America, cannot repeat the European experience (Shariati 1980b, pp. 22-23). If Shariati had a vague notion of the theory of dependency, the

Mojahedin were explicit in pointing out that the path toward capitalist development has come to a dead end with the domination of imperialism on the world economic order. They declared that economic development in the periphery can be accomplished only after eradicating dependency and pursuing a strategy that would negate class relations (Mojahedin 1980a, #6, pp. 16-17). This anti-dependency position of the Mojahedin was indeed in line with the dominant view among the intellectuals in the Third World (and many on the left in the West) and the political strategy of most important factions of the Marxist movement in Iran.[21]

Shariati calls for a "genuine socialist" society. Shariati's *genuine* socialism was to be unlike the dictatorial-bureaucratic socialism. It was not to be unidimensional in its materialism. Instead, it was to be a society for human liberation, where the essentiality of humanism is recognized (Shariati 1982, p. 222, 1978a, pp. 83-85). Does Shariati see a contradiction between his Islam, and socialism and Marxism? He clearly does not. The resolution of the contradiction between historical materialism of Marxism and metaphysics of Islam is more essential. Riffat Hassan, in an effort to explain the same issue about Muhammad Iqbal, points out that in the contrapositioning of Marxist materialism and Hegelian idealism we are led to the dichotomy of matter and spirit. However, Hassan points out, in materialism "the reality of thought and other mental phenomena is not denied, only their primacy" (1971, p. 154). Thus, to Shariati historical materialism and Islam are not necessarily contradictory. The resolution of socialism and Islam is much simpler for Shariati. The life and statements of Muhammad, Ali, and Husayn, to Shariati are all clear proofs that collectivism and egalitarianism are integral parts of Islam. To Shariati, Abu Dharr, Muhammad's Companion, who dared to say in condemnation of private property what even Proudhon did not (Shariati 1980b, p. 29), was simply following Muhammad's teaching and Tradition. Mojahedin brought Shariati's thought into the Iranian revolution.

Vision II. Islamic Capitalism

Seyyed Abul Ala Maududi, the Pakistani Muslim revivalist, has been regarded by some as a "Muslim Adam Smith" and a champion of "laissez faire economy" (Malik 1971b, p. 75). Maududi (1903-1979) founded the Jama'at-i Islami in 1941, and following the independence of Pakistan as an Islamic state, carried on a campaign for the formulation of an Islamic constitution, based on the Shari'a (Nasr 1994; Maududi 1969). Maududi believed that the "Islamic Movement...has to change the whole pattern of living: it has to change the politics, ethics, economics, and the civilization of the world" (S.R. Ahmad 1976, pp. 26-27).

Maududi elaborates on his political theory in a speech delivered at a meeting of the Inter-Collegiate Muslim Brotherhood in Lahore in 1939. He states that sovereignty in the state belongs to God, and no "person, class or group, not even the entire population of the state as a whole, can lay claim to sovereignty." Furthermore, the real lawgiver is God and "the authority of absolute legislation vests in him." From this it follows that the state must "in all respects, be founded upon the laws laid down by God through his Prophet," and no independent legislation can be made, nor can God's laws be modified, "even if the desire to effect such legislation or change in Divine laws is unanimous" (Maududi 1969, p. 132).

Maududi, however, suggests a redefinition of the Shari'a. In a speech delivered at the Law College in Lahore in 1948 he points out that one is not compelled to "accept any and every saying or expression of opinion by an authority on *Fiqh*, or anything and every thing written in a book of *Fiqh*" (Maududi 1969, p. 104). This is an obvious fact to any student of Islamic jurisprudence (fiqh) because of the existence of many contradictory positions that have been taken by various jurists (*fuqaha*) on many minor and some major issues over the centuries of Islamic rule in various historical, social, and geographical circumstances. According to Maududi, only the following can constitute elements of Islamic law: (1) "an explicit commandment of God" in the Quran; (2) Muhammad's Tradition (explanations of Quranic verses or his explicit order or prohibition); (3) juristic interpretations of the Shari'a "on which there has been a consensus (*ijma'*)" of the umma, or "the majority decision of the *ulama* which has been accepted by an overwhelming majority of our own people;" and (4) a consensus reached by "our own men of learning and authority" (Maududi 1969, pp. 104-105). Here Maududi reaffirms the tenets of Islamic law accepted by the majority of jurists in various schools of jurisprudence, except placing "our own men of learning and authority" at par with the ulema. This implies that jurisprudential determinations are not in the monopoly of the ulema. The category ulema (or ulama) is, however, a vague concept. It is a plural noun meaning "the learned," as a distinction from lay persons. But it also means a cleric and it is in this context that the word is generally used. However, a cleric may be the lowly mullah of the village officiating at weddings and burials, or a grand mufti or ayatollah with the authority to make jurisprudential judgments and with millions of followers. On the other hand, anyone who goes through the training of Islamic jurisprudence (fiqh) can attain the authority of a jurist, whether he (only men qualify to be jurists) is a member of clergy or not. However, generally, only the clergy go through the training necessary for becoming a jurist, and they generally reject those outside of their rank attempting to break into the circle. Maududi himself is an example, which may explain some of the antagonisms between Maududi and the Muslim clergy in India and Pakistan. But aside from this issue, there is little difference between Maududi and the Islamic orthodoxy on any matter of jurisprudential significance.

According to Maududi (1969, p. 48), the Shari'a is "a complete scheme of life and an all-embracing social order where nothing is superfluous, and nothing is lacking." This follows the presumption that it is an eternal and universal system of law, which may be applied with some minor variations in different circumstances, but can never be modified in terms of its principle. Moreover, as Maududi (1969, pp. 48-49) asserts, "the scheme of life envisaged by the [Shari'a]" constitutes "an organic whole." Thus, no judgment can be passed on any parts of this system before the entire scheme of Islamic life has been established. He attributes many of the "misunderstandings" about the Shari'a to the "faulty attitude" of judging its parts separately (Maududi 1969, p. 49). As an example, he cites the Islamic penalty of amputating the hand of a thief. He argues that "this injunction is meant to be promulgated in a full-fledged Islamic society," where the Islamic tax of *zakat* is paid, everyone is provided with "equal privileges and opportunities" and people are "God fearing," love their neighbors and provide for the needy, and so forth. This injunction is not for the present society where, Maududi says, "the economic system only leads to enrichment of the few at the cost of crushing poverty and intolerable misery of the many," and where the government promotes "injustice, class-privileges and distressing economic disparities." He then even goes so far as to say that in these circumstances "it is doubtful if theft should be penalized at all, not to speak of cutting of the thief's hand" (Maududi 1969, p. 50).

This is the patented response in the revivalists' debate with their secular opposition. A similar point is generally made about stoning to death for adultery, and Maududi finds himself compelled to address that issue, too, on the same pages. On these issues, and also on the issue of women's rights, Muslim reformers, like Shariati, find themselves in serious difficulty. They do not believe that such injunctions are viable in the modern circumstances, however, many of these injunctions are either explicitly stated in the Quran or are established as Muhammad's Tradition, and therefore, their negation (*naskh*) is not permissible. In this situation, Muslim reformists attempt to present a positive interpretation of these injunctions, and to view them similar to Islam's injunction about trade in, and treatment of, slaves, an issue no longer relevant. But Maududi, true to his revivalist position, does not question the viability of these injunctions. Instead, he suggests that such injunctions may not be carried out before the complete Islamization of society. This position, however, raises a theoretically relevant, and a politically crucial issue in the newly established Islamic states. That is, at what point may the society be regarded Islamic, when the various injunctions of Islam can be carried out? Or rather, at what pace and order should the Islamization of society take place? Clearly Islamization would be a process, and society may approach its utopian ideal only, and at best, asymptotically. Maududi, too, believes that Islamic law may be introduced gradually, as

Muhammad himself did not implement it all at once (S.R. Ahmad 1976, p. 110). It is, however, a curious fact that the implementation of the penal code of Islam, with flogging, amputation and stoning to death, and coercive imposition of the veil on women have been among the first acts of Islamization in the newly formed revivalist states. Sudan, Pakistan, Iran, and Afghanistan are the most vivid examples.

Maududi has been quite explicit on the economic features of an Islamic society. In an address delivered in 1941 at the Muslim University in Aligarh, he states that the economic problems of human society begin when "natural selfishness of man exceeds the limits of moderation,...with the aid of certain other immoral habits and...further support from an inherently defective political system" (Maududi 1978, p. 13). Then, Maududi proposes his Islamic solution by delineating the line of demarcation of the Islamic economic system with those of capitalism and socialism. To him, the "uncontrolled economy of capitalism," which allows individuals to pursue their gains in any ways that they wish, gives rise to evils of capitalism, such as the unemployment of "hundreds of thousands of workers" and in the continuation of "the curse of artificial soaring prices, and regularly planned scarcities of commodities," or, in short, in the persistence of the "disease of Trade Cycle" (Maududi 1977, pp. 49-50, 55). The excessive accumulation of wealth by the rich "would injure" the society, as they neglect the deprived individuals and engage in "artificial and self-created requirements of their self-indulgence;" as they get engaged in adultery, which necessitates "an army of prostitutes" and "another army of...musicians, dancing girls, drum beaters" and many other wasteful uses of society's resources by the "pleasure-loving" wealthy people (Maududi 1978, pp. 15-16). Maududi asserts that it is true that socialism may eliminate some of the evils of capitalism "but at the cost of huge losses of life and property and open rebellion against morals and religion," and will also "suppress individual freedom totally" (Maududi 1977, p. 55).

While Maududi tries to distinguish Islamic order from capitalism and socialism by criticizing both, his opposition to socialism and his affinity with capitalism are fundamental. In a treatise on *Interest*, written in the 1950s, Maududi articulates his epistemological and ideological commitment to individualism. He states as one of the fundamentals of the "Islamic Solution:"

> ...Islamic point of view attaches real importance to the individual and not to any party, nation or society...No party, nation or society in its entirety is accountable before God, rather every individual singly and in his personal capacity is answerable to Him for his deeds....Collective life here does not necessarily mean the collective prosperity but individual welfare....[T]he criterion of goodness...of a social order is the degree to which it can help individuals prosper and develop their potentialities....Islam does not approve of any form of social organization or any measures aimed at Collective welfare (Maududi 1977, p. 55).

Maududi points out explicitly that economic freedom is the central tenet of the Islamic order, which converts the "oppressive" and "uncontrolled economy" of capitalism into a "free economy" in accordance with the limits Islam defines for such freedom. The most controversial issue in the arena of economics is the limits on property rights. He states that

> Islam recognizes, like other proprietary rights, the right of individual ownership of land....And there is no limit placed on such ownership; it may be thousands of acres. If it has come in his possession legally, it is his lawful property (Maududi 1977, p. 61).

Maududi's acceptance of unlimited private property rights is in striking contrast to the view expressed by Shariati and the Mojahedin in Iran some years later. But more significant is Maududi's acceptance of unlimited land ownership in the Zamindari system (landlordism) of Pakistan in the 1950s, when many Muslim reformers were promoting a land reform (Hakim 1953; Parwez 1968; Sheikh 1961). But Maududi adds that the problems of "the feudal system" in Pakistan "are neither purely the outcome of the Zamindari system nor call for the drastic measures of abolition of the individual ownership of land, or need such limitations as are being advocated by 'lay-practitioners' of the day." In his view, application of the Islamic law of inheritance and restriction on land allowed to remain fallow would be sufficient to take care of the land problem in Pakistan (Maududi 1977, p. 61, 1994).

Thus, to Maududi ownership of private property is certainly legitimate as long as the subject of ownership has been acquired legitimately. Moreover, by rejecting revolutionary acts he suggests that once an Islamic state is established all the wealth that had been previously acquired would be safe from confiscation, even if the wealth had been acquired "unlawfully." Maududi claims this to have been the policy of the Prophet and the victorious Muslims who took "no cognizance...of the past deeds of once usurers, brothel-keepers and highway robbers" and raised no question about "previous hoardings or that all the unlawfully earned wealth ought to be confiscated" (Maududi 1977, pp. 59-60). Great Islamic jurists, Abu Yusuf in *Kitab al-Kharaj*, Abu Ubaid al-Qasim in *Kitab al-Amwal*, and Ibn al-Qayyim in *Zad al-Muad,* provide proof for the above assertion (Maududi 1994, pp. 135-136). Maududi, however, points out the restrictions that Islam imposes on ownership, even on the lawfully earned property. One may not "spend his wealth on illicit and dissolute pastimes." Drinking of alcoholic beverages, gambling, adultery, making free persons slaves and selling them to the wealthy to "fill their harems with female slaves," are all prohibited. But "if a man desires to utilise his surplus wealth to earn more wealth, he can do so only by practicing a lawful business" (Maududi 1994, p. 112).

Maududi spells out the general organizational principles of an Islamic economy. In such an economy interest on loans, hoarding, speculation and fraud are prohibited, and "as far as possible" industry and commerce "should be open to free competition." Maududi believes that the application of the law of inheritance of Islam (dividing the assets of the deceased among the heirs in specific proportions), imposition of 2.5 percent tax on "annual assets" of "industrialists and businessmen" (*zakat*), 5-10 percent tax on agricultural products (*kharaj*), 20 percent tax on "the wealth obtained from mines and buried treasures" and "of the war spoils," and the charity of "the businessmen, industrial magnets and other traders," and "landlords and cultivators" would take care of extreme inequalities in society (Maududi 1977, pp. 59-60).

The role of the state in society is the guardianship of the Shari'a, to ensure compliance of the members of the community with its restrictions and to facilitate the virtuous activity of the pious. In the economic domain the state would safeguard freedom of economic activities and would guarantee that the principles enumerated above, and those necessary for the Islamic conduct of affairs in any specific circumstances, prevail. Here, Maududi expresses certain suspicion about the activity of the industrial sector. While he condemns the government's control over the economy in the "Nazi fashion," he gives the state the role of "guidance and co-ordination, so that the industries and commerce may not enter wrong channels and there may be complete harmony among various phases of economic life." Even more specifically, he states that no application of labor-replacing technological changes by "industries and other businesses" should be allowed before "thorough stock has been taken as to how many hands it is going to render idle and that ample provisions has been made for them" (Maududi 1977, pp. 62-63). While the state has no responsibility to provide employment for "each and every one of its citizens," the state "has to provide dole" to those who are out of work or are disabled, and to children with no guardians (Maududi 1977, pp. 64-65).

On July 5, 1977 General Zia-ul Haq staged a coup d'etat against Prime Minister Zulfiqar Ali Bhutto and his People's Party of Pakistan. In his program for "Islamic justice," Bhutto's government had nationalized some large enterprises, carried out a land reform, and promoted unionization of workers (Nomani and Rahnema 1994, pp. 121-126). Bhutto's government had come under severe criticism for trying to impose socialism on Pakistan. He also caused an uproar when he confessed that he drank wine (A. Iqbal 1984, pp. 104-105). Immediately following the coup d'etat, Zia imposed martial law. Worker's strikes were broken, some of the striking workers were shot dead; and "thousands of tenants were forcibly evicted" from the land that Bhutto had distributed (A. Iqbal 1984, p. 107). Jama'at-i Islami applauded the coup and was delighted to join Zia's government which committed itself to set up the Islamic system (*Nizam-i Mustafa*). Zia's Islamization effort entailed imposition of zakat and *ushr* (tax on trade), elimination of interest and enforcement of the Islamic penal code (*hudud*). The imposition of zakat and ushr confronted many practical difficulties (Mayer 1986; Clark 1986; Nienhaus this volume), and the elimination of interest was no more than cosmetic changes, renaming of banking practices and application of "legal tricks" (Nomani and Rahnema 1994, pp. 126-129; Nomani this volume).

The enforcement of the Islamic penal code was, however, swift. In 1979 the laws dealing with Offenses Against Property, Offenses of Zina, Offenses of Qazf, and Prohibition were put into effect (A. Iqbal 1984, p. 113). The law on offenses against property provides for amputation of the right hand of the offender for the first offense, the left foot for the second offense, and life imprisonment for the third offense. The offense of zina (extramarital sex) can result in stoning to death or one hundred stripes, and qazf is the penalty for false accusation of zina, in cases of adultery and rape, punishable by eighty stripes. Ironically, women victims of rape can be charged under the qazf provision for not being able to convince the court of their claim of rape, and then being themselves charged and punished for zina. Finally, the Prohibition Ordinance makes drinking intoxicants punishable by thirty stripes and three years in prison (A. Iqbal 1984, p. 113).

All these penalties were enforced in spite of Maududi's earlier statement that the penal codes would not be enforced until the Islamization of society is completed and "injustice, class-privileges and distressing economic disparities" have been rectified. Perhaps, as far as the Jama'at-i Islami is concerned, the Islamization program in Pakistan had progressed sufficiently by 1979. Perhaps the impositions of penal codes for zina and the consumption of intoxicants and promotion of Islamic religious practices are themselves indications of Islamization of society and elimination of injustice, class privileges, and economic disparities. After all, that is how for centuries caliphs, kings and governors and their qadis and muhtasibs proceeded to rule what everyone viewed as an Islamic society, except occasional voices of reformist dissidents who, following Abu Dharr and Husayn, asked for return to the "true Islam" of Muhammad. The stories of Abu Dharr and Husayn have become legends in the liberation movement of the oppressed Muslims, and those who dared follow their examples met the fate of their heroes. The Iranian revolution of 1979 appeared to be the first victory of an Abu Dharrian Islam over the Mu'awiyas and Yazids of the time. Alas, it, too, proved to be a failed utopia.

THE REVOLUTIONARY ISLAM (The Rule of the Oppressed)

The 1979 Iranian revolution claimed to establish the rule of the oppressed. The slogans, songs, posters and the resolutions of the revolution, Islamic or not, were all on the high tempo of social justice and liberation of the masses. The revolution demanded "the guarantee of the rights of workers and peasants to the full benefit from the product of their labor," an end to "the exploitation of man by man," and elimination of "exploitive profiteering and economic domination which will result in accumulation of great wealth, on the one hand and depravation and poverty, on the other" (Khalili 1981).

The revolutionary movement enjoyed powerful popular support and was deeply influenced by the radical agenda of the left, both secular and Islamic (Behdad 1995a). The views of Shariati and the Mojahedin overshadowed those of the conservative clergy. The revolutionary movement soon accepted the slogans demanding elimination of the "system of dependent capitalism." Fighting "dependency," "dependent capitalism," and "dependent capitalists" soon became a part of the revolutionary agenda of nearly all Islamic and non-Islamic groups. Khomeini, who opposed the materialist interpretation and the classless society of Shariati and the Mojahedin, nevertheless asked the seminary students, prior to the revolution, not to denounce those who have such beliefs, because "Islam is for justice" and "for class balance" (Khomeini 1982, vol. I, p. 177). It was after the revolution that the Mojahedin were banished as hypocrites (monafeqin)[22] and Shariati received severe condemnation.[23]

An Islamic Republic was established as the Pahlavi regime fell in the insurrection of February 1979. The various "revolutionary organizations" of the Islamic Republic championed the expropriation movement that accompanied the revolution and was on the top of the agenda of Marxist organizations. Jehad-e Sazandegi (the Construction Crusade) in the rural areas helped, and in some places led, the peasants to confiscate and cultivate agricultural land, and Bonyad-e Maskan (the Housing Foundation) helped the "oppressed" take over vacant apartments or to build in the vacant lots inside and outside the city limits. Within a few months, most of the large manufacturing establishments and all the banks and insurance companies were declared nationalized by the Revolutionary Council. The Revolutionary Islamic Courts also condemned "lackeys of imperialism and of the Pahlavi regime" and those who had "embezzled public funds" as "Corrupt on Earth" and confiscated all of their properties. The Revolutionary Islamic Courts acted swiftly in the Muhammadan spirit in confronting the enemies of Islam and collected a grand treasure of war booty, possibly unparalleled in the history of Islam. It constituted a significant part of the national wealth, including the entire assets of the Pahlavi Foundation (the largest conglomerate in Iran), many factories, farms and a huge collection of highly valuable real estate properties belonging to the pre-revolutionary Iranian oligarchy, at the national and local levels. This large booty of the war of Islam against its "enemies" was placed, with the order of Ayatollah Khomeini, at the custody of the Bonyad-e Mustaz'afan (the Foundation for the Oppressed), which was formed for this purpose, independent from the structure of the Islamic state. Khomeini, in a posture not unlike that of Muhammad in Medina, proclaimed in the early months of the Revolution that "We will deal with those capitalists, whose capital and wealth could not have become so large from legitimate sources."

And "even if we assume that someone has legitimate properties but the Islamic judge or *vali-ye faqih* realizes that an individual having so much will adversely affect the welfare of Muslims, he can expropriate those properties" (Khomeini 1982, vol. III, p. 126). Khomeini had stated in his treatise on *Islamic Government* (1971) that Islam "must not allow masses to remain hungry and deprived while plundering oppressors usurp the sources of wealth and live in opulence" (Khomeini 1981, p. 49). The Abu Dharrian followers of Khomeini, many of whom were the young students of the Shariati and the Mojahedin school of Islam, proceeded to "conquer," with a revolutionary fervor, every fort and fortress of the arrogant (*taghoti*) enemy. Later on, some of these young activists became known as the Students of the Imam's Path and formed the more radical segment of the populist faction of the Islamic Republic.

The Constitution of the Islamic Republic reflects this ebullient spirit. It asserts that "the Iranian Revolution...has been a movement aimed at the triumph of all oppressed and deprived persons over the oppressors," and condemns "concentration and accumulation of wealth and maximization of profit" (*Constitution* 1980:19, 21). The Constitution diminishes the place of private sector in the economy as the residual of that will supplement the state and cooperative sectors (Article 44) and puts the private sector completely out of certain economic activities, most importantly, foreign trade. Although the Constitution recognizes "legitimately acquired" property (Article 47), it sets such a vague set of criteria (Article 49) that any property, especially those acquired prior to the revolution, may be declared illegitimate.

In 1981, a clear conflict surfaced when the newly established Islamic Consultative Assembly (Majlis) attempted to establish the legal structure of the Islamic Republic. The confrontation was between the populist forces gathered around an Abu Dharrian interpretation of Islam, promoting an interventionist state and a widespread social welfare policy, and a conservative alliance of privileged classes, particularly the bazaar merchants, upholding Islam's recognition of the sanctity of private property. The latter alliance had the strong support of a large segment of the high ranking clergy, including all Sources of Imitation (*Maraje' Taqlid*), except Khomeini, who tried to balance the two factions.

With the Mojahedin banished as hypocrites, the radicalized young activists of the Islamic movement adhered to some variations of the Mojahedin's view. Habibollah Peyman's radical interpretation of Islam on property ownership was the most attractive alternative for these Islamic activists.[24] Peyman, in a scholastic interpretation of the Islamic notion of God's ultimate ownership argues that since all natural resources belong to God, everyone has the right to take advantage of them, and, therefore, individuals may not be allowed to possess any more of the fruits of nature than they need (Peyman 1979, p. 32). Furthermore, he argues that since tools are products of advances in human civilization and are an inseparable part of the labor process, and since everyone has the natural and social right to engage in productive work, everyone has the right to own tools of production (Peyman 1979, p. 66) On this basis, Peyman sees capital accumulation as the means for exploitation of workers, which is only possible when many workers do not own their tools of production or do not have access to the natural resources (Peyman 1979, p. 71) Therefore, he argues that only what is received as the result of one's direct labor is legitimate. This leads Peyman to a Sismondian utopia of small family firms, supplemented by the state or cooperative ownership of those industries that have to be large because of the technological necessities of the modern age (Peyman 1979, pp. 137-138, 273)

Many populist leaders of the Islamic Republic adhered to a less radical interpretation of the Islamic social order by Muhammad Baqir Sadr. To Sadr, the contemporary Islamic economic system would be a market economy with mixed ownership, where an interventionist state imposes some limits on economic activities and maintains the Islamic "social balance" (Sadr 1968/1971, vol. I, p. 354). The interventionist activity of the state is legitimized by Sadr on the basis of God's ultimate ownership. That implies that private property rights are subject to fiduciary principles, and individuals can exercise "their" property right as long as they look after society's welfare, as determined by the Shari'a. Since the Islamic state is the guardian of the divine law, it can limit the exercise of private property rights to maintain the social balance required by Islam (Sadr 1968/1978, vol. II, pp. 32-33). Sadr, then, argues that technological changes now necessitate the existence of an interventionist state to maintain the social balance mandated by Islam. This is necessary because, in Sadr's opinion, with modern technology there is no limits on the amount of resources that individuals can bring under their control. Thus, the Islamic social balance may be easily disturbed, and, therefore, the state may have to limit the scale of certain activities that were not limited by Muhammad (Sadr 1968/1978, vol. II, pp. 34-44, 57-58, 209-210).[25]

It is important to note that Murtaza Mutahhari (d. 1979) had apparently reached a similar conclusion toward the end of his life.[26] Mutahhari (1985, 1986), a student and disciple of Khomeini, and one of the principal theoreticians and ideologues of the Islamic Republic, was the major defender of Islamic orthodoxy in opposition to Shariati.[27] Mutahhari relied on the conservative Islamic tradition to refute Shariati's radical interpretation as a dangerous ploy that threatens Islam (Mutahhari 1978, p. 34). In the early 1980s some manuscripts of Mutahhari were published posthumously. In these manuscripts Mutahhari states that "new capitalism is a separate, independent and unprecedented phenomenon and requires separate and independent jurisprudential considerations" (Mutahhari 1983, p. 57). Here, Mutahhari, similar to Sadr, argues that advanced machinery has changed the character of relations of production, and since, as Peyman points out, technology is the product of human civilization, no individual can claim the "surplus" that it creates. Hence, Mutahhari suggests that with modern technology, production is not the outcome of individual's actions and its output must be owned collectively (Mutahhari 1983, pp. 191-192).[28] This was strong support from a reputable source for the proponents of an interventionist state, an issue of intense debate in the Islamic Republic. The publication of Mutahhari caused an uproar in the Seminaries and the bazaar. Since his critics could not challenge the authority of Mutahhari, they questioned the authenticity of the manuscripts. Some claimed they are counterfeits, others, when Mutahhari's handwriting was published, regarded them as Mutahhari's notes to himself and therefore inconsequential. The book was soon banned and taken out of the market.[29]

The orthodoxy on the Islamic social order was presented by a group of seminary teachers (*modarressin*) who were officially charged, as a part of the Cultural Revolution, with the responsibility for Islamization of various academic disciplines (Behdad 1995b). They constituted the Center for Cooperation of Seminaries and Universities (Daftar-e Hamkari-ye Hozeh va Daneshgah) and published their principal views on the economic system of Islam in 1984. The Modarressin reject, unequivocally, the radical views of Shariati and the Mojahedin, the populist-statist interpretations of Peyman and Sadr, and the newly revealed ideas of Mutahhari about private property rights and market activities. The Modarressin, relying on the traditional jurisprudential scholarship and juristic practices, argued that there is no limitation on accumulation of capital and wealth acquired from legitimate economic enterprises (Daftar 1984, pp. 186-187). The domain of legitimacy is limited only by prohibition of riba (interest), production of goods that are forbidden in Islam (mainly intoxicants and pork), speculation, monopoly, fraud, deception and similar practices that are banned by Islam. In this way, the Modarressin's vision of an Islamic economy is similar to Maududi's outline presented some forty years earlier. The Modarressins, however, find the discourse of modern mainstream economics quite useful for justification of private property rights and for promoting limited government intervention in the economy (Behdad 1995b).

The scholarly and intellectual debate on the Islamic economic order was only an extension of the political disputes in the Islamic Republic and the actual skirmishes on the frontiers of property rights in the society. Khomeini's attempt to play a balancing act, by calling for the unity of the whole (*vahdat-e kalameh*] did not resolve the problem, it simply postponed its resolution. To avoid a jurisprudential confrontation between himself and the other Grand Ayatollahs, who opposed the nationalization policies, Khomeini packed the Council of Guardians with proponents of Islamic orthodoxy and let them fight a populist-statist Majlis. But the Council of Guardians kept rejecting the laws passed by the Majlis. Even Khomeini's delegation of the power of the *vali-ye faqih* (the just jurist) for recognition of urgency (*darura, zarorat* in Persian) to the Majlis in October 1981 (which enabled this body to pass laws that may be in opposition to the Primary Rules of Islam) did not settle the issue. Finally, in December 1988 Khomeini established the Council of Expediency to intervene in the conflict between the Majlis and the Council of Guardians. Khomeini died in June 1989 without resolving the conflict between the populist-statist proponents of an Islam social order and the uncompromising front of the Islamic orthodoxy. This conservative front was firmly constituted by the alliance of the Grand Ayatollahs, their religious seminaries, and the propertied class, especially the bazaar merchants, who have had traditional ties with the religious establishment and have served as its concentrated source of financial contributions.

With the death of Khomeini, the Islamic Republic began a process of transformation. Frustrated from the destructive impacts of a prolonged war with Iraq, a sharp glut in the oil market between 1985 and 1989, which wiped out much of the Iranian source of livelihood, and an unyielding orthodox opposition, and faced with an intense economic crisis, the populist-statist faction began abandoning its revolutionary slogans. The search for an ideal Islamic society, the struggle for establishing the rule of the mustaz'afin had come to its dead end. Instead, a pragmatic coalition of conservatives and reformed populists was formed. The experts of the World Bank and the International Monetary Fund were invited to serve as the source of guidance, which they were happy to oblige. They offered to the Islamic Republic their usual remedy, a strong dose of a economic liberalization, which entails de-nationalization and privatization of state enterprises, elimination of many social programs and subsidies, and "rationalization" of the markets, including decontrolling prices and a free floating foreign exchange rate. This policy is a clear repudiation of all the revolutionary claims of the Islamic Republic. That is an issue that has not been easily accepted by those who had formed the popular base of the regime, and as the warriors of Islam, very much in the tradition of Muhammad, had received an enticing share of the booty. They find it hard to see that the sons of Abu Sufyan and Hind are placed in positions of consultation and power. The major proponent of the liberalization policy, President Hashemi Rafsanjani, found himself, similar to Uthman, in a precarious situation and tried to reach the end of his tenure without meeting Uthman's fate. His newly elector successor, President Khatami faces the same dilemma. Nearly two decades after the revolution, and almost a decade since the implementation of the liberalization policy, the Islamic Republic has not been able define its public policy framework (Behdad 1995a). It is clear, however, that the heyday of the revolutionary Islam of Abu Dharr and the mustaz'afin is over in Iran. Pax Islamicus, Islamic orthodoxy, is being established within a social order that exhibits little difference with what existed prior to the revolution, except the coercive imposition of certain Islamic cultural norms and rules of social conduct.

CONCLUSION

Islamic revivalism is a movement for renewal of Islamic social order. It is a utopian struggle to return to a morally upright, politically successful, and economically prosperous golden ages of Islam. The experience of Islamic revivalism in the twentieth century reveals two clearly distinct and completely opposite presentations of Islam. One is the popular image of an Abu Dharrian social order, in which an egalitarian system of social justice would prevail, where there would be neither any "excessive" accumulation of wealth nor any suffering from deprivation. This utopia, this social order of Ali and Husayn, is the heaven on earth for the mass of urban and rural populations of Muslim societies. These are the people who in this century have suffered social and economic dislocation in the face of acute social inequalities.

The story of Abu Dharr's opposition to the luxurious life of Mu'awiya, and Husayn's martyrdom at the hands of Mu'awiya's impious and arrogant son, Yazid, are vivid representations of the contradiction between justice and oppression, "social balance" and disparity. This is the vision of Islam, the rebellious, the idol smasher, the hope of the mustaz'afin (oppressed), the banner of the mobilized masses. This is the Islamic populism that conquered the temples of the wealthy, arrogant and corrupt; and the shrines of polytheism (shirk). It is the Islam that its towhid (monotheism) means not only the one-ness of God, but also one-ness of the umma. It was the collectivism of the band of warriors of Islam, who distributed the booty of the war among their ranks and left some for the newly formed state of Islam only when the Prophet presented them with God's words (Sura Al-Anfal). This is the Islamic populism that has the power of mobilizing the poor and destitute, the unemployed and the low-wage urban workers in Tehran, Istanbul, and Cairo. These followers of Islam are also delighted to see a return to Islamic morality and rules of social conduct, so that corruption, fraud, addiction and prostitution would disappear, so that their sons and daughters could live a decent and moral life, so that they would have a job, a house, a family and social respect. This is the vision that populist theorists of Islamic revivalism present, from Parwez, Hakim, and Sheikh in Pakistan to Shariati, Mojahedin, and Peyman in Iran. The case of Islamization of Iran is of special interest to us because it presents the most radical and the most articulate expression of the Abu Dharrian Islam. Moreover, the Iranian Islamic movement grew out of the 1979 popular revolution, which provided it with a forceful momentum for establishing a new social order.

The experience of Islamization in Iran and Pakistan shows that the Abu Dharrian vision of Islam soon confronts another interpretation of Islam, one that is a vision of order in an Islamic state, a class society, Pax Islamicus. This is the Islam that protects property rights and its muhtasib maintains order in the suq, where a brisk market would offer merchandise of competing entrepreneurs to desirous buyers. The presence of the muhtasib guarantees that contracts are honored, transactions are honest, and weights and measures are accurate. Fourteen centuries of Islamic jurisprudence is the legal means for the conduct of affairs in the class societies of the Islamic world in a long historical span that began during the rule of the Umayyads and extended to those of the Ottoman Empire and the Saudi Kingdom. In this vision, too, adherence to the moral values of Islam, chastity and thriftiness, modesty and veracity are valued. They are valued for the hard working, well disciplined and productive members of a society who do not expect, nor demand, any more than what they work for and the sustenance that God has set aside for them. This vision, whether presented by Maududi or the Modarressin, is a call for returning to the moral values and norms of conduct of the Islamic societies of the past, while supporting the status quo in the social relations of production.

The thundering power of Islamic revivalism is in its populism. It derives its power from mobilizing masses under the green flag of Islam raised to fight injustice, oppression, and exploitation. As vague as these terms are, they are even appealing to Maududi and the Modarressin who subscribe to them to attract the poor and destitute. The conservative vision of Islam that Maududi and the Modarressin propagate corresponds closely to the traditional jurisprudence of Islam. This is the vision of Islam that receives the seal of approval from the mujtahids or muftis who are the guardians of the legal structure developed by generations of great theorists such as Abu Yusuf and Imam Ghazali. This vision is also supported by an alliance of propertied class, particularly its traditional segments, the bazaaris, who find the vision of suq and muhtasib especially appealing since it tends to limit, at least temporarily, the freedom of action of the large modern oligopolies, who have enjoyed, for decades in some Muslim countries, the support of the secular state.

Any attempt toward Islamization of society, whether by a newly established Islamic state, as in Iran, or by a secular state attempting to accommodate some of the demands of a revivalist movement, as in Egypt or Algeria, would provide an arena of contestation between these two visions of Islam. Clearly there are sufficient ambiguities in the Quran and Muhammad's Tradition to inspire the believers to design a wide range of models for the social order. It is also understod that the "true Islam" that would be established in any Muslim society would be most fundamentally a reflection of the social balance existing in that society. It must be recognized, however, that Islamic hierarchy, the guardian of the jurisprudencial tradition, in its alliance with the privileged classes, forms an incredibly strong force in determination of the outcome of the struggle within the revivalist movement. In the absence of a thundering opposition to the Islamic hierarchy and the traditional jurisprudence, there is little to expect from the Islamization efforts in terms of changes in the social order, except those that have to do with appearance. Even Ayatollah Khomeini, riding strong on the waves of a popular revolutionary movement, did not mange to win over this fourteen-century-old fortress. The failure of the Islamic revolutionary movement in Iran, undoubtedly, has weakened the populist dimension of the revivalist movement. This would imply weakening of the social welfare orientation of the revivalist movement and its proposed public policy agenda. The shift in the Islamic movement is toward cultural and symbolic dimensions of Islamization, and incorporation of modern concept and categories into the system of Islamic jurisprudence. The problem would be more in dealing with issues such as segregation of the work and educational space of men and women, organization of an interest-free banking system and a speculative-free stock market, control of foreign satellite transmission of un-Islamic television programs in the national space, and redefining the categories and forms of Islamic taxation. In this change of emphasis, there is little promise for improvement of the lot of the Muslim masses, while limitation of human rights, particularly for women, religious minorities and secularists appears to be the main outcome.

ACKNOWLEDGMENTS

This study was supported by a Faculty Development Grant from Denison University. I would like to express my appreciation for this financial support. I would also like to thank Bahram Tavakolian and Farhad Nomani for their comments and corrections. I am responsible for the opinions expressed and errors remaining.

NOTES

1. From the English translation of *Tajdid-o-Ihya-i-Din* published in Urdu. See Maududi (1972, pp. 17-18).
2. See Hakim (1953, pp. x-xi).
3. See also Rahnema and Behdad (1995, pp. 1-18).
4. The Quran is quoted from Pikthall (nd).
5. As quoted in Cameron (1973, pp. 51-52).
6. Abu Sufyan was a leader of Meccan opposition to Muhammad. He was murdered by Abdallah ibn Unays, who received a prize when he threw Abu Sufyan's head at Muhammad's feet. Hind is known for expressing her hatred for Islam by tearing out and eating the liver of Hamza, Muhammad's uncle, in the battle of Uhud (Rodinson 1971, pp. 189, 295). Umar appointed Mu'awiya to the governorship of Syria in an effort to "associate the Meccans with the cause" (Lewis 1960, p. 59).
7. Hasan died in 669.
8. Aharon Ben Shemesh in his translation of Yahya Ben Adam's *Kitab al-Khraj* (1967, pp.4-6) notes there are references to twenty-one treatises titled *Kitab al-Kharaj* or *Kitab Risala fi al-Kharaj*.
9. He had a number of other books which did not survive: *Kitab Siyar Muluk al-Ajam* (Chronicle of Kings of Persia), *Kitab al-A'in* (on customs of the court) *Kitab al-Taj* (on the life of Anoshirawan Sassanid). Only the last of these titles survived. Others are quoted in segments in the works of others.

10. Yassine Essid (1995, Part II). This is an excellent source on the study of hisba. It also provides references to the original Arabic treatises. See also Khan (1982).

11. On fabrication of hadith see Kamali (1991, pp. 65-68) and on the sources of hadith in general see Siddiqi (1993).

12. On Ibn Taymiya see Islahi (1988), Lewis et al. (1971, pp. 951-955) and Maududi (1972, pp. 63-69).

13. The original source for this statement is Ibn Taymiya (1963, vol. 8, p. 583).

14. Some scholars of Islamic jurisprudence view the closing of the gate of ijtihad as a fiction. See for example Aghnides (1969, p. 124) and Hallaq (1984).

15. Among the four major schools Hanafis, Malikis and Hanbalis have validated and practiced one or these methods, but Shafi'is have refused to accept such juristic practices (Kamali 1991, pp. 245-282).

16. The other possibilities are "need" (*hajiyyat*) and "improvement" (*tahsiniyyat*). Since there is less "urgency" in these cases, it is more difficult to invoke maslaha in dealing with them (Khadduri 1989, p. 739).

17. For example Abduh suggested including modern sciences in the curriculum of al-Azhar (the center of religious learning in Egypt) (Ramadan 1993, p. 153) and Kasravi (1990) condemned most religious ceremonies.

18. For a discussion of the necessity of tajdid by a leading revivalist see Maududi (1972) and for an analysis and review of literature in the Arab world about this topic see Haddad (1986).

19. The pamphlet does not have a date of publication. The "Explanatory and Discursive" introduction of the author is dated 1912. I learned about the existence of this pamphlet from Ansari (1986).

20. They ruled from 1502 to 1736, and made Shi'ism the official religion of the court.

21. For an example see Jazani (1980).

22. Starting in the summer of 1981, many were killed in armed confrontations or were executed in prison; many others fled to exile.

23. Although the extreme popularity of Shariati prevented the Islamic Republic from criticizing him sharply in the mass media, some strong condemnations were issued by the clergy. See, for example, Abolhasani (1983).

24. On property ownership in Islam see Behdad (1989).

25. Sadr's main point is about *taqrir* in Tradition, meaning Muhammad's silence about actions taking place around him with his knowledge. For a discussion of the issue see Sadr (1978, vol. II, pp. 44-49) and Behdad (1994).

26. Mutahhari was assassinated in May 1979 by members of Forqan, a group which claimed to be followers of Ali Shariati.

27. On the debate between Mutahhari and Shariati see Behdad (1994), Nomani and Rahnema (1994), Dabashi (1993), and Abolhasani (1983).

28. For a more detailed discussion of the issue see Behdad (1994).

29. Occasionally parts of the manuscripts have been published in the newspapers or magazines (e.g., *Kayhan Hava'i* in December 1986) supporting the populist faction of the Islamic Republic.

REFERENCES

Abdul-Rauf, M. 1979. *The Islamic Doctrine of Economics and Contemporary Economic Thought; Highlight of a Conference on Theological Inquiry into Capitalism and Socialism*. Washington, DC: American Enterprise Institute for Public Policy.

Abolhasani (Monzer), A. 1983. *Shahid Mutahhari*. Qum: Entesharat-e Eslami.

Abrahamian, E. 1989. *Radical Islam: The Iranian Mojahedin*. London: I.B. Tauris.

Abu Yusuf, Y. 1969. *Kitab al-Kharaj*. Translated, with an introduction and notes, by A.B. Shemesh. Published as *Taxation in Islam, Volume III*. Leiden: E.J. Brill.

Aghnides, N.P. 1969. *Mohammedan Theories of Finance with an Introduction to Mohammedan Law and a Bibliography*. New York: AMS Press.

Ahmad, K. 1982. "Preface." Pp. 5-11 in *Public Duties in Islam: The Institution of the Hisba*, by Ibn Taymiya. Leicester: Islamic Foundation.

Ahmad, S.R. 1976. *Maulana Maududi and the Islamic State*. Lahore: People's Publishing House.

Ahmed, I. 1987. *The Concept of an Islamic State: An Analysis of the Ideological Controversy in Pakistan*. New York: St. Martin's Press.

Al-Azmeh, A. 1993. *Islam and Modernities*. London: Verso.

Arjomand, S.A., ed. 1988. *Authority and Political Culture in Shi'ism*. Albany: State University of New York.

Ansari, K.H. 1986. "Pan-Islam and the Making of the Early Indian Muslim Socialists." *Modern Asian Studies* 20(July): 509-537.

Behdad, S. 1989. "Property Rights in Contemporary Islamic Economic Thought: A Critical Perspective." In *Review of Social Economy* XLVII(Summer): 185-211.

________. 1992. "Islamic Economics: A Utopian-Scholastic-Neoclassical-Keynesian Synthesis!" Pp. 221-232 in *Research in the History of Economic Thought and Methodology*, Vol. 9, edited by W. Samuels. Greenwich, CT: JAI Press.

________. 1994. "A Disputed Utopia: Islamic Economics in Revolutionary Iran." *Comparative Studies in Society and History* 36(October): 775-813.

________. 1995a. "The Post Revolutionary Economic Crisis." In *Iran After the Revolution: Crisis of an Islamic State*, edited by S. Rahnema and S. Behdad. London: I.B. Tauris.

________. 1995b. "Islamization of Economics in Iranian Universities." *International Journal of Middle East Studies* 27(May): 193-217.

________. 1997. "Islamic Utopia in Pre-Revolutionary Iran: Navvab Safavi and the Fada'ian-e Eslam." *Middle Eastern Studies* 33(January): 40-65.

Bosworth, C.E. et al., eds. 1989. *The Encyclopedia of Islam, New Edition, Vol. VI*. Leiden: E.J. Brill.

Cameron, A. 1973. *Abu Dharr al-Ghifari: An Examination of His Image in the Hagiography of Islam*. London: Royal Asiatic Society/Liuzac.

Clark, G. 1986. "Pakistan's Zakat and Ushr as a Welfare System." In *Islamic Reassertion in Pakistan: The Application of Islamic Law in a Modern State*, edited by A.M. Weiss. Syracuse, NY: Syracuse University Press.

Constitution of the Islamic Republic of Iran. 1980. Translated from Persian by H. Algar. Berkeley, CA: Mizan Press.

Dabashi, H. 1993. *The Theology of Discontent: The Ideological Foundation of the Islamic Revolution in Iran.* New York: New York University Press.

Daftar-e Hamkari-ye Hoze va Daneshgah. 1984. *Daramadi bar Eqtesad-e Eslami.* n.p.: Salman-e Farsi.

Donner, F. McG. 1981. *The Early Islamic Conquests.* Princeton, NJ: Princeton University Press.

Essid, Y. 1995. *A Critique of the Origins of Islamic Economic Thought.* Leiden: E.J. Brill.

Gibb, H.A.R. et al., eds. 1960. *The Encyclopedia of Islam, New Edition, Vol. I.* Leiden: E.J. Brill.

Gordon-Polonskaya, L.R. 1971. "Ideology of Muslim Nationalism." In *Iqbal: Poet-Philosopher of Pakistan*, edited by H. Malik. New York: Columbia University Press.

Haddad, Y.Y. 1986. "Muslim Revivalist Thought in the Arab World: An Overview." *The Muslim World* LXXVI(July/October): 143-167.

Hakim, K.A. 1953. *Islam and Communism.* Lahore: Institute of Islamic Culture.

Halaq, W.B. 1984. "Was the Gate of Ijtihad Closed?" *International Journal of Middle East Studies* 16(March): 3-41.

Hassan, R. 1971. "The Development of Political Philosophy." In *Iqbal: Poet-Philosopher of Pakistan*, edited by H. Malik. New York: Columbia University Press.

Ibn al Muqaffa', R. 1960. *Al-Adab al-Saghir wa l-Adab al-Kabir.* Beirut.

Ibn Taymiya. 1982. *Public Duties in Islam: The Institution of the Hisba.* Translated by M. Holland. Leicester: Islamic Foundation.

________. 1963. *Majmu' Fatwa Shaik al-Islam.* Riyad: Matabi' al-Riad.

Ibn Umar, A.Z. 1975. *Ahkam al-Suq*, edited by Abdelwahab. Tunis: STD.

Iqbal, A. 1984. *Islamisation of Pakistan.* Delhi: Idarah-i Adabiyat-i Delli.

Iqbal, M. 1955. *Poems from Iqbal.* Translated by V.G. Kiernan. London: John Murray.

Islahi, A.A. 1988. *Economic Concepts of Ibn Taimiyah.* Leicester: Islamic Foundation.

IWW (Industrial Workers of the World). 1976. *Songs of the Workers.* Chicago: IWW.

Jazani, B. 1980. *Capitalism and Revolution in Iran.* London: Zed Press.

Kamali, M.H. 1991. *Principles of Islamic Jurisprudence.* Cambridge: The Islamic Text Society.

________. 1996. "Fiqh and Adaptation to Social Reality." *The Muslim World* LXXXVI(January): 62-84.

Kasravi, A. 1990. *On Islam and Shi'ism.* Translated from Persian by M.R. Ghanoonparvar. Costa Mesa, CA: Mazda.

Keddie, N.R. 1994. "Sayyid Jamal al-Din 'al-Afghani'." Pp. 11-29 in *Pioneers of Islamic Revival*, edited by A. Rahnema. London: Zed Books.

Kerr, M.H. 1966. *Islamic Reform: The Political and Legal Theories of Muhammad Abduh and Rashid Riza.* Berkeley: University of California Press.

Khadduri, M. 1989. "Maslaha." Pp. 738-740 in *The Encyclopedia of Islam*, New Edition, Vol. VI, edited by C.E. Bosworth et al. Leiden: E.J. Brill.

Khalili, A. 1981. *Gam be Gam ba Enqelab.* Tehran: Soroush.

Khan, M.A. 1982. "Appendix: Al-Hisba and the Islamic Economy." Pp. 135-150 in *Public Duties in Islam: The Institution of the Hisba*, by Ibn Taymiya. Leicester: Islamic Foundation.

Khomeini, R. 1971. *Hokomat-e Eslami.* np.: n. pub. [An English translation of this treaties appears as "Islamic Government" in *Islam and Revolution: Writings and Declarations of Imam Khomeini*, translated from Persian and annotated by H. Algar. Berkeley, CA: Mizan Press, 1981.]

________. 1982. *Payam-e Enqelab: Majmo'eh-ye Payamha va Bayanat-e Emam Khomeini*, Vol. I-IV. Tehran: Payam-e Azadi.

Kidwai, S.M.H. 1912. *Islam & Socialism.* London: Luzac & Co.

Lewis, B. 1960. *The Arabs in History.* New York: Harper and Row.

________. 1974. *Islam from the Prophet Muhammad to the Capture of Constantinople, Volume II: Religion and Society*, edited and translated by B. Lewis. New York: Harper & Row.

Lewis, B. et al. 1971. *The Encyclopedia of Islam, New Edition,* Vol. III. Leiden: E.J. Brill.

Malik, H., ed. 1971a. *Iqbal: Poet-Philosopher of Pakistan.* New York: Columbia University Press.

________. 1971b. "The Spirit of Capitalism and Pakistani Islam." In *Contributions to Asian Studies* (The Canadian Association for South Asian Studies). Leiden: E.J. Brill.

Marty, M.E. and R.S. Appleby, eds. 1993. *Fundamentalism and the State, The Fundamentalism Project, Vol. 3.* Chicago: The University of Chicago Press.

Maududi, A.A. 1969. *The Islamic Law and Constitution.* Translated and edited by K. Ahmad. Lahore: Islamic Publications Ltd.

________. 1972. *A Short History of the Revivalist Movement in Islam.* Translated by al-Ashari. Lahore: Islamic Publications.

________. 1977. *Capitalism, Socialism and Islam.* Translated by S.A. Khan. Kuwait: Islamic Book Publishers.

________. 1978. *The Economic Problem of Man and its Islamic Solution.* Lahore: Islamic Publications.

________. 1994. *Economic System of Islam.* Edited by K. Ahmad, Translated by R. Husain. Lahore: Islamic Publication.

Mayer, A.E. 1986. "Islamization of Taxation in Pakistan." In *Islamic Reassertion in Pakistan: The Application of Islamic Laws in a Modern State*, edited by A.M. Weiss. Syracuse, NY: Syracuse University Press.

Mojahedin-e Kalq-e Iran. 1972. *Shenakht [Metodolozhi], Ideolozhi, Part 1.* n.p.: Mojahedin-e Kalq-e Iran (September/October).

________. 1980a. *Tabyin-e Jahan [Qava'ed-e va Mafhoom-e Takamol]; Amoozesha-ye Ideolozhik.* n.p.: Entesharat-e Sazman-e Mojahedin-e Khalq-e Iran.

________. 1980b. *Elteqat va Ideolozhiha-ye Elteqati; Majmo'eh-ye Seh Sokhanrani.* n.p.: Anjoman-e Daneshjoyan-e Mosalman-e Kompiyoter.

Momen, M. 1985. *An Introduction to Shi'i Islam.* New Haven, CT: Yale University Press.

Mutahhari, M. 1978. *Elal-e Gerayesh be Maddigari.* Qum: Entesharat-e Sadra.

________. 1983. *Barresi-ye Ejmali-ye Mabani-ye Eqtesad-e Eslami.* n.p.: Entesharat-e Hekmat.

________. 1985. *Fundamentals of Islamic Thought: God, Man and the Universe.* Translated from Persian by R. Campbell. Berkeley, CA: Mizan Press.

———. 1986. *Social and Historical Change: An Islamic Perspective*. Translated from Persian by R. Campbell. Berkeley, CA: Mizan Press.

Nasr, S.V.R. 1994. *The Vanguard of the Islamic Revolution; The Jama'at-i Islami of Pakistan*. Berkeley: University of California Press.

Nizam-al Mulk. 1960. *The Book of Government or, Rules for Kings: The Siyasat-Nama or Siyar al-Muluk*. Translated from Persian by H. Darke. New Haven, CT: Yale University Press.

Nomani, F. and A. Rahnema. 1994. *Islamic Economic Systems*. London: Zed Books.

Paret, R. 1978. "Istihsan and Istislah." Pp. 255-259 in *The Encyclopedia of Islam, New Edition, Vol. IV*, edited by E. van Donzel et al. Leiden: E.J. Brill.

Parwez, G.A. 1968. *Islam: A Challenge to Religion*. Lahore: Idara-e-Tulu-e-Islam.

[Peyman] Paydar, H. Circa 1979. *Bardashtha'i dar Bareh-ye Malekiyyat, Sarmayeh va Kar az Didgah-e Eslam*. n.p.: Daftar-e Nashr-e Eslami.

Pikthall, M.M. n.d. *The Meaning of the Glorious Koran; An Explanatory Translation*. New York: New American Library.

Rahnema, A., ed. 1994. *Pioneers of Islamic Revival*. London: Zed Books.

Rahnema, S. and S. Behdad, eds., 1995. *Iran After the Revolution: Crisis of an Islamic State*. London: I.B. Tauris.

Ramadan, A.A. 1993. "Fundamentalist Influence in Egypt: The Startegy of the Muslim Brotherhood and the Takfir Group." Pp. 152-183 in *Fundamentalism and the State, The Fundamentalism Project, Vol. 3*, edited by M.E. Marty and R.S. Appleby. Chicago: The University of Chicago Press.

Richard, Y. 1988. "Shari'at Sangalaji: A Reformist Theologian of the Rida Shah Period." Pp. 159-177 in *Authority and Political Culture in Shi'ism*, edited by S.A. Arjomand. Albany: State University of New York.

Rodinson, M. 1971 *Mohammed*. Harmondsworth: Penguin.

———. 1981. *Marxism and the Muslim World*. New York: Monthly Review.

Sadr, M.B. 1968. *Iqtisaduna*. Beirut: Dar al-Fikr. [The references in the text are from the Persian translation: *Eqtesad-e Ma*, Vol. I. 1971, translated by M.K. Mosavi-Bojnordi; Vol. II, 1978, translated by A. Espahbodi. Tehran: Entesharat-e Eslami.]

Schacht, J. 1982. *An Introduction to Islamic Law*. Oxford: Clarendon Press.

Shariati, A. n.d. Circa 1978a. "Chegooneh Mandan." In *Majmo'eh-ye Asar*, Vol. 2. Daftar-e Tadvin va Enteshar-e Asar-e Bradar-e Shahid Doktor Ali Shariati dar Oropa.

———. n.d. Circa 1978b. "Khodsazi-ye Enqelabi." In *Majmo'eh-ye Asar*, Vol. 2, Daftar-e Tadvin va Enteshar-e Asar-e Bradar-e Shahid Doktor Ali Shariati dar Oropa.

———. 1979. *On the Sociology of Islam*. Translated from Persian by H. Algar. Berkeley, CA: Mizan Press. [Exerpts from *Eslamshenasi*, reprinted as Shariati, 1981b, 1981c.]

———. 1980a. *Marxism and Other Western Fallacies: An Islamic Critique*. Translated from Persian (*Ensan, Marksism va Eslam*. Qum: n.pub., 1976) by R. Campbell. Berkeley, CA: Mizan Press.

———. 1980b. *From Where Shall We Begin? and The Machine in Captivity of Machinism*. Translated from Persian by F. Marjani. Houston, TX: Free Islamic Literature.

________. 1981a. *Tarikh va Shenakht-e Adyan,* (1) *Majmo'eh-ye Asar*, Vol 14. Tehran: Sherkat-e Sehami-ye Enteshar.

________. 1981b. *Eslamshenasi,* [*1*] *Majmo'eh-ye Asar,* Vol. 16. Tehran: Entesharat-e Qalam.

________. 1981c. *Eslamshenasi,* [*2*] *Majmo'eh-ye Asar,* Vol. 17. Tehran: Entesharat-e Qalam.

________. 1982. "Mazhab Aleyh-e Mazhab." In *Mazhab Aleyh-e Mazhab, Majmo'eh-ye Asar*, Vol. 22. Tehran: Enesharat-e Sabz.

Sheikh, N.A. 1961. *Some Aspects of the Constitution and the Economics of Islam.* Surrey: Working Muslim Mission and Literary Trust.

Siddiqi, M.Z. 1993. *Hadith Literature; Its Origin, Development & Special Features.* Cambridge: The Islamic Text Society.

Spuler, B. 1960. *The Muslim World: A Historical Survey; Part I: The Age of the Caliphs.* Leaden:. Braille.

van Donzel, E. et al., eds. 1978. *The Encyclopedia of Islam, New Edition. Vol. IV.* Leiden: E.J. Brill.

Voll, J.O. 1986. "Rivivalism and Social Transformation in Islamic History." *The Muslim World* LXXVI(July/October): 168-180.

Weiss, A.M., ed. 1986. *Islamic Reassertion in Pakistan: The Application of Islamic Laws in a Modern State.* Syracuse, NY: Syracuse University Press.

Yahya, B.A. 1967. *Kitab al-Kharaj.* Edited and translated, with an introduction and notes, by A.B. Shemesh. Published as *Taxation in Islam, Vol. I.* Leiden: E.J. Brill.

IDEOLOGY AND INSTITUTIONS IN ISLAMIST APPROACHES TO PUBLIC POLICY

S.V.R. Nasr

INTRODUCTION

Much has been written over the past fifteen years about Islamist movements. The principal aim of these studies has been to identify the ideological directives and political imperatives that guide the Islamist challenge to established orders from Morocco to Malaysia. In so doing, the studies to date have focused mainly on the revolutionary and utopian rhetoric of Islamism. Comparatively little attention has been paid to what Islamists promise in terms of public policy, or how their programs may constitute a public policy platform. Public policy here is defined broadly as those policies of the state, or those aspirations of the political opposition, that intend to mold the individual in an intended image; and more specifically, to those policies that are designed to transform society and contend with the needs of its citizenry in general, and address the problems of the poor in particular.

International Review of Comparative Public Policy, Volume 9, pages 43-69.

ISBN: 0-7623-0268-2

The question of Islamists' approach to public policy has become more pertinent as Islamists have occupied positions of power in Iran and the Sudan at the center, and in Afghanistan, Algeria, Turkey, and Malaysia at the municipal and provincial levels. Even where Islamists remain out of power—which is the primary focus of this paper—their social control has made public policy central to their politics. Their greater inclusion in the political process in recent years has only increased this tendency. Islamist strategies in Bangladesh, Egypt, India, Indonesia, Malaysia, Sudan, Turkey, Pakistan, and among the Shi'i of Lebanon bears ample testimony in this regard.

This paper seeks to elucidate the Islamist approaches to public policy: how they have encountered the issue of public policy, which aspects of it have featured in their ideologies, and which have become important to their political strategies as they have deepened their roles in Muslim societies, and finally what institutional arrangements and mechanisms have embodied their public policy choices.

ISLAMIST APPROACHES TO THE STATE, POLITICS, AND THE ECONOMY

Both the state and its Islamist opposition in the Muslim world have been concerned with molding the individual as a prelude to carrying out successful social change. This has been the avowed policy of Kemalist states from Turkey to Pahlavi Iran, and Indonesia to the revolutionary Arab states.[1] To a lesser extent this has also been the case in Malaysia, Pakistan, Bangladesh, and Arab monarchies. The state's concern with music, dress, popular beliefs, and cultural outlook of Muslims has per force made the issue of public policy, and its implications for the social role of religion and the extent of state penetration of society central to politics (Chehabi 1993).

Islamism was from inception nothing short of a rejection of both the reality and intent of the secular state's penetration of society and, as such, it was a clear attempt at undoing the impact of public policy in Muslim societies to date. In this regard Islamism serves as an important source of social resistance to expansion of state capacity—what has elsewhere been termed as the "infrastructural power" of the state (Mann 1994, pp. 189-190)—and as the focal point for the rallying of those social forces that resist state domination (Migdal 1988; Migdal, Kohli, and Shue 1994). This function began in Turkey, Iran, and Egypt in the early part of this century, and since the 1950s it has gained momentum as ulema and Islamist groups across the Muslim world have been at the forefront of opposition to land reform and nationalization of industries and, of course, to state domination of education, commerce, charities, and religious endowments. The scope of Islamist resistance to state power is so broad that it portends to float an alternate paradigm for social organization and political change. In effect, Islamism itself, for all practical purposes, is a grand public policy project, in that it too seeks to mold the individual, and to regulate his or her music, dress, private beliefs, and cultural outlook—all as a prelude to its vision of development—and to change distribution of resources in society. One can therefore conclude that, although Islamism is concerned with the "infrastructural power" of the state, it is not inherently opposed to it; its opposition to that power emanates from the ability of the secular state to use it to further its secularist agenda. In effect, resisting expansion of state capacity is a consequence of Islamism's resistance of state-sponsored secularization, and not to *étatisme* itself.

In essence, the public debate in the Muslim world—between secular and Islamist forces—has focused on ideology alone, and has largely ignored the problematic of the growing size of the state and the reach of its public policies, which is embedded in the programs of both the state and its Islamist opposition. Muslim societies—as have many other Third World societies—have become stuck with the notion of an intrusive state, and the ideal of defining Muslims has become central to all political discourse. The bone of contention is not the state's right to formulate public policy (Ayubi 1995; Norton 1995), but the aim and content of those policies. In this sense, the Islamist project does not purport to some form of liberalism—as it does not engage the problematic of the dominant state—but by sticking to *étatiste* assumptions promises to continue the state-centered authoritarian tendencies of the regimes it challenges.

The Islamist project is, however, not limited to rejecting the general thrust of the secular state's public policy, but is also concerned with the direction and outcomes of economic development strategies to date. Islamist activism across the Muslim world has been animated, to varying degrees, by what the Islamists see as failures of development strategies pursued by the secular state (Ayubi 1991). As a result, Islamism has evolved as a serious critique of existing economic development regimes and their impact on Muslim societies, and as the main advocate of formulating social policy programs to contend with the results of those failures.

Islamism is first and foremost a political ideology. It promises a utopian order that would follow the radical change in the established order. That change is necessitated by the apparent corruption and injustice of the established order, its inherent moral decay, and its negative impact on the Islamic character of society. Islamism as such, is a project for change in which social concerns find historical roots and cultural responses, and solutions that can be found only in the order that Islamism will harbinger. The Islamist perspective is therefore premised on two separate although interrelated discourses: the first concerns the Islamist view of the existing sociopolitical orders and economic systems in Muslim countries; and the second, the ideal order that will replace them. Both are directly concerned with undoing the impact of the modern state on

Muslim life and thought, and seek instead to mold Muslims anew in the image of the teachings of Islam. Both discourses also, directly as well as indirectly, involve questions of institutional change in Muslim societies. It is the contention of this paper that, although the Islamist ideology and plan of action in various social settings do not refer to public policy directly, nor do they openly view it as a central preoccupation, both of their discourses are concerned with public policy concerns. It is for this very reason that public policy has become an important dimension of Islamist strategies, especially as their ideological visions have been put into practice.

As a direct challenge to political authority in secular Muslim states, Islamist forces have challenged ruling regimes across the Muslim world on the political front in armed confrontations, through civil unrest, and at the ballot box. They have also challenged political authority by restricting the extent of the state's social control as they have created spaces in the public sphere where from the state is excluded. In so doing they have presented themselves as the "foil" to state authority, the reality of which undermines state legitimacy and rejects the established order's promise of progress.

The Islamists' ability to restrict the state's social control and to eliminate its influence from certain public domains is perhaps far more threatening to the state than armed confrontations, civil unrest, or electoral competition. For, it weakens state institutions and their control of the society just as it erects a virtual "counter-state." Also important for our purposes is that the competition over social control between the state and the Islamists must happen over public policy, the management of which now becomes the central focus of both protagonists. The greater sophistication of the Islamist challenge has per force made public policy a central aspect of their political agenda.

To understand the Islamists' plans of action one must examine how Islamists interpret existing public policy ideals and programs, and the nature of the decision-making process; what do they put forward in the way of public policy platforms, and what kinds of social programs have they put in practice so far, where and when they have had the ability to do so. This paper will examine these three areas of concern with reference to diverse cases from across the Muslim world with the aim of elucidating the essence of Islamists' conception of public policy.

THE ISLAMIST CRITIQUE OF EXISTING PUBLIC PROGRAMS

Islamists share the vision of ruling elites that states in Muslim countries are teleogical (Kohli 1987, pp. 15-50), namely that their principal aim and function is to transform society—to develop it (Evans 1995, pp. 5-6). Although the Islamists' ideal political order is inspired by early Islamic history, and is defined using Islamic symbolisms and ideals (Mousalli 1992; El-Effendi 1991;

Mohamed 1994; Arjomand 1988, pp. 177-188; Nasr 1996; Enayat 1982), it still strives to merely complete the task of development. The Islamic state, although different in each of its elaborations as it has been discussed in the works of Islamist thinkers from Mawdudi and al-Banna' to Rashid al-Ghannoushi and Abbasi Madani, therefore, is forward-looking, an agent of modernization, just as it accepts and promotes values of earlier times. The notion of development is in fact deeply embedded in the notion of the Islamic state; for the Islamists' views on this subject have been inspired by what they perceive to be the failures of the secular developmental state. The idea of the Islamic state is therefore floated as an alternative to the secular developmental state, and is defined in contradistinction to it. This is clearly apparent in the manner in which the idea of the Islamic state has served as a discursive tool in the struggle against the secular state, most notably, in Iran by Ayatollah Khomeini and in Egypt by Sayyid Qutb and the Takfir groups that have followed his lead.

Beyond their denunciation of the secular nature of modern states, Islamists most clearly challenge the teleological aim of those states, and their records to date. The Islamist perspective questions the gains made by the secular states' developmental agendas, drawing attention to the discrepancies between the elites' promises and the realities that confront the Muslim masses.[2] It is for this reason that justice plays such an important role in questioning the legitimacy of the state as well as in convincing Muslims of the promise of the Islamic state.

It is this challenge to the teleological aim of the secular state that makes Islamism a threat to the established order. For, on par with questioning the religious credentials of the secular state, the Islamist discourse questions the ability of the secular state to achieve veritable development—the secular state's avowed *raison d'être*. Ayatollah Khomeini's success in 1978-79 in Iran owed as much to his ability to depict the Pahlavi state as "irreligious" as to his convincing rejection of its developmental gains and ability to realize veritable development in the future.[3] It is this dimension of Khomeini's discourse that appealed to nationalist sentiments, and attracted those among the intelligentsia, leftist activists, urban professionals, and middle classes who viewed development as the central function and primary duty of the state. Similarly, the Refah party's rapid growth in Turkey owes to its successful questioning of the direction of socioeconomic developments under Kemalism, and more specifically, since the consequences of economic reform initiated by Turgut Özal. In short, the Islamist challenge to the teleological foundations of the secular state serves to make its religious message relevant to a broader spectrum of the population, and it is by so doing that it can pose a serious threat. It may be concluded that in fact the Islamists' ability to question the teleological aim of the state is as important to the success of their discourse than its purely religious dimension.

The Islamist rejection of the secular state—and the assertion that these states are not genuinely developmental—builds on the leftist and nationalist arguments that have laid the foundations of oppositional politics in Muslim countries for most of this century. The Islamists argue that the secular state is merely an extension of the colonial state and serves the aims of imperialism. The staying power of the state is owed to foreign assistance and intervention. The upshot of this line of attack is to suggest that the principal aim of the state is not the genuine uplifting of the Muslim nation, but the continuation of patron-client relationship with the West, plunder of national resources, enrichment of the political elite, and the disenfranchisement, cultural subjugation, and weakening of the nation. The echo of nationalist and leftist arguments are clearly evident here (O'Donnell 1975; Evans 1981). In short, the popular view of public policy-making, on which the Islamist critique is based and which Islamism has in turn helped shape, is premised on the assumption that public policy has been surrendered to international capital, whose aim is to further its interests at the cost of all other national and class interests (Bates and Lien 1985).

From this it follows that the developmental aims and policies of the state are detrimental to national interests. While the Islamists' rhetoric suggests that it is the state's secularism and subservience to the West that incapacitate it, it is in the crises born of development strategies that the Islamists find their evidence. The rise in income inequality, urban and rural poverty, reliance on foreign capital, and inability to defend Muslim interests in war all attest to the ultimate failure of the secular state (Ajami 1981; Ayubi 1991; Nasr 1994, pp. 170-171). In this regard the leftist critique of the capitalist experiments with development have reinforced the Islamist claim and made a wider audience receptive to its arguments. The case of Iran in the 1970s is instructive in this regard (Arjomand 1988, pp. 92-96).

The Islamist ideology operates in the gap between the lofty ideals of the state and the reality of Muslim life. Clive Kessler has found this to be particularly true in the Malaysian state of Kelantan where Islamists have been strongest (Kessler 1978, pp. 220-230). Similarly, in Iran and Turkey Islamists have argued that the cultural compromises of the Pahlavi and the Kemalist states failed to eradicate poverty, and merely compromised the nation's political and cultural integrity without any real gains in return (Arjomand 1988; Abrahamian 1994; Milani 1988; Toprak 1981). Similar arguments have been put forward by Islamists in denouncing Suharto's regime in Indonesia, the National Liberation Front (FLN) years in Algeria, Z.A. Bhutto's populist policies in Pakistan, Sadat's *infitah* in Egypt, and King Hasan II and Habib Bourghiba's developmental agendas in Morocco and Tunisia.

There is little doubt that development strategies across the Muslim world have faced serious problems. They have, more often than not, produced bloated bureaucracies that are incapable of delivering adequate social services but create bottlenecks and red tape; inefficient public sectors; and confused public policies (Richards and Waterbury 1990; Sullivan 1994). Elsewhere, rapid economic change has entrenched urban and rural poverty, encouraged exodus to urban areas, and still failed to bring about veritable economic development (Keddie 1981; Kazemi 1980; Sayeed 1980; Kepel 1985). The concomitant centralization of authority that has followed development initiatives in all these cases has led to opposition. In Pahlavi Iran, and in Ayub Khan and Bhutto's Pakistan, Islamists opposed the expansion of the state capacity and reach even when the objectives of the state were acceptable. In short, state capacity itself became a bone of contention. In Iran, the state's land reform policies of the early 1960s led to Ayatollah Khomeini's first uprising against the Shah. In Pakistan the state take over of industries and land sparked opposition among Islamist forces (Nasr 1994, pp. 147-187). More recently, opposition to the expansion of state capacity and reach led to the mobilization of tribal and Islamist forces against the communist government of Afghanistan which culminated in the decade-long Afghan war (Rubin 1994).

The Islamist ideological perspective from the works of Khomeini in Iran, Sayyid Qutb in Egypt, or Mawlana Mawdudi in Pakistan addressed the consequences of expansion of state capacity (Dabashi 1993; Mousalli 1992; Nasr 1996; Rahnema 1994; Esposito 1991; Sivan 1985). Although their works were often motivated by direct challenges to the autonomy of religious institutions such as endowments, schools, courts, and mosques, in their proclamations they provided a systematic rejection of state domination of and intervention in society. However, the state that featured in the works of these ideologues was not a "generic" state, but the secular state. As a result, although by implication their ideological position challenged the role of the state in Muslim society, more directly and immediately, their attack remained focused on only the type of the state that was in control. Consequently, Islamism never fully developed its critique of the intrusive state—remaining content with merely denouncing its secular variant—and was unable to guard against the domineering and intrusive state creeping into its own utopian vision.

The failure to satisfactorily contend with the problematic of the scope of state authority in Muslim societies is a serious shortcoming of the Islamist discourse. It limits the scope of its dissent, which in turn has direct implications for growth of civil societies, pluralism, and democracy. In addition, the Islamist vision of the state is ultimately molded by the very model that it seeks to reject. It also limits the effectiveness of the Islamist opposition which falls short of addressing fundamental questions of authority, legitimacy, freedom, and state power.

Still the Islamist challenge to the secular state is a serious one. Its scope goes beyond politics to involve the nature of the state's role in public life and in defining the identity, mode of social behavior, cultural setting, and commercial and political transactions of individual Muslims. The Islamist challenge therefore begins with defining a Muslim anew, and proceeds to reform their social life, and ultimately political relations. At each level the

Islamist project both aims to undo the secular state's social control and vision of public policy, and to construct alternate Islamist ones in their place. Public policy issues therefore surface at different levels in the Islamist discourse, as facets of discussions of purifying and spreading Islam, and constructing an Islamic state as well as an important component of Islamist activism in communities across the Muslim world. At each level, the Islamist vision finds manifestation in institutional arrangements, some already present in Muslim communities—but now appropriated or reinterpreted—and some reformulated or newly introduced by Islamists. The role of each will be considered in the remainder of this paper.

INSTITUTIONS: *DA'WAH*, THE ISLAMIC STATE, MOSQUES, AND VOLUNTARY ORGANIZATIONS

It was mentioned earlier that the Islamist ideology and plan of action is by nature a public policy approach. Islamism not only rejects the secular state for its vision of public policy and the outcome of its policies to date, but by seeking to shape the lives of individual Muslims based on preconceived notions of ethics, religiosity, and proper cultural and social behavior, it too has put forward a grand public policy scheme. Moreover, by looking to social and political institutions to implement its vision, Islamism, above and beyond its political implications, operates in the realm of public policy. The institutions of da'wah (call to Islam) or the Islamic state serve as the principle vehicles for expression of Islamist perspectives on public policy. Also important, are those institutions whose origins are not found the ideology of Islamism—and in fact have always been an integral part of Muslim life and piety—but that have been used, or whose functions have been reinterpreted, by Islamists. Most notable, among these are mosques and voluntary organizations. These institutions are less important to shaping individual Muslims, and are instead more central to regulating their daily affairs, and to providing social services. In effect, they are primarily of significance to social policy, and only as such to discussions of public policy.

Da'wah

Da'wah as an institution is directly associated with Islamism. It has its roots in Islamic theological and juridical doctrines; still, its prominence in discussions surrounding the place of Islam in the modern world owes to its central role in the Islamist ideology and discourse, and to the manner in which Islamists have appropriated and even reinterpreted the concept to define their plan of action. Da'wah is a clearly proselytical institution; however, its aim is not to convert the non-Muslim to Islam, but to call the "fallen" or "nominal" Muslims back to their faith. Conceptually, it is an extension of Islam's doctrine of *al-amr bi'l-ma'ruf wa nahy an al-munkar* (enjoining the good and forbidding the reprehensible). In practice, however, it is nothing short of putting forward the Islamist interpretation of Islam, committing Muslims to follow its lead, and instituting it as true Islam.

Da'wah, therefore, purports to redefine Muslims, not only in terms of their piety, but more importantly, in terms of whether or not they follow "true" Islam—which is not orthopraxy per sé, but the Islamist interpretation of it. Insofar as Islamism distinguishes between true faith and nominal Islam,[4] and it seeks to eradicate the latter in order to promote the former, da'wah is a tool of molding individuals in the image of predetermined notions of right and wrong, proper behavior and their lack of, and to commit them to social action as an incumbent and necessary aspect of their piety.

Not all da'wah movements have associated their agendas with politics (Eickelman and Piscatori 1996, pp. 35-36). Some like the South Asian Tablighi Jama'at (Propagation Society) (Ahmad 1991; Metcalf 1994) or the Malaysian Daru'l-Arqam (Abode of Numbers) (Nagata 1984)—which may not even be called Islamist—are concerned only with orthopraxy and shun politics. Others, have reiterated the by now famous Islamist dictum, "there is no separation between religion and politics in Islam," to conclude that orthopraxy can only be fostered if social and ultimately political transactions are brought in line with the Islamist vision. Mawlana Mawdudi, one of the most influential writers on the notion of da'wah was of the belief that so long as governments are hostile to the cause of Islam all efforts to protect and promote Islamic values would be in vein (Nasr 1996, p. 69). The conclusion that he and his followers reached was that successful da'wah begins with taking over the state. As we will see later on, many Islamists came to view the takeover of the state as necessary in that it would provide Islamism with the coercive tools and infrastructural features of the state that would facilitate efficient da'wah. Mawdudi, however, initially justified the takeover of the state as a means of denying control of state institutions to secularist forces, who were likely to oppose da'wah. He was therefore less interested in using the state to promote da'wah than to deny its powers to oppnents of da'wah.

These conceptions of da'wah would surface in the Arab world in the form of the radical Takfir wa'l-hijra (Condemnation and Migration) groups, where the Islamists, claiming to emulate the example of the Prophet in early Islamic history, withdraw from the society as a whole, viewing it as completely un-Islamic, and begin an actual war against the state. Here the concern with the state has clearly dominated the da'wah function of Islamism, but the attitude toward the society at large, no doubt, suggests that the underlying justification remains that of the da'wah. For the withdrawal from the larger society is justified by depicting it as "pagan" (the pre-Islamic *jahiliya*). The relation between the "Islamic" and the "jahiliya" will be as they were under the Prophet in the early years of Islam: the former rejects the latter, inviting it to accept the truth of Islam through da'wah. If this were to fail the da'wah is extended to armed struggle. Jihad (struggle in the path of truth) is thus the logical next step in the Islamists' conception of da'wah, a choice which is sanctioned by the example of early Islamic history, and which is justified as a necessary vehicle for spreading the truth and resisting domination by the "jahiliya."

So central is the notion of da'wah in contemporary Islam that while they may differ over political and doctrinal matters, the apolitical Tablighi Jama'at or the Arqam, and the militant Takfir groups all share in their binary view of society, divided into Islam and un-Islam. Although they disagree over what construes "true" Islam, and what they mean by "un-Islam" or "jahiliya," and what ought to be the relation between "true" Islam and "jahiliya" (Troll 1982), there is no doubt that their religious and social mission is guided by the belief in the existence of such a division. The paradigm of da'wah leads to the consolidation of the notion of a "holy community"—which in South Asia has taken the form of *jama'at* (societies or parties), and in the Arab world the form of urban guerilla cells. The holy community claims to represent "true" Islam and by example, as well as by serving as the only bastion of truth, will spearhead intellectual and social change, and lead the way in redefining the individual. The holy community expects to eventually expand its boundaries to include the whole of society. Since, the aim of the "holy community" runs counter to the public policy vision of the state, it will ultimately find itself at loggerheads with the state.

It is for this reason that from the Arqam in Malaysia, to the Iranian Hujatiyah, to the Takfir groups in Egypt, Islamist groups not only challenge the state's political practices and legitimacy, but seek to engage the state's vision of "new Muslims" suited for the task of development. Hence, while the post-colonial state, following the lead of Kemalist Turkey, sought to change the attire, social mores, and even private habits of Muslims as a prelude for modernization, Islamists seek to change again these very aspects of Muslim lives to prepare them for an Islamic state. The preoccupation with cultural issues is a direct response to the preoccupation of the existing secular states with these very issues. The battle is over control of culture, and the importance of culture to religiopolitical action lies in the fact that the secular state has tied its social and economic agenda to changing Muslim culture.

The Islamic State

The Islamic state has become the most distinctive feature of the Islamist ideological discourse. It is a utopian vision, one which has inspired and justified all religiopolitical activism. Islamist struggles are in all cases directed at creating an Islamic state, without which they have no meaning and aim. The idea of an Islamic state is a modern construction. It is based on Islamic notions of good government and the incumbency of the rule of the Shari'a, and it mobilizes Islamic symbols and historical paradigms—most notably the period of the Rightly Guided Caliphs (632-661). Above and beyond its source of legitimacy and political values, however, the Islamic state is a modern state, whose very idea was conceived of and operationalized in contradistinction and as an alternative to modern conceptions of the state.

Mawlana Mawdudi, one of the first Islamist thinkers to elaborate on the idea of the Islamic state—which has since become the basis for all discussions of the idea—formulated his views on the subject with Western models in mind (Nasr 1996, pp. 80-106). His writings on the Islamic state were surprisingly devoid of any references to the "mirrors of princes" literature, principles of the caliphate, or constitutions of Muslim monarchies of India, which should have served as the basis for his discourse. Mawdudi's Islamic state instead incorporated Western constitutional and administrative ideals, and boasted a modern machinery of government, functioned on the basis of relations of power that have been institutionalized in the modern conceptions of the state; and even replicated procedures, codes of conduct, relations of power, and institutions of authority that comprise Western notions of the state (Nasr 1996; Ahmed 1987; Mujeeb 1967, pp. 402-403). What is true of Mawdudi's model of the Islamic state is also evident in other formulations of the concept from Malaysia to North Africa. Moreover, where and when an Islamic state has emerged, as in Iran after 1979, it has functioned as a modern bureaucratic state with all the trappings of such a state, and with little relation to caliphates and sultanates of earlier eras.

The Islamic state was envisioned as a legal-rational entity, politically controlled by the three branches of the executive, the legislature, and the judiciary, whose powers are constitutionally delimited. The relations of the officeholders to the citizenry is also outlined in the constitution of the state. The leadership will be elected to their offices and will be accountable to the citizenry, although concern with the Islamicity of the state will limit pluralism. In fact, the Islamic state has more in common with the various incarnations of the modern mobilizational state which have surfaced in Europe and the Third World since the 1930s—and which dominated conceptualizations of the state among Muslim intellectuals at the time when Mawdudi or Hasan al-Banna' were writing on the Islamic state—than with the liberal notions of the state that have dominated in the West since World War II. It is for this reason that the Islamic state is a largely state-centered model with very little room for pluralism, civil societies, and individual expression. In the end, it is important to note that in trying to make sense of the Islamic state, it is important to pay equal attention to the Western intellectual roots of the Islamic state as to the role of the Shari'a in its working. The Shari'a, in fact, has little in it in terms of a coherent model of state, and has more to do with the details of laws and procedures. That those laws and procedures are used to conceptualize the Islamic state does not mean that the blueprint for such a state is readily available in the Shari'a, or that it is possible to think of the Islamic state in terms of the Shari'a alone.

While the modernizing impetus of the Islamic state is, in and of itself, an important issue for consideration, the implications that it has for relations between Islamists and existing states is what will concern this paper. The modernizing impetus of the Islamic state, and its attempt to incorporate procedures and apparatuses from modern conceptions of the state, are perhaps more significant in making the Islamic state a "foil" to the developmental state in the Muslim world than its claim to stand for Islamic values. It is in the claim that the Islamic state is not simply a rule of the Shari'a, but an efficient institutional arrangement that can compete with the modern state in every sense, that makes its challenge to the fledgling secular states serious.

Mawdudi, and those who followed his lead in elaborating on the Islamic state adopted, modern ideas in articulating their views in response to suggestions that religion cannot serve as the basis for modern statecraft. Their attempt to make the Islamic state equally as modern as Islamic has made their challenge to the modern state, its public policy platforms and developmental agenda even more potent. In short, while the Islamic state may have been conceived as a religious ideal—"not a form of state so much as a form of Islam" (Smith 1977, p. 213)—in practice it has become an alternative to the developmental state; and as such, it forms the foundation of the Islamist public policy perspective.

The principal aim of the Islamic state as a public policy tool is to "reconvert" Muslims to "true" Islam, and to establish and maintain a sociopolitical order in which they would be able to live according to the Shari'a. The Islamicity of the citizenry is, in turn, the guarantor of the institutionalization and continuity in the working of the Islamic state. In a word, Islamicity of the citizenry and that of the state reinforce and support each other.

The task of "reconverting" Muslims to Islam is one of da'wah. The foundations of the Islamic state are, therefore, to be found in calling Muslims back to their faith, which will eventually culminate in Islamizing their social transactions and political orders. The Islamic state is at the same time itself important to successful da'wah in that Islamization cannot be carried through without neutralizing the resistance of the secular state, and yet having recourse to the authority of the state. Education of the masses in "true" Islam therefore is the harbinger of the Islamic state and, still, cannot be successfully completed without the support of that state.

This has created much debate among Islamists as to which path must be pursued more vigorously: should Islamists seek to educate the masses, as was argued by Mawdudi; or should they focus on taking over the state first, as have the Takfir groups in Egypt. In either case, the Islamists have nevertheless continued to strive in the other path as well. For instance, in Pakistan Mawdudi continued to keep the Jama'at-i Islami involved in politics and to push for an Islamic state while an equal share of his party's efforts were dedicated to "reconverting" Pakistanis. Similarly, in Egypt the Takfir groups have continued to distribute their literature, and to carry out da'wah in tandem with their armed struggle. In many cases, as in Egypt, the presence of several Islamist tendencies, each of which may emphasize one or the other of the aforementioned paths, has had the effect of providing for simultaneous emphasis on both education and political struggle. In Egypt today, groups such as the Muslim Brotherhood tend to be more involved in disseminating Islamist views, whereas the radical Takfir groups are more directly involved in creating an Islamic state.

Islamists tend to view the priority that is given to da'wah or political activism as one of choice of strategy, namely which of the two pillars of successful Islamization—da'wah or the Islamic state—must be pursued most vigorously at first. The confusion arises from the fact that the object of the Islamization endeavor, namely, the Islamic state, is at the same time important to its own realization. As mentioned above, Islamists have sought to combine the pursuit of Islamization of the individual with political activism. What separates various Islamist groups from one another—the "militant" from the "accomodationist"—is the degree of emphasis that is placed on one of the two, da'wah and religious works on the one hand, and political activity on the other. Those Islamists who emphasize education and da'wah, and view involvement in politics as a lesser concern, are none other than the "moderate" Islamists. Obversely, those Islamist groups that see political activism as more important than da'wah tend to be the more political and "militant." Although the issue of the use of violence and attitude toward participation in the political process are also a determining factor in ascribing the titles "moderate" or "militant," the division of labor between education, da'wah, and religious works on the one hand, and politics on the other, is also an important indicator. One could even conceive of a spectrum in Islamism, which is enveloped by the two extremes, the first completely devoted to da'wah, as is the case with the South Asian Tablighi Jama'at or the Malaysian Arqam, and the second completely immersed in political activism, as is true of the Egyptian Takfir groups. The majority of Islamists can be placed in between the two extremes in this spectrum.

The debate over what ought to constitute the primary concern of Islamism has also been an important force in deciding continuity and change within this tendency, and especially within individual Islamist movements. For instance, in 1956-57 a large number of Jama'at-i Islami's leadership left that party in protest of the greater importance that political activity—electioneering at the time—was accorded at the cost of religious work. Again, in 1995 fifteen of the party's most senior members resigned in protest over "excessive" attention that was paid to political activity. It is interesting to note that this latter schism arose after the lectures of a senior Jama'at leader were interrupted by march music that was being prepared for the party's election rallies (*Herald* 1995, pp. 52-53). In post-revolutionary Iran similar tensions developed between the Islamist ulema that have controlled the machinery of the state and the Hujatiyah group—which was generally opposed to the Islamic Republic's role in the economy—over the use of coercion as opposed to education in Islamizing Iranians and their society. For instance, the Hujatiyah opposed the policy of physical eradication of the Baha'is and would have preferred converting them to Islam through da'wah.[5]

Such direct clashes over choice of strategy, and ultimately the meaning and intent of Islamism, have forced Islamist groups to rationalize and reformulate Islamist ideology and plan of action. Islamism is defined in terms of da'wah and political activism, and it is the dilemma of successfully pursuing both goals that serves as the problematic that decides the trajectories along which Islamism is evolving.

The importance of education to the success of Islamism in the political arena has also greatly confused Islamists' understanding of the roots and function of revolution. The first advocates of Islamic revolution, and Mawlana Mawdudi who coined the term in particular, in fact viewed revolution as an evolutionary process in which education rather than class conflict served as the principal agent of change (Nasr 1996, pp. 69-79). In Mawdudi's conception, education would bring Muslims back into the fold of their faith, and once this was accomplished the Islamic state would emerge without the need for a struggle. More to the point, Mawdudi directed Islamizing education at the society's elite—the educated and the political leadership. As such, Islamic revolution was to be propelled by education, and then from the top of the society downward. It was for this reason that Pakistani Islamists, and later Southeast Asian and Egyptian Islamists, organized students—the future bureaucratic and political elite. The Islami Jam'iat-i Tulabah in Pakistan, the Refah party's student cadres in Turkey, ABIM (Angkatan Belia Islam Malaysia; Muslim Youth Movement of Malaysia) in Malaysia, and Jama'at groups in Egypt have their roots in the "revolutionary" use of education in the Islamist conception. The rise of many of Jam'iat and ABIM's alumni to positions of power in Pakistan and Malaysia—and in particular, the rise of Anwar Ibrahim, ABIM's founder, to the top in the ruling UMNO party—can be seen as signs of success of the strategy.

In other cases, Islamist movements in Iran and Egypt developed far more Marxian conceptions of revolution, where revolution was understood as a cataclysmic process that would follow some form of social struggle and lead to overhauling of the social structure. Still, Mawdudi's view of revolution, and the role of education in materializing it has remained a large part of the Islamist view. In particular, Mawdudi tied education to politics and the rise of the Islamic state, and as such made education an important tool of public policy.

He turned the secular state's strategy of changing Muslims as a prelude to development on its head by arguing that Muslims ought to be re-Islamized as a prelude to the Islamization of their public lives.

While education is meant to foster conversion and conformity, the Islamic state—which will benefit from that conversion and conformity—is charged with the task of guaranteeing Islamicity of the society and the state and bringing about genuine development. The Islamic state will use its institutions of power to enforce Islamicity, although Islamist thinkers such as Mawdudi believed that success in education would make the use of coercion unnecessary. Islamists have disagreed over whether or not the Islamic state should rely on force to bring about Islamization. In fact, not all Islamists have supported the "Islamization of the state first" strategies of General Zia ul-Haq in Pakistan or Ayatollah Khomeini in Iran, arguing that without prior Islamization of society, the Islamic state would face insurmountable difficulties that would force the state to resort to coercive measures that would, in turn, doom it to failure. Here too, the debate over the proper degree of attention to religious works and education, politics, and the Islamic state is an important aspect of the evolution of Islamist ideology.

While the Iranian revolution seemed to have resolved the dispute, and even Mawdudi, who advocated pursuing education first or at least in tandem with political activity, defended the Iranian case and later the Zia model in Pakistan, by the 1990s the situation had changed and the debate was once again waging. The apparent failure of the Iranian Islamic state, the unpopularity of the Zia period in Pakistan, and the collapse of similar experiments in Bangladesh and Sudan, created much doubt about the "Islamization of the state first" strategy. Most important, the failure of the Islamic Salvation Front (FIS) in Algeria to topple the Algerian state was proof that without broad popular support the political dimension of the Islamist agenda was unlikely to succeed in capturing power and, even if that was done, it would fail to establish a viable Islamic state. Overreliance on coercion in enforcing Islamization of society fell far short of the lofty ideals of the Islamist thinkers.

In recent years, Islamists have become once again more convinced of the importance of da'wah and education in successful Islamization. The proper balance between education and the Islamic state continues to elude Islamist thinkers; the two, however, remain the most important tools of public policy among Islamists.

Mosques

In contemporary Islamist discourse mosques are not only physical spaces wherein Muslims pray or congregate, but they have an important symbolic role (Gaffney 1994, pp. 13-26). The term mosque has come to symbolize the righteous Islamic order—articulating the Islamist agenda in terms of a highly charged religious institution and sacred space.[6] Islamists often associate construction or preservation of mosques with their political platforms. In Turkey, the call for reconverting the Hagia Sophia to a mosque was among the first demands of the Islamists. Moreover, when the Refah party won the municipal elections in Istanbul in 1994 it promised to construct a new mosque complex at the center of the city so as to restore Istanbul to its forsaken position of prominence in Islam.

At a different level, the Islamist literature is replete with references to the term "mosque" when talking about the Islamic order and Islamization of society and state. On more than one occasion Mawdudi referred to the term "mosque" when talking about the Islamic state; for instance, when he asked: "Will the architects who are well-versed in building bars and cinemas spend their energies in erecting a mosque? If the answer is in the affirmative, it will indeed be a unique experiment of its kind in human history; godlessness fostering godliness to dethrone itself!" (Maudoodi 1955, p. 53). More recently, the Tunisian Islamist leader, Rashid al-Ghannoushi, in his trial before the State Security Court in Tunisia remarked, "If God wishes me to become a martyr of the mosques, then let it be so" (Gaffney 1994, p. 26).

Beyond its symbolism, the mosque has also been important to the Islamist project in more immediate ways. Mosques have long been an area from which the secular state has been excluded. While some Muslim states like Malaysia and Turkey—and to some extent Egypt (Gaffney 1994, p. 50)—were able to gain control of mosques by dominating the education and licensing of preachers, and others like Saudi Arabia were able to exert influence on mosques outside the Kingdom—notably in Europe and the United States—through controlling the funding and appointment of *imams*, more often than not, Muslim countries from Morocco to Pahlavi Iran have had little control over mosques. As a result, mosques have become important arenas for congregation of dissidents and propagation of antiregime ideas. In Pahlavi Iran, and even in Egypt since the mid-1970s, the political value of an independent network of mosques to Islamist oppositional activity has been quite clear (Gaffney 1994; Bakhash 1984).

The importance of mosques to Islamism is in fact more fundamental than its logistical contribution to the struggle against the state. Insofar as the Islamist vision has been floated as a form of rejection of the secular state's intrusion into the various spheres of Muslim life, those arenas from which the state is excluded possess momentous symbolic value in constructing a vision of "counter-state"—an Islamist alternative to the secular state. As such, the network of mosques serve as the nucleus for the creation of state apparatuses, political authority, and social control. They are the first platforms for imagining a sociopolitical order without the secular state, and one that is Islamic in character. Hence, the logistical uses of the mosque in an actual battle against the state only amplify the importance of the mosque to the entire Islamist thinking, and its posturing as an alternative to the secular state.

More immediately, the network of mosques are important in furthering the cause of the Islamic state by enabling Islamist activists to recruit members, disseminate their ideas, weaken support for the secular state, and legitimate resistance to and rejection of the state. In short, they serve as appropriate arenas for conducting da'wah—educating the masses in Islamism—while also mobilizng and organizing them. For Islamists, mosques are not sacred places in that it is there that Muslims pray and perform rituals, but that it is in mosques that the masses are "reconverted" to Islam, become politically active, and are mobilized into well-knit and cohesive political units. As such, mosques attain direct institutional status in the Islamist scheme of events, and become pivotal to the Islamist notions of shaping Muslim identities and political choices. Mosques thus become political institutions, and a tool of implementing policies.

It is for this reason that from Sheikh Omar Abdu'l-Rahman's mosque in Jersey City in the United States to the mosque of Amr ibn As in Cairo to al-Jawad or Hedayat mosques in Tehran in the 1970s, mosques have played a central role in Islamism. It is also important to note that Islamists have often created their own networks of mosques. In Pakistan or Egypt traditional mosques remain under the control of ulema, traditional religious authorities, or the state. Islamism tends to flourish in new, and often smaller, mosques. Here the rise of "holy communities" goes hand-in-hand with the construction of new networks of mosques—some mere *musalla*s (spaces for prayer)—which often operate outside of the traditional networks of established mosques. In Turkey this development was necessitated by the state's control of the network of mosques. The rivalry between the secular state and its Islamist challengers is symbolically captured in the competition between the two over the building of mosques. From King Hasan II of Morocco, to General Zia of Pakistan, King Fahd of Saudi Arabia, the Sultan of Brunei, and Prime Minister Mahathir of Malaysia, political leaders have invested large sums in building lavish state mosques at home and donating mosques to states and communities abroad, efforts which mirror the Islamist-sponsored small scale neighborhood mosques. The symbolism and logistical utility of mosque networks have also encouraged ruling establishments to attempt gaining control of mosques and their preachers. Under Hosni Mubarak, the Egyptian government monitors mosque preaching more carefully, and those whose sermons are viewed as a challenge, such as the blind preacher Sheikh Abdu'l-Hamid Kishk, will be prevented from delivering public sermons.

The cassette and video revolutions in recent years have made mosques somewhat less essential to the successful dissemination of Islamist views; still, Islamism, its conception of "holy community," its rejection of the expansion of state capacity and reach, and its emphasis on da'wah all make mosques central to the Islamist agenda, and its evolving public policy platforms.

Voluntary Organizations

It was mentioned earlier that Islamism, as an ideology of dissent, has objected to the socioeconomic problems that developmentalism has brought about in the Muslim world. In Iran, Ali Shariati or the Ayatollah Khomeini clearly viewed Islamism as an ideology of empowerment for the poor, and styled the Islamic revolution as a means for reversing the balance of power between the social classes. In Malaysia, PAS (Partai Islam Se-Malaysia; Islamic Party of Malaysia) has opposed rapid economic development (Mohamed 1994, pp. 92-93), and the Arqam has advocated the cause of rural development (Anwar 1987, p. 37); and in Turkey, the National Salvation Party—the precursor of the Refah party—criticized industrialization strategies of the government for failing to eradicate poverty (Toprak 1981, pp. 94-95, 98). In the words of Mehmet Ozay, "the roots of Islamic resurgence in Turkey are...in the failure of the nationalist development strategy to deliver social justice" (Ozay 1990, p. 126). The idea of the "just order" (*adil duzen*) has been central to the Refah party's political platform. In 1991 the party's leader, Necmeddin Erbakan wrote a book titled, *Adil Ekonomik Duzen* (The Just Economic Order), in which he declared elimination of economic crises, and development of the relatively backward Southeastern Turkey, to be a priority of his party (*FBIS* 1995, pp. 1-15).

Still, it is not clear whether Islamism as an intellectual trendency has a clear position on social issues, economic justice, and economic development. At times, Islamist parties may espouse opposing views on these themes, confounding the aim of their political posturing. For instance, the National Salvation Party of Turkey, which was critical of the state's record in addressing the problem of poverty, nevertheless firmly supported industrialization, which it believed would give Turkey its due place in history (Toprak 1981, p. 102). The Refah party has been even more openly pro-business. In Iran too, soon after the revolution, the populist and class-oriented rhetoric of the revolution was greatly tempered by those Islamist leaders who favored sanctity of private property, and supported the bazaar and commercial interests against public sector control (Bakhash 1987, 1989; Moaddel 1991).

This issue brings to the fore an aspect of Islamism that is often omitted in the generalizations of its ideology and aim in the Western literature; namely, that when it comes to economic issues there is no one Islamist position (Nomani and Rahnema 1994; Kuran 1993; Nasr 1987). Over social justice and economic concerns Islamist movements can be separated into two groups: the "left," consisting of those who favor populism, nationalization of property, and egaliterianism; and the "right," which includes those who favor protection of property rights, are supportive of commercial interests, and view social justice as an issue of secondary concern to their struggle against the cultural aspects of state policy (Behdad 1989). Factions of Iranian and Egyptian Islamists generally fit the first category, and the Pakistani, Malaysian, and Turkish

Islamists the second. In Iraq, Sayyid Baqir al-Sadr, and in Iran, Ali Shariati, Abu'l-Hasan Bani Sadr, or the Majma'-i Ruhaniyun-i Mubariz (Association of Militant Clergy) capture the essence of the populist and Marxian trends (Katouzian 1983; Behdad 1994). Conversely, the Refah party, Mawdudi, the PAS, or the Iranian Jami'ih-i Ruhaniyat-i Mubariz (Society of Militant Clergy) best represent the more conservative group. Mawdudi, for instane, was opposed to land reform in Pakistan, and throughout the 1960s, when rapid industrialization mobilized leftist forces, he charged the Jama'at to avoid populism and instead to confront the Left (Nasr 1994, pp. 170-187). While in Iran the leftist challenge to the Shah led Islamists to borrow heavily from leftist ideologies, in Pakistan Islamism resisted the lure of populism. In Turkey, the Refah party presents perhaps the most openly economically liberal and pro-business face of Islamism. The party promises that in its ideal Islamic order it will be the business sector and not the government that will control the economy, and the government will merely guard against the rise of monopolies (*FBIS* 1995, pp. 1-15).

From its inception, Islamism was not concerned with issues of social justice, and as has been elucidated above, would not develop a uniform position on the subject. Islamism would not therefore present a vision of social policy, although it was quite articulate in rejecting the secular state's developmental claims, all along pointing to rising income discrepancies and poverty as symptoms of its failure. A number of Islamist groups, such as the Pakistani Jama'at-i Islami or the Malaysian PAS, would remain focused on questioning the state's cultural policies. For instance, although PAS has criticized the Malaysian government's rapid growth strategy and encouragement of foreign investment, the party's 1986 and 1995 election manifestoes have been singularly lacking in social policy initiatives (Mohamed 1994, pp. 192-93). In fact, PAS or the National Salvation Party in Turkey were concerned with state economic policies insofar as they had a bearing on the Islamist cultural perspectives, namely, these parties opposed foreign investment as a sign of national weakness and as a gateway for foreign intrusion into the local cultural sphere (Mohamed 1994, pp. 192-93; Toprak 1981, p. 98).

There is a complex relationship between the social base of Islamism and its views on social policy. Islamism has been generally associated with the lower middle classes—the petite-beorgeoisie (Binder 1988, pp. 328-441; Laroui 1967; Nasr 1994, pp. 81-100; Arjomand 1988; Munson 1988, pp. 98-104; Ayubi 1991, pp. 158-177; Ibrahim 1980, pp. 423-453; Fischer 1982, pp. 101-125; Eickelman and Piscatori 1996, pp. 109-121). The social base has much to do with the Islamists' self-perception and agenda. In Malaysia Islamism represented the interests of the upwardly mobile rural classes against the upper-class background of the leadership of the ruling UMNO party (Kessler 1978, p. 230ff). In Turkey, the Refah party is perhaps unique in recruiting its leadership cadre from outside of the established political elite, and especially from the backward regions of Anatolia. In Pakistan too, the Jama'at-i Islami has served as a vehicle for the empowerment of upwardly mobile lower middle classes who are frustrated with the limits to progress in a feudal society (Nasr 1994, pp. 81-100). Even where society is more permissive of upward mobility, as in Iran of the 1970s, still Islamism represents the declassé elements and those whom modernization has left behind, or has not fully integerated into the modern sectors of the economy. More important, it represented those who had received the necessary education to join the modern economic sector but found elite culture as unreceptive to their Islamic orientation. The lower middle-class base has resulted in the highly text-based orietation of Islamism—for this class has been largely literate, and has been associated with the expansion of what Benedict Anderson has called print capitalism (Anderson 1991). The consequence of the reliance on texts has been to limit the downward spread of Islamism among lower classes that are generally less literate and more steeply immersed in oral traditions.

In recent years, however, Islamism has begun to break out of its lower middle-class base, and to spread downward into the lower classes. In Iran and Egypt where Islamism was more left of center in orientation, and therefore possessed greater social consciousness, this happened more readily. It was also in Iran and Egypt that Islamism first departed from its primary reliance on texts, and through the use of sermons, and more importantly, cassettes began to relay its message more directly to the plebian masses. The impact of this strategy on Iran is well-known, and in the case of Egypt it has given Islamism a base in poorer districts of Cairo, such as Imbaba.

This has in turn forced Islamism to take the problems of the poor more seriously. Where they failed to do so they lost support. For instance, soon after the creation of Pakistan the Jama'at-i Islami found a base of support among the recent migrants from India, mainly because of the social services that the Jama'at performed in refugee camps at a time when the governmernt was unable to address the needs of the migrants. This base of support began to vanish when the Jama'at focused on demand for Islamization alone and failed to address the needs of the migrant settlers in the 1980s. It was with this lesson in mind that the Jama'at made social works the principal means of creating a base of support among Afghan refugees who entered Pakistan following the Soviet invasion of that country, and generally began to invest more in its orphanges, hospitals, schools, youth clubs, and its division of Shu'baha-i Khidmat-i Khalq (Division of Service to the People).

Islamists in Egypt have found that it is not possible to impose effective social control over the poor without advocating their cause; Islam alone is not enough. In a similar vein in Turkey, the Refah party's administrations in Istanbul and Ankara pay lip-service to Islamic idealism, but their power base is guaranteed by subsidizing bread and fuel in poor neighborhoods, maintaining a populist approach to politics, and promising clean, efficient, and open government. In fact, they have done very little in terms of enforcing any Islamic codes of conduct so far (*FBIS* 1994, p. 58). In the occupied territories the political power of Islamists is greatly dependant on their network of clinics and schools, and their long-standing tradition of providing social services to the needy (Roy 1995).

The Islamists realized the importance of social policy initially on university campuses in Malaysia, Pakistan, or Egypt, where for the first time they gained control of a territorially defined population, and were forced to impose social control on it as well as to govern it. It was on these campuses that they understood that their political and religious rhetoric must be complimented with the development of coherent public policy platforms and provision of services. In Pakistan, Islamists have controlled and managed housing on campuses, resolved student problems, helped with administrative and academic issues, and in effect entered into a patronage deal with the student population, according to which they provide services and in turn receive support. The details of this deal and the public policy that manages it are more important in election outcomes on campuses than the cultural purification policies of the Islamists. This fact is evident in countries like Pakistan where various Islamist student groups fight for the control of campuses, and Islamism in itself ceases to be a defining issue in campus elections (Nasr 1992).

In the course of time, therefore, the Islamist rhetoric has gravitated toward populism. The famed Egyptian Islamist slogan and battle cry *al-Islam huwa'l-hall* (Islam is the solution) clearly looks to Islam to resolve socioeconomic problems, poverty, shortages, and unemployment (Sullivan 1994, p. 58). Even in Pakistan where Islamism has always shunned populism, since 1988 it has increasingly advocated the cause of the poor (Nasr 1994, pp. 81-100, 206-218). Islamists have also found out that it is not enough to adopt a populist rhetoric, but to effectively dominate the politics of the poor, to control their neighborhoods, and most importantly to exclude the state from poor neighborhoods, they have to devise and implement social policies—to provide social services that the state has failed to provide, and manage municipal affairs that the state has neglected (Kotob 1995, p. 327). It is this realization that has turned Islamism as a militant challenge to the state into a social force that is rapidly creating an alternate normative order and a shadow state. The defining characteristic of this challenge to state authority and legitimacy is social policy. The new Islamist posture is based on the failure of the state's social policy and the effective provision of Islamist alternatives to it. As a consequence, social policy has become the new battleground between Islamism and the secular state. Islamists may not as yet present a clear view of social policy, but increasingly their religious edicts and cultural programs are being cast in the image of social policy. For instance, Islamist students in Egypt justify the promotion of the use of *hijab* among female students as a way of alleviating the problem of shortage of space by regulating the use of space based on gender (Kepel 1985, p. 129).

The principal vehicle for effecting an Islamist social policy is the "voluntary organization," a form of civil society based non-governmental organization that mobilizes funds from the society in the name of Islamic charitable activity or as *zakat* collection, and invests it in various social services. The individual organizations do not put forward a social or public policy position—they view their work as charitable and voluntary assistance—but put together, the network of these organizations amounts to an alternate civil society-based normative order, and a clear-cut social policy agenda. For, they aim to fill the gaps left by the state, and as such both underscore the failure of the state, and begin to restrict the extent of its social control. The scope of these activities across the Muslim world have been quite significant. Even in Islamist Iran, voluntary organizations such as the Kahrizak charitable project, which provides services to orphans, the disabled, and war veterans, have begun to dominate social services in specific neighborhoods.

However, nowhere is the scope of this form of social activism more extensive, or its implications for state authority and legitimacy more far-reaching than in Egypt, where both the failures of the state, and the investment of Islamists in social services is most noticeable. The Cairo earthquake of 1992 perhaps best elucidates this. Most of the immediate relief after the quake was provided by Islamic as well as Islamist agencies and mosques, whereas the government relief agencies were unable to begin their work in time to have a bearing on the crisis (Sullivan 1994, p. xiii; "Mosques Scoop Up Support in Safety Nets" 1993, p. H/7).

Increasingly, Islamist organizations—and their apolitical Islamic sister institutions—have become the principal providers of social services. Denis Sullivan writes of the growing popularity of Islamic clinics in Egypt in the following manner: "many patients come to an Islamic hospital because....[it] has the concern for the poor that the public sector is supposed to have and has the efficiency and quality that the private sector is supposed to have" (Sullivan 1994, p. 79). Estimates have placed the number of such clinics in Cairo in 1988 at 300 to 350 (Morsy 1988, p. 356). These clinics vary in size; most have 10 or 15 doctors, but perform extensive and sophisticated medical services (Clark 1995, p. 174). They are located close to mosques with which they are affiliated, and receive their funding from the locality that they serve (Clark 1995, p. 175). The clinics therefore serve as community institutions, and encourage support and participation from the community. This has made them social institutions with broad political implications (Clark 1995, pp. 176-179), the most obvious of which is that the citizenary begin to develop initiative, and to assert their rights through institutions over which the state has no control, and which exist to the embarrassment of the state.

The Islamic clinics model exists in the case of schools and other social services as well.These networks then compliment, reinforce, and interact with each other to create a web that only reinforces the notion of an all-encompassing Islamist social policy, and in effect restricts the reach of the state, challenges its public policy platform, and perhaps undermines the state in a far more fundamental fashion than the political challenge of Islamism.

The advent of Islamist social policies also has more immediate political ramifications. In Turkey, for instance, since the 1980s Islamists have formed cells that work in neighborhoods, establishing close ties with individual families. These cells discuss social concerns, and learn about the needs of the families. The result, as was witnessed in the general elections of December 1995 wherein the Islamist Refah party won the plurality, has been to provide Islamists with an effective organizational structure that not only is able to mobilize support at the grass roots level, but has the means to more acurately gauge the demands and aspirations of the voters. Hence, even short of challenging the state, Islamist social policy has strengthened Islamism where and when it has become active in the political process.

CONCLUSION

The foregoing discussion suggests that Islamism never consciously produced a public policy platform. Its discourse, however, was never divorced from it either. The manner in which Islamism has challenged the legitimacy of the secular state, its goal of development, and the increase in its capacity, has per force made public policy the focus of its challenge to the state. In fact, Islamism as an ideological perspective developed largely in the image of the secular state, but in rejection of its agenda. Hence, Islamism too, gave rise to a developmental vision that was premised on reorganizing every aspect of the individual's social existence, not to mention his identity. The Islamist perspective was anchored in a number of institutional arrangements, the da'wah, the Islamic state, the mosque, and the network of voluntary organizations. The latter has pushed the scope of the Islamist public policy position further as it has amounted to a thoroughgoing social policy position, one which more than any other political position or institution of Islamism purports to construct an Islamist "counter-state."

ACKNOWLEDGMENT

I am grateful to the editors for their careful reading of an earlier version of this paper, and for their suggestions.

NOTES

1. Alan Richards and John Waterbury argue that although of different ideologies the cultural outlook and economic function of modern Middle East states is essentially Kemalist (Richards and Waterbury 1990, pp. 187-218).

2. Evans argues that the questions of growth and distribution are ineluctably tied together and impact state legitimacy (Evans 1995, pp. 5-6).

3. It is important to note that decades of leftist opposition to the Shah's regime had ensconced doubts about the promise of his modernization strategy in the minds of Iranians. As a result, the Islamist rejection of the Pahlavi modernization programs merely built on existing criticisms, and tapped directly into the popular beliefs that had been established by the leftist opponents of the regime.

4. This distinction has been evident in the works of Mawlana Mawdudi and Sayyid Qutb, but has most clearly emerged in the works of Ali Shariati, who distinguished between Alavi and Safavid (or red and black) Shi'ism, and of Ayatollah Khomeini who dubbed traditional and established Islam as "American Islam" (Keddie 1981, pp. 218-220; Algar 1981).

5. Interviews with members of Hujatiyah group, Washington, DC, May 1993.

6. Campo provides a thorough picture of the place of sacred space in Islamic piety (Campo 1991).

REFERENCES

Abrahamian, E. 1994. *Khomeinism.* Berkeley: University of California Press.

Ahmad, M. 1991. "Islamic Fundamentalism in South Asia: The Jama'at-i-Islami and the Tablighi Jama'at." Pp. 457-530 in *Fundamentalisms Observed*, edited by M. Marty and R.S. Appelby. Chicago: University of Chicago Press.

Ahmed, I. 1987. *The Concept of An Islamic State: An Analysis of the Ideological Controversy in Pakistan.* New York: St. Martin's Press.

Ajami, F. 1981. *The Arab Predicament: Arab Political Thought and Practice Since 1967.* New York: Cambridge University Press.

Algar, H., ed. 1981. *Islam and Revolution: Writings and Declarations of Imam Khomeini.* Berkeley, CA: Mizan Press.

Anderson, B. 1991. *Imagined Communities: Reflections on the Origin and Spread of Nationalism*, 2nd ed. New York: Verso.

Anwar, Z. 1987. *Islamic Revivalism in Malaysia: Dakwah among the Students.* Petalingjaya: Peladunk Publications.

Arjomand, S.A. 1988. *Turban for the Crown: The Islamic Revolution in Iran.* New York: Oxford University Press.

Ayubi, N. 1991. *Political Islam: Religion and Politics in the Arab World.* New York: Routledge.

———. 1995. *Over-Stating the Arab State: Politics and Society in the Middle East.* London: I.B. Tauris.

Bakhash, S. 1984. "Sermons, Revolutionary Pamphleteering and Mobilization: Iran, 1978." Pp. 174-194 in *From Nationalism to Revolutionary Islam*, edited by S.A. Arjomand. Albany: SUNY Press.

________. 1987. "Islam and Social Justice in Iran." Pp. 95-115 in *Shi'ism: Resistance and Revolution*, edited by M. Kramer. Boulder, CO: Westview Press.

________. 1989. "The Politics of Land, Law, and Social Justice in Iran." *Middle East Journal* 43(Spring): 186-201.

Bates, R.H. and D.-H.D. Lien. 1985. "A Note on Taxation, Development, and Representative Government." *Politics and Society* 14(1): 53-70.

Behdad, S. 1989. "Property Rights in Contemporary Islamic Economic Thought: A Critical Perspective." *Review of Social Economic* 47(Summer): 185-211.

________. 1994. "A Disputed Utopia: Islamic Economics in Revolutionary Iran." *Comparative Studies in Society and History* 36(October): 775-811.

Binder, L. 1988. *Islamic Liberalism: A Critique of Development Ideologies*. Chicago: University of Chicago Press.

Campo, J.R. 1991. *The Other Sides of Paradise: Explorations into the Religious Meanings of Domestic Space in Islam*. Columbia: University of South Carolina Press.

Chehabi, H.E. 1993. "Staging the Emperor's New Clothes: Dress Codes and Nation-Building Under Reza Shah." *Iranian Studies* 26(Summer/Fall): 209-229.

Clark, J.A. 1995. "Democratization and Social Islam: A Case Study of the Islamic Health Clinics in Cairo." Pp. 167-186 in *Political Liberalization and Democratization in the Arab World, Volume 1: Theoretical Perspectives*, edited by R. Brynen, B. Korany, and P. Noble. Boulder, CO: Lynne Rienner.

Dabashi, H. 1993. *Theology of Discontent: The Ideological Foundations of the Islamic Revolution in Iran*. New York: New York University Press.

El-Effendi, A. 1991. *Turabi's Revolution: Islam and Power in Sudan*. London: Grey Seal.

Eickelman, D.F. and J. Piscatori. 1996. *Muslim Politics*. Princeton, NJ: Princeton University Press.

Enayat, H. 1982. *Modern Islamic Political Thought*. Austin: University of Texas Press.

Esposito, J.L. 1991. *Islam and Politics*, 3rd ed. Syracuse, NY: Syracuse University Press.

Evans, P. 1981. "Recent Research on Multinational Corporations." *Annual Review of Sociology* 7: 199-223.

________. 1995. *Embedded Autonomy: States and Industrial Transformation*. Princeton, NJ: Princeton University Press.

Fischer, M.M.J. 1982. "Islam and the Revolt of the Petit Bourgeoisie." *Dædalus* 111(Winter): 101-125.

Foreign Broadcast Information Service: WE. (August 9, 1994).

________. (December 5, 1995).

Gaffney, P. 1994. *The Prophet's Pulpit: Islamic Preaching in Contemporary Egypt*. Berkeley: University of California Press.

Herald. (Karachi) (April 1995).

Ibrahim, S.E. 1980. "Anatomy of Egypt's Militant Islamic Groups: Methodological Note and Preliminary Findings." *International Journal of Middle East Studies* 12(December): 423-453.

Katouzian, H. 1983. "Shi'ism and Islamic Economics." Pp. 145-165 in *Religion and Politics in Iran: Shi'ism from Quietism to Revolution*, edited by N. Keddie. New Haven, CT: Yale University Press.

Kazemi, F. 1980. *Poverty and Revolution in Iran.* New York: New York University Press.

Keddie, N. 1981. *Roots of Revolution.* New Haven, CT: Yale University Press.

Kepel, G. 1985. *Muslim Extremism in Egypt: The Prophet and the Pharaoh.* Berkeley: University of California Press.

Kessler, C. 1978. *Islam and Politics in a Malay State: Kelantan, 1838-1969.* Ithaca, NY: Cornell University Press.

Kohli, A. 1987. *The State and Poverty in India.* Cambridge: Cambridge University Press.

Kotob, S.A. 1995. "The Accommodationists Speak: Goals and Strategies of the Muslim Brotherhood." *International Journal of Middle East Studies* 27(August): 321-339.

Kuran, T. 1993. "The Economic Impact of Islamic Fundamentalism." Pp. 175-196 in *Fundamentalisms and the State*, edited by M. Marty and R.S. Appelby. Chicago: University of Chicago Press.

Laroui, A.. 1967. *L'ideologie arabe contemporaine.* Paris: Françoise Maspero.

Mann, M. 1984. "The Autonomous Power of the State: Its Origins, Mechanisms, and Results." *Archives Europeénnes de Sociologie* 25: 185-213.

Maudoodi, S.A.A. 1955. *Islamic Law and Constitution.* Karachi: Jamaat-e-Islami Publications.

Metcalf, B.D. 1994. "'Remaking Ourselves': Islamic Self-Fashioning in a Global Movement of Spiritual Renewal." Pp. 706-726 in *Accounting for Fundamentalisms*, edited by M. Marty and R.S. Appelby. Chicago: University of Chicago Press.

"Mosques Scoop Up Support in Safety Nets." 1993. *Los Angeles Times*, April 6, p. H/7.

Migdal, J.S. 1988. *Strong Societies and Weak States: State-Society Relations and State Capabilities in the Third World.* Princeton, NJ: Princeton University Press.

Migdal, J.S., A. Kohli, and V. Shue, eds. 1994. *State Power and Social Forces: Domination and Transformation in the Third World.* New York: Cambridge University Press.

Milani, M. 1988. *The Making of Iran's Islamic Revolution: From Monarchy to Islamic Republic.* Boulder, CO: Westview Press.

Moaddel, M. 1991. "Class Struggle in Post-Revolutionary Iran." *International Journal of Middle East Studies* 23(August): 317-343.

Mohamed, A. 1994. *PAS' Platform: Development and Change, 1951-1986.* Petaling Jaya, Malaysia: Gateway.

Morsy, S. 1988. "Islamic Clinics in Egypt: The Cultural Elaboration of Biomedical Hegemony." *Medical Anthropology Quarterly* 2(December): 355-369.

Moussalli, A.S. 1992. *Radical Islamic Fundamentalism: The Ideological and Political Discourse of Sayyid Qutb.* Beirut: The American University of Beirut.

Mujeeb, M. 1967. *The Indian Muslims.* Montreal: McGill University Press.

Munson, H., Jr. 1988. *Islam and Revolution in the Middle East.* New Haven, CT: Yale University Press.

Nagata, J. 1984. *The Reflowering of Malaysian Islam: Religious Radicals and their Roots.* Vancouver: University of British Columbia Press.

Nasr, S.V.R. 1987. "Towards A Philosophy of Islamic Economics." *The Muslim World* 87(July-October): 175-196.

________. 1992. "Students, Islam and Politics: Islami Jami'at-i Tulaba in Pakistan." *Middle East Journal* 46(Winter): 59-76.

________. 1994. *The Vanguard of the Islamic Revolution: The Jama'at-i Islami of Pakistan*. Berkeley: University of California Press.

________. 1996. *Mawdudi and the Making of Islamic Revivalism*. New York: Oxford University Press.

Nomani, F. and A. Rahnema. 1994. *Islamic Economic Systems*. London: Zed Books.

Norton, A.R., ed. 1995. *Civil Society in the Middle East*. New York: E.J. Brill.

O'Donnell, G.A. 1975. *Modernization and Bureaucratic Authoritarianism*. Berkeley: University of California Press.

Ozay, M. 1990. *Islamic Identity and Development: Studies of the Islamic Periphery*. London: Routledge.

Rahnema, A., ed. 1994. *The Pioneers of Islamic Revival*. London: Zed Books.

Richards, A. and J. Waterbury. 1990 *A Political Economy of the Middle East*. Boulder, CO: Westview Press.

Roy, S. 1995. "Beyond Hamas: Islamic Activism in the Gaza Strip." *Harvard Middle Eastern and Islamic Review* 2(Spring): 1-39.

Rubin, B. 1994. "Redistribution and the State in Afghanistan: The Red Revolution Turns Green." Pp. 187-227 in *The Politics of Social Transformation in Afghanistan, Iran, and Pakistan*, edited by M. Weiner and A. Banuazizi. Syracuse, NY: Syracuse University Press.

Sayeed, K.B. 1980. *Politics in Pakistan: The Nature and Direction of Change*. New York: Praeger.

Sivan, E. 1985. *Radical Islam: Medieval Theology and Modern Politics*. New Haven, CT: Yale University Press.

Smith, W.C. 1977. *Islam in Modern History*. Princeton, NJ: Princeton University Press.

Sullivan, D. 1994. *Private Voluntary Organizations in Egypt: Islamic Development, Private Initiative, and State Control*. Gainsville: University Press of Florida.

Toprak, B. 1981 *Islam and Political Development in Turkey*. Leiden: E.J. Brill.

Troll, C.W. 1982. "The Meaning of Din: Recent Views of Three Eminent Indian Ulama." Pp. 168-177 in *Islam in India: Studies and Commentaries*, Vol. 1, edited by C.W. Troll. Delhi: Vikas Publishing House.

Voll, J.O. 1991. "Fundamentalism in the Sunni Arab World: Egypt and the Sudan." Pp. 366-395 in *Fundamentalisms Observed*, edited by M. Marty and R.S. Appelby. Chicago: University of Chicago Press.

ISLAMISM AND ECONOMICS:
POLICY IMPLICATIONS FOR A FREE SOCIETY

Timur Kuran

INTRODUCTION

One of the most visible triumphs of the ongoing global movement known as Islamic fundamentalism, Islamic revivalism, or Islamism[1] has been the spread of Islamic banks. These are banks that claim an Islamic identity on the grounds that their operations are free of interest. They have growing clienteles, and many have proven profitable.

Throughout the Islamic world, the successes of Islamic banking have alarmed many supporters of secularization, modernization, and economic development. One outspoken worrier was Uğur Mumcu (1987, pp. 171-196, 1988, pp. 48-50, 151-153), a widely-read Turkish columnist who was assassinated in January 1993, probably on orders from the Islamic Republic of Iran.[2] Mumcu saw the advent of Islamic banking as part of a sinister ploy to advance Islamic fundamentalism, isolate Muslims from the evolving global civilization, and force Muslim nations into a despotic political union established on Medieval principles.

International Review of Comparative Public Policy, Volume 9, pages 71-102.

ISBN: 0-7623-0268-2

To readers familiar with the actual practices of Islamic banks, Mumcu's fears may seem hysterical. The typical Islamic bank is a commercial enterprise whose choices are sensitive to market pressures. Notwithstanding the utopian claims of Islamist ideologists, its practices differ only cosmetically from those of conventional banks. Where an ordinary bank charges interest openly and unabashedly, the Islamic bank charges a commensurate "commission." A Malaysian observer of Islamic banking puts it more bluntly: "The only difference is whether the man behind the counter is wearing a religious hat or a bow tie" (*Economist*, August 7, 1993, p. 57).

The notion that Islamic banking differs minimally from its conventional counterpart is not inconsistent, however, with the assertion that the spread of Islamic banking is socially disruptive. Even if inherently beneficial to commerce and investment, it may serve a potentially harmful political and cultural mission. Going beyond Islamic banking, this paper offers reasons to be concerned about the spread of Islamist economic teachings and practices. And the paper's ultimate objective is to propose a set of responses to the challenge of Islamism. Following an analysis of the economic effects of Islamism, both actual and potential, I offer three policy prescriptions. The first is to identify and disseminate the flaws in the Islamist economic agenda and the second to show that Islamist leaders tend to overstate their popular support. The last prescription, to which secularists of Mumcu's persuasion might object, is that we should pay close attention to Islamist views on social problems. Many Islamist complaints about modernity, including various socioeconomic complaints, stem from genuine policy failures.

To prescribe public policies one must have a sense of what kind of society one wants to create and maintain. Let me make explicit my own ideals and biases. In my view, the organizing principles of an economy should be individual liberty and limited government. I believe that our material aspirations, not to mention many of our nonmaterial needs, are best served by a free economic order—one that accords individuals broad freedoms to produce, consume, negotiate, and exchange according to their own preferences, expectations, and abilities. In a free economy the state's central economic role is to protect property rights. It may regulate commerce, communication, and innovation only to forestall demonstrated inefficiencies. Though it may provide certain forms of social insurance, it must avoid arbitrary attempts at redistribution.[3]

THE SOCIAL SIGNIFICANCE OF ISLAMIC BANKING

In and of themselves, Islamic banks pose no challenge to a free economic order. As long as they are sustained by private capital, the government gives them no special privileges, and their clients are free to switch to conventional banks, their operations will not violate economic freedoms. Moreover, at least where these conditions are met, the very success of Islamic banking implies that it is meeting an apparent social need; so it should be tolerated, given the full protection of the law, and even welcomed.

It has been argued that, in its ideal form, Islamic banking encourages certain classes of depositors, such as the retired, to carry risks better borne by others. It has been argued, too, that Islamic banking necessitates excessive monitoring of commercial and industrial borrowers.[4] However justified, such criticisms are beside the point. If a septuagenarian chooses to risk her savings by placing them in a variable-return account, third parties have no reason to object—provided they will not be forced to support her in the event her investment turns sour. Likewise, if a private bank chooses to tie the returns on its industrial loans to the profits from the financed projects, third parties need not be concerned—provided, again, they are not obligated to keep the bank solvent.

Every society gives its members wide latitude to make decisions that observers might characterize as wasteful or silly. As a case in point, Americans are free to purchase nutritionally worthless, even harmful, food. Granting Muslims, not to mention non-Muslims, the liberty to "misspend" resources in trying to run their finances in accordance with an alleged Islamic tenet is no different, in principle, from allowing Americans to "misspend" resources on unhealthy food. Even if Islamic banks are inefficient, in itself that is no reason to oppose their right to exist.

The alarmists would agree that the practice of Islamic banking presents no reason for concern;[5] many recognize that a religious hat is no more threatening than a bow tie. They would hasten to add, however, that Islamic banking is motivated by more than the provision of a neglected financial service. A common theme in the alarmist literature is that Islamic banking represents not simply a new financial instrument but also a conspicuous, lucrative, and symbolically potent instrument of Islamism. Mumcu considered Islamic banking an affront to Atatürk's vision of a secular and Westernized Turkish Republic. In the same vein, Islamist Egyptian writers view it as both a source of financial support for Islamist activities and a vehicle for creating financial ties among individual Islamic activists (Ayubi 1991, ch. 8, esp. p. 192).

The link between Islamic banking and Islamism is anything but imaginary. The founders and executives of Islamic banks see themselves as contributing to the re-establishment of the primacy of Islam in the lives of Muslims. Some want the entire social order restructured according to Islamic criteria, not only economic relations but also gender roles, education, the mass media, government, and much else. An appliance manufacturer might launch a new line of refrigerators or enlarge his workforce without any intention of changing the basic framework of the economy, to say nothing of revising the high school curriculum or reordering male-female relations. By contrast, Islamic bankers commonly view themselves both as profit seekers and as agents of social renewal. They do not ask merely to be left alone, to be permitted to withdraw from the economic mainstream. They want to be noticed and to inspire. They wish to make their preferred financial practices the norm. And they consider themselves in the vanguard of a struggle to cleanse the economy, indeed the whole social system, from harmful secular influences.

The vast literature that falls under the rubric of "Islamic economics"[6] confirms that, unlike ordinary firms, Islamic banks are intended to serve a broad social mission. The literature holds that the prevailing capitalist and socialist economies generate injustice, corruption, inequality, poverty, and discontent. A major source of all these problems is interest, which pushes distressed borrowers into poverty, allows lenders to make risk-free gains without exerting any effort, and weakens social ties. By offering an alternative to interest, Islamic banking will contribute, so claims Islamic economics, to making the economy fairer, more egalitarian, and less harmful to the social fabric. But the objective is not simply to make Islamic banking more accessible; it is to make all banking Islamic. Certain campaigns against conventional banking have been successful, at least in the sense that they have made "interest-laden" banking illegal. In 1979, all banks in Pakistan were ordered to purge interest from their operations within five years, and in 1992 the Shari'a court removed various critical exemptions (*Economist*, January 18, 1992, p. 33). Bans on interest have gone into effect also in Iran and the Sudan.[7]

The religious justification for prohibiting interest has been that the Quran bans all its forms, irrespective of the prevailing institutional framework and the nature of the involved loan. Actually, what the Quran bans is riba, an ancient Arabian lending practice whereby defaulters saw their debts grow exponentially. The notion that all interest is equivalent to riba has never enjoyed unanimous acceptance among Islamic authorities, to say nothing of skepticism on the part of laypeople (Rodinson 1966/1973). Nevertheless, Islamic economists[8] treat the issue as settled, and they consider anyone who questions the wisdom or feasibility of prohibiting interest as misguided, corrupt, or ignorant. They feel justified, moreover, in compelling everyone, including Muslims who reject the precepts of Islamic economics, and even non-Muslims, to abide by a ban.

For many proponents of Islamic banking, its religious rationale is far more important than its economic justification. Revealingly, some respond to criticisms of the methods of Islamic banking by saying that its economic merits are secondary to its contribution to the revival of Islam. For them, the overriding objectives are the reassertion of Muslim identity, the reaffirmation of Islam's relevance to the modern world, and the restoration of Islamic authority. Islamic banking defies the separation between economics and religion. It invokes religious authority in a domain that modern civilization has secularized. Moreover, by promoting the distinctness of Islamic economic behavior, it counters the absorption of Islamic civilization into Western civilization. Many leaders of Pakistan's Jama'at-i Islami support Islamic banking as a vehicle for replacing the cautious and limited Islam of Iqbal and Jinnah with one that is more assertive and more intrusive (Ahmed 1994, pp. 688-689).

THE BROADER ISLAMIC SUBECONOMY

"True Muslims," wrote Sayyid Abul-Ala Maududi (1940/1985), the Pakistani thinker who founded Islamic economics in the 1940s, "... merge their personalities and existences into Islam. They subordinate all their roles to the one role of being Muslims. As fathers, sons, husbands or wives, businessmen, landlords, labourers, and employers, they live as Muslims" (translation edited). Following in Maududi's footsteps, contemporary Islamists hold that Islam is a "complete way of life," a source of guidance and inspiration in every personal and social domain. This position obviously clashes with the secularist view that religion is separable from public realms such as science, law, politics, and economics.

To discredit the notion that Islam's relevance is limited to matters of personal faith, Islamists have needed to demonstrate its comprehensiveness as a way of life. Islamic economics emerged, through the writings of Maududi and others, as one component of a broad philosophical attempt to meet this challenge.[9] With the literature growing rapidly, concrete applications were quick to follow. Major efforts to exhibit the practicality, distinctness, and wisdom of Islamic economics got underway in the 1960s, and the world's first Islamic bank went into operation in 1975.[10] Subsequently, Islamic banking spread to more than sixty countries, putting it at the center of the campaign to demonstrate the superiority of Islamic economics over secular economic traditions. The practical demonstrations of Islamic economics, which go far beyond banking, may be grouped in four categories.

In the first category are efforts to give an Islamic identity, mainly through symbolism, to existing economic activities. Islamic banking falls into this category. From a substantive standpoint, the Islamic banks do not operate very differently from their conventional counterparts. In making loans, they adhere mainly to financial criteria; like a conventional bank, they pay little attention to nonfinancial considerations. Contrary to the stipulations of Islamic economics, they tend not to give priority to long-term development projects over projects aimed at quick profits. They do not limit their business to pious Muslims, or even to Muslims. And they show no appreciable bias against projects with heavy social costs, such as high-polluting factories. They differ from conventional banks primarily in appearance. Their operations are formally under the control of a religious council. Their quarters feature various signs of Islamic piety: verses from the Quran on the walls, veiled female and bearded male tellers, prayer rooms, and breaks during times of prayer. Finally, and as already noted, they give and take interest under various religious disguises.

Cosmetically Islamic variants have emerged for other types of businesses, too. There are now Islamic money utilization companies, Islamic retail outlets, Islamic investment companies, even Islamic conglomerates. What makes all such enterprises Islamic, observes Nazih Ayubi (1991, pp. 188-189), is essentially that their owners have taken to "wearing white gowns, growing thick beards, and holding long rosaries." Indeed, the substantive differences between Islamic and ordinary firms are minor. Self-consciously Islamic firms tend not to deal in commodities considered un-Islamic. Thus, an Islamic grocery store does not sell pork or liquor, and an Islamic boutique does not cater to the needs of discotheque patrons. Another difference involves the beneficiaries of the firm's charitable contributions. Where an ordinary firm makes donations to secular causes—an art show, a scholarship program, a sports team—the Islamic firm tends to favor explicitly Islamic causes.

If the employees of Islamic firms were asked what makes their operations Islamic, their responses would include the point that they do not deal in interest. In fact, most Islamic enterprises do so routinely, though usually under Islamic garb. In no essential respect do their business practices differ systematically from their non-Islamic counterparts. Though Islamic enterprises have not been studied adequately,[11] casual observation suggests that they are not exceptionally honest in their dealings with clients unlikely to be repeat buyers. Nor do they appear any less prone than ordinary firms to tax evasion, or especially generous toward their employees. Though Islamic economics is highly critical of the prevailing standards of honesty, trustworthiness, and fairness in contemporary economies (Siddiqi 1972; Naqvi 1981), there is nothing to suggest that established Islamic enterprises behave substantially differently from ordinary enterprises.

The efforts that form our first category of Islamic economic activity do no great harm. In refusing to stock dancing outfits, an Islamic boutique does not keep women who want them from taking their business elsewhere. Islamists also try, however, to prevent economic choices they find objectionable, sometimes by force. Such efforts form our second category. They include attacks on establishments deemed to foster moral laxity: video stores, theaters, bookstores, dance clubs, bars, and hotels that cater to "naked" tourists, and restaurants that serve food during fasting times (Rashwan and Gad 1993). The attacks are aimed at destroying certain economic sectors, and the consequences have hardly been insignificant. In Egypt, attacks against tourists and tourism establishments have lowered foreign exchange earnings, thus harming investment and growth (Murphy 1992). Such attacks serve to differentiate Islamic economic values from non-Islamic ones. They also make it known that the goals of preserving and advancing Islam take precedence over such objectives as individual liberty, financial opportunity, hospitality toward visitors, and economic development.[12]

The economically harmful efforts to establish the primacy of Islam have not been limited to violent attacks against morally controversial sectors. In various countries, it is now common for employees to be pressured to attend communal prayers several times a day. During Ramadan, employees have been forced to fast, often with adverse effects on productivity, especially in years when Ramadan falls in the summer. Communities have been discouraged from spending resources on nonreligious social services, lest these reduce the amount available for the construction of mosques and Quran schools. In one case that has received scholarly attention, Islamists seeking space for a new mosque secured the removal of a Turkish community's only health center, without bothering to find it a new location (Gürsoy-Tezcan 1991). Still another example of giving primacy to religious needs lies in the pressures to make governments devote more resources to religious education. In Turkey between 1976 and 1984, the number of government-run religious highschools (*imam-hatip okulu*) rose from 73 to 374, largely because of Islamist pressures. For a comparison, over the same period the number of secular highschools (*lise*) rose from 927 to 1190 (Reed 1988, pp. 2-4).

The third category of economic instruments for advancing Islamism consists of nongovernmental Islamic social services. Such services include hospitals, clinics, legal-aid societies, schools, dormitories for college students, youth centers, summer camps, and sports facilities. Benefiting primarily the poor, they demonstrate that organizations infused with an Islamic spirit can tackle various social problems more adequately than the government. Also, by giving a religious dimension to traditionally secular services they lend credibility to the claim that Islam is a complete way of life.

In Egypt, mosque-affiliated clinics are delivering health services at rock-bottom prices, generally more efficiently than the overburdened health facilities run by the government. Where a poor Egyptian might wait three months for an X-Ray at a government facility, at an Islamic clinic access to the procedure typically takes much less.[13] For another example, thousands of underprivileged Turkish teenagers have attended Islamic summer camps, their expenses paid by Islamist organizations (*Cumhuriyet*, July 10, 1993, pp. 1, 15, August 7, 1993, pp. 1, 17). These camps put great emphasis on religious instruction. In addition to earning general goodwill, therefore, they serve as Islamist training grounds.

It is costly to run services like camps and hospitals. By contrast, the final category of economic instruments consists of services that bring Islamists net financial gains. The most important such instrument is *zakat*, a religious tax on wealth and income. By tradition, the proceeds from zakat serve partly to deliver relief to the poor and partly to finance "praiseworthy activities," defined to include religious education and pilgrimage. In several countries, including

Pakistan, zakat has become a legal obligation, and the government now organizes both its collection and its disbursement. In most parts of the Islamic world, however, zakat is collected and disbursed in a decentralized manner by local religious organizations (Kuran 1993b, pp. 318-325). Whatever the pattern, substantial shares of the raised funds end up financing Islamist activities. Another source of income for Islamist causes is the hides of sacrificial animals. In Turkey, such hides traditionally went to the Turkish Aviation Society. Now, a large share, in some localities the lion's share, goes to Islamist collection agencies, which have taken to spreading the word that it is sinful to make donations to a secular authority. One estimate puts the Islamist income from a recent year's collection at around US $90 million (*Cumhuriyet*, July 19, 1993, pp. 1, 15).

No one knows the total income of Islamist organizations. What is known is that Islamists derive revenue from many channels, including not only zakat and hides but also donations from Islamic businesses, individuals sympathetic to Islamic causes, and unsympathetic individuals hedging their bets. There is also reason to believe that many Islamist organizations receive funds from foreign governments. The governments of Saudi Arabia, Libya, and Iran, among other countries, deliver much help to Islamist organizations throughout the world.[14] Some of the diffuse income supports Islamic services. Much of the rest is used to support an array of political activities. In Muslim countries that have elections, the candidates of Islamic parties regularly receive financial support from Islamist organizations, including ones whose primary purpose is nonpolitical.

All told, the economic activities of the Islamists hardly amount to a radical reordering of economic relations. They either serve to raise resources for Islamist advancement or to demonstrate that Islamists are capable of solving entrenched social problems. It is true, of course, that Islamists, including ones with socialist leanings, talk of radical change. Yet the existing economic activities associated with Islamism generally do not represent major departures from the economic status quo.

It is essential here to remember that Islamic economics did not originate from a need to reorder economic relations. Rather, it arose out of an impulse to demonstrate the distinctness, continuing social relevance, and priority of Islam. From this perspective, Islamists have been remarkably successful in pursuing their economic objectives. They would have been less successful, perhaps, had they advanced a detailed, coherent, and comprehensive plan for truly revolutionary economic restructuring. This is because a complete and fully explicit plan would have identified many potential losers. As matters stand, the vagueness of Islamic economics allows Islamists to appeal to a broad audience without fueling widespread anxieties about the economic consequences of an Islamist takeover. It also gives Islamists the flexibility to adjust the content of Islamic economics to changing needs and opportunities.

THE DOMINO EFFECT

Successful Islamization in one domain lends credibility to Islamization attempts in other domains. Therefore, a significant consequence of the economic activities undertaken in the name of Islam is the support they give to the broader Islamist agenda. The economic accomplishments of Islamism encourage activists to press new demands. They also discourage resistance from opponents and embolden Islamist movements elsewhere. It follows that the economic advances of Islamism may lead, through a domino effect, to further victories in both economic and noneconomic domains.

When Islamist writers insist, then, that economics, family structure, dress, and education are all inseparable, they do not mean that it is only in the minds of individual believers that these domains are interlinked. They mean also that every move toward Islamization facilitates further moves. The spread of Islamic commerce discourages women from dressing immodestly in public settings. Conversely, as veiling becomes more common, Islamic economics is taken more seriously, and its opponents become less vocal.

In pursuing power, any political movement will rely on two complementary strategies. The first is to make people understand and accept its objectives, the second to pressure individuals of all persuasions, including skeptics and potential opponents, into supporting the objectives. Each of these strategies is apparent in the economic activities of Islamists. A philosophical contribution to Islamic economics serves primarily to persuade. By contrast, a profitable Islamic bank and a well managed clinic serve both to persuade and to build political pressure. On the one hand, they show that Islam can make positive contributions to economic justice and efficiency. On the other, they boost the perceived power of Islamism, thus reinforcing the pressure to support Islamist positions.

Bolstering a movement's perceived strength is not riskless: it can induce a major reaction. The military response to the imminent victory of the Islamic Salvation Front in Algeria's legislative election of 1991-92 illustrates the point (Roberts 1994). Yet, under the right set of contingencies a pro-Islamist turn in public opinion will make fear change sides: people who were reluctant to appear too sympathetic to Islamism will become reluctant to appear too unsympathetic. Thus, a shift in public opinion may weaken the anti-Islamist resistance, make it equivocate, and eventually crush it. Islamist leaders can never be certain about their hidden support or about their opponents' susceptibility to social pressure. Like all political players, they face uncertainty regarding the appropriate timing for action. Nevertheless, they all recognize that their chances of breaking the resistance to their agendas are likely to grow with increases in Islamism's perceived strength.[15]

In some places Islamism in general and Islamic economics in particular already derive critical support from fears of being stigmatized as un- or anti-

Islamic. Though many leaders of Pakistan's People's Party have serious misgivings about Islamic economics, few have gone on record with their reservations. In fact, they generally make a point of claiming that their own economic programs are consistent with, if not derived from, the economic teachings of Islam. Zulfikar Ali Bhutto, who once questioned the feasibility of organizing a modern society according to Islamic principles, began asserting in the mid-1970s that the inspiration for his socialist ideas had come from Islam (Ahmad 1991, pp. 474-479). A decade later, his daughter, Benazir Bhutto, refrained from denouncing the ban on interest, even as the Jama'at-i Islami was conspiring to topple her government. Such episodes are at once a manifestation of, and a contributor to, social pressures to appear supportive of Islamism. All fundamentalist objectives, not simply those associated with attire, gain legitimacy and immunity to criticism when Benazir Bhutto appears in an election poster with her head fully covered and reading the Quran (Ahmed 1994, p. 687). So, too, they all benefit when Pakistani statesmen, bureaucrats, intellectuals, and professionals exhibit reticence to take issue with the claims and practices of Islamic economics.

To sum up thus far, Mumcu's point that Islamic economics is part of a broad mission is unassailable. Islamic economics serves to differentiate the Islamic social order from prevailing social orders, to signal the strength of Islamism, and to make potential opponents practice self-censorship. As such, schemes such as zakat and Islamic banking are both objectives in their own right and instruments of the wider Islamist cause. Policymakers who reject the Islamist agenda are justified, then, in treating Islamic economics as more than a quaint doctrine.

If this point is granted, the obvious next question is, "How should non-fundamentalist policymakers respond to the economic activities undertaken in the name of Islam?" I will offer three suggestions in due course, but only after making two additional points. I will note that Islamism has been grappling, not always unsuccessfully, with some deep social problems. And I will then suggest that Islamism—like other fundamentalisms—carries the potential of harming the global economy.

ISLAMISM AS ECONOMIC INSTRUMENT

Given that the rise of Islamic economics has been driven largely by noneconomic factors, it is not surprising that it has solved no major economic problem. A half-century after the publication of Maududi's seminal writings, there is no evidence that Islamic economics has made Muslim economies more equal, more productive, or more innovative. Nevertheless, for reasons essentially unrelated to the principles of Islamic economics, Islamism has helped alleviate some of the discontents of rapid urbanization. In particular, it has enabled metropolises like Cairo, Istanbul, and Karachi to cope with historically unprecedented demographic changes.

The Islamic world has been urbanizing against the backdrop of rampant government corruption and widespread dishonesty in business relations. Explosive population growth, by making it increasingly difficult to monitor government officials and enforce private contracts, has aggravated both problems.[16] Under the circumstances, businessmen find it difficult to succeed without bribing government officials and deceiving their suppliers and clients. Their moral compromises give rise to guilt, a condition for which Islamism offers an array of cures. Specifically, Islamist activities offer guilt-ridden Muslim businessmen opportunities to cleanse their consciences. By holding his savings in an Islamic account and making a point of shopping at Islamic stores, a businessman can achieve the feeling that he is doing his best to live as a good Muslim. Likewise, by donating money to a mosque and time to a Quran school, he can satisfy himself that, despite the adverse conditions, he is doing his share to improve the moral climate. He can alleviate his guilt also by assuming an Islamic identity for *his own* business.

In view of these observations, it would be simplistic to portray the businessmen who contribute to Islamic causes and participate in the Islamic subeconomy as mere tools of Islamist leaders—although there are undoubtedly Muslim, non-Muslim, and agnostic businessmen who take such actions under real or imagined social pressures. Many of the businessmen who have fueled the growth of the Islamic subeconomy derive important personal benefits from their activities. To them, Islamism is not just a social force to be accommodated. It is also a vehicle for emotional support and a means of fitting into a hostile economic culture.

A related consequence of rampant dishonesty is that it makes it advantageous to join a subeconomy that fosters trustworthiness. Relative to businessmen who must constantly guard against being cheated, ones who can count on timely deliveries and payments incur lower costs of operation. In particular, they spend less time, energy, and money on negotiating, drafting, monitoring, and enforcing agreements (Williamson 1985; Etzioni 1988, ch. 12). Newcomers to the cities of the Islamic world have immediate access to networks built on ties of kinship and regional origin. As a rule, however, these networks provide fewer benefits than those comprised of longtime residents, because members of the former tend to be poorer, less experienced, and politically powerless. The newcomers would like to break into the most lucrative networks. They seldom can, however, if only because they lack the proper education and the requisite social etiquette. The Islamic subeconomy offers a second-best alternative. Within it, they are able to establish business ties with ambitious but culturally handicapped people who, like themselves, are excluded from the economic mainstream. Their shared commitment to Islam, even if partly feigned, provides a basis for mutual trust. Their costs of doing business are thus lower than they are for the typical newcomer who chooses to operate outside the Islamic subeconomy.

The networks of the Islamic subeconomy are not limited to the private sector. Throughout the Islamic world they extend into various levels of government. Islamists elected or appointed to positions of government authority are known to grant financially valuable political favors to those outside government. They make it cheaper for Islamist businessmen to learn about regulations, obtain licenses, avoid tax audits, and pass inspections.

Hence, low standards of honesty fuel Islamism through two channels: by creating a need for guilt relief and by making the economically insecure seek a vehicle for forming networks based on trust. One reason why the Islamic world displays low standards of honesty is that overregulation and misregulation provide officials, high and low, with vast opportunities for earning bribes.[17] And a related reason is that the government has been ineffective at its most essential task, namely, providing the legal foundations of a complex, dynamic, and increasingly impersonal economy. In most parts of the Islamic world, the legal system fails to provide traders low-cost instruments for collecting money owed to them. Traders thus have strong incentives to join private networks for protection against being cheated.[18] Explosive urbanization has compounded the incentives by stretching government services, including legal services, to the breaking point.

In high-population countries like Egypt and Turkey, there was a time when the government could employ every high school graduate in need of a job. Now it cannot even hire every applicant with a university degree. In Cairo, a university graduate may have to wait a decade for a low-paying, entry-level position. At the same time, today's young job seekers have much higher aspirations than ones of earlier generations. Populist ideologies, schooling, and the mass media have all raised their expected living standards way above what they are likely to achieve. Yet, many must wait until they are around thirty to get a stable job, rent a place, and get married. In the meantime they remain sexually and emotionally frustrated, and they feel robbed of honor and dignity.

This brings us to still another way in which contemporary Islamism helps marginalized segments of the urban population cope with frustrations. In promoting a puritanical lifestyle, it gives the deprived a justification for accepting a bad situation they are powerless to change. In effect, it offers both sexes the means for adjusting their aspirations to their possibilities—much like the fox who decided the grapes he could not reach had to be sour.[19] Islamists communicate to their adherents that to get married and have children one need not wait to achieve financial security. One can live as a "good Muslim," they say, even without fancy clothes, a private home, and modern amenities. And, they add, the affluent lifestyles they may have dreamed of emulating are in any case immoral and un-Islamic. In some Egyptian Islamist circles, a prospective bride demands nothing from her future husband but a deep commitment to Islam. Moreover, before she accepts and her family agrees to the marriage, the groom must prove, usually through an oral quiz, that he has sufficient religious knowledge.[20]

Against this background, the Islamist obsession with veiling may be seen as another effort to alleviate the frustrations of marginalized urban youths. The veil reduces women's need to compete socially by keeping up with changing fashions. And it makes a statement against the materialism of modern civilization. Yet another of its functions is to protect the dignity of men who have been socialized into measuring their manhood by how well they protect their women. City life offers great opportunities for the mixing of the sexes, and overcrowding in poor urban communities drastically limits family privacy. Under the circumstances, the veil attenuates the vulnerability of culturally conservative men, both young and old (Ayubi 1991, ch. 2).

By no means do all women who veil themselves belong to economically deprived segments of the urban population. Some come from upwardly mobile families, who live in better neighborhoods and have the financial means to keep up with fashions. For such women, the motivation for veiling lies primarily in its political and religious symbolism. This observation is supported by the emergence of an Islamic fashion industry, which even stages fashion shows that display the latest styles in Islamic dress (Saraçgil 1993; Sullivan 1994, pp. 58-63). Relative to poor Islamist women, prosperous ones tend to wear more expensive and more stylish veils. They thus affirm their religious identity and express opposition to modern sex relations *without* protesting the materialism of modern civilization.

THREATS TO THE GLOBAL ECONOMY

The encouragement of veiling is a shared characteristic of Islamist movements everywhere. Likewise, the promotion of Islamic banking is a standard element of every Islamist economic agenda. Notwithstanding such commonalities, the economic orientations of Islamists exhibit potentially significant variations. Most critically, Islamists, including Islamic economists, have always been divided on the merits of the market. The religious opposition that toppled the Iranian monarchy harbored several rival economic agendas. Schools led by Morteza Motahhari and Navab-Safavi put great faith in the market system and saw the private accumulation of capital as essential for healthy economic development. A rival school led by Ali Shariati rejected private ownership and insisted on endowing the government with vast redistributive powers (Rahnema and Nomani 1990, ch. 3; Behdad 1989). In Pakistan, Maududi and many of his followers have maintained that Islam is sympathetic to private property and the market mechanism. But in the 1970s prominent Pakistani leaders interpreted Islam as favoring planned economic development and massive redistribution.[21]

Contemporary Islamism is capable of supporting both pro- and anti-market ideologies, because this is true of Islam itself, as it is of every other major religion.[22] In the fundamental sources of Islam one can find justifications for respecting market outcomes along with ones for restricting them; and precepts favoring interventions on behalf of the poor along with ones that preach tolerance of inequality. Such variations provide immense flexibility to anyone who wants to put Islam in the service of an economic agenda or to use economic policies grounded in Islam for noneconomic ends. They make it possible to ground any number of arguments, strategies, and policies in venerable traditions or timeless commandments. They enable public figures whose rhetoric may make them appear solidly attached to a distant past to select among various economic positions without rejecting the authority of religion. As a case in point, they allowed the Ayatollah Khomeini and his associates to adapt to various exigencies without ever stepping outside Islamic discourse. And they have permitted Iran's theocracy to accommodate new needs, yield to social pressures, and make pragmatic adaptations—all while continuing to claim allegiance to a timeless, unchanging, and well-defined economic agenda (Rahnema and Nomani 1990, ch. 4).

Modern Islamist movements possess, then, the ideological capacity and flexibility to sustain a liberal economic agenda. Even if they promote illiberal policies while in opposition, they may be able, once in power, to assume a liberal orientation. In any case, to pursue de facto liberal policies they need not make deliberate or explicit ideological adaptations. By giving low priority to economic issues, they may end up promoting private investment, self-management, private ownership, and free trade by default. Such unintended liberalism is all the more likely where illiberal economic goals are overshadowed by objectives concerning family, sexuality, manners, and education. Though a prominent theme in Khomeini's pre-revolutionary rhetoric was the elimination of poverty and exploitation, once he rose to Iran's helm he subordinated his stated economic objectives to the general goal of restoring the centrality of Islam in public life—even to such particular objectives as eliminating the consumption of alcohol, veiling women, banning Western music, and severing Iran from its pre-Islamic heritage. After the revolution, when pressed to speed up the economic transformation, he quipped that the revolution was not made to make watermelons more plentiful (Bakhash 1989, p. 16). On another occasion, he dismissed demands for concrete economic reforms on the grounds that economic well-being is worthy of the donkey (Rahnema and Nomani 1990, p. 296).

None of this means that the Islamists' market-constraining enactments and prescriptions, whether involving interest or planning or redistribution, should be viewed as gimmicks without long-term significance. Although there exists no necessary connection between Islamism and economic illiberalism, Islamism is capable of supporting restrictions on economic liberties. Islamist leaders who, for whatever reason, desire to restrict international trade or regulate the composition of private investment will have little difficulty justifying their actions in religious terms. Also, leaders who whip up anti-market feelings as a tactical move to gain power may be unable, once in power, to control the forces they have unleashed.

Illiberal economic policies can harm the countries implementing them. Protectionism causes the misallocation of local resources, and it insulates inefficient firms from external competition (Bhagwati 1988). Measures to direct an economy's evolution can produce excessive bureaucratic growth, thus reorienting individual enterprise from wealth-creating economic activities to wealth-distributing political ones (Rowley, Tollison, and Tullock 1988). The colossal failures of the centrally planned economies and the persistent disappointments of countries that have pursued inward-oriented and highly interventionist development strategies bear testimony to these adverse consequences (Chenery 1979; Papageorgiou, Michaely, and Choksi 1991). The burden of illiberal policies do not fall entirely, of course, on the countries pursuing them. Trade restrictions generate misallocations in countries forced to produce goods that, under free global trade, they would have obtained more cheaply from abroad. Likewise, excessive bureaucratic meddling in one country imposes costs on anyone who consumes its products, directly or indirectly.

When a new regime starts pursuing an anti-market agenda, the consequent costs need not be evident immediately. Part of the reason is that the new regime's apologists will be endeavoring to make its failures look like successes. Another reason is that the hardships and inefficiencies may initially appear transitional. Even when the costs gain wide recognition, however, it will not be easy to switch gears. For one thing, certain leaders may lose authority if they repudiate the policies they have championed. For another, the government bureacuracy and protected segments of the private sector will have developed a vested interest in the status quo. As a case in point, many Iranian leaders now recognize that the statist and nationalistic measures taken after the revolution of 1978-79 have imposed great burdens on the Iranian people. Accordingly, they have begun to change course: recently nationalized industries are now being re-privatized, and measures have been taken to encourage foreign trade, foreign investment, and joint ventures with foreign firms. Nonetheless, the Iranian economy remains more regulated and more inward-oriented than it would have been in the absence of illiberal post-revolutionary policies (Behdad 1994; Amuzegar 1992). Even though leaders who would prefer a radical change in direction could easily concoct an Islamic rationale for broad liberalization, they find it politically infeasible to undo all the policies they now consider unwise. The principle at work here is "path dependency:" current possibilities are shaped by what has occurred in the past. Of two societies, one market-oriented and the other anti-market, the former can retain liberal policies more easily than the latter can adopt them.

Government policies influence economic performance, but they are not the sole determinant; education, culture, expressive freedoms, and the mass media all play important roles. Islamism seeks to reshape each of these factors—mostly for reasons other than the creation and distribution of wealth. The relevant political efforts carry implications, then, for economic variables. A culture that insists on the inerrancy of traditional sources of wisdom, treats religious teachings as unambiguous, and holds certain issues as beyond controversy may leave the societies it touches unprepared to face the challenges of a dynamic global economy. It is likely to encourage risk aversion as opposed to risk taking, cautious imitation as opposed to daring innovation, and the production of established commodities as opposed to new, and thus highly profitable, ones. In a world where some societies are inquisitive and expressively free, therefore, an Islamist-controlled society may find itself economically disadvantaged. It is worth noting that the long economic descent of the Middle East, from about the eleventh century to the twentieth, coincided with the spread of a mindset that equates innovation with heresy and success with stability (Rahman 1982; Lewis 1982). Technological creativity has always been a major source of relative economic success, and creative societies are generally those tolerant of political dissent, distrustful of conventional wisdom, and open to social change (Mokyr 1990).

If a society starts pursuing economically unwise policies, should outsiders raise objections? As long as they respect the society's right to make bad choices, they should not hold back. Just as an individual has a right to eat foods high in cholesterol, so a society has every right to restrict its economic growth for other objectives, be it environmental preservation, political stability, or spiritual harmony. By the same token, anyone bothered by a society's economic choices, including an outsider, has a right to be heard and to offer alternatives. There is no reason, from a liberal point of view, to prevent a society from choosing to make its economy unproductive. Nor is there a reason, however, for people disturbed by its inefficiencies to keep their misgivings hidden.

When one part of the world falls substantially behind another, there can be troubling consequences for global political stability. The economically backward societies will accumulate resentments against the global economic system, against advanced societies, and even against similarly disadvantaged others. Such resentments can become a source of international friction. Whatever their source, international conflicts threaten global trade flows and transfers of capital. They also impart uncertainty to property rights, as when an ethnic group held responsible for another's economic frustrations faces the possibility of expropriation. If, then, a society run by Islamists—or, for that matter, by anyone else—performs poorly in an economic sense, other societies, too, may suffer.

Islamist movements need not sow social tensions only after achieving power as a by-product of their economic failures. Because they thrive on drawing boundaries between Muslims and others, they may deliberately antagonize outsiders in the hope of benefiting from the consequent social and political polarization. Agitated outsiders may then get organized in self-defense, thus fueling protracted intercommunal struggles. Such a process is already underway in Egypt, where Islamic militancy has prompted Copts to organize in self-defense. Intercommunal struggles can throw property rights into jeopardy, discourage local investment, and cause capital flight, thus impairing economic growth. In the long run, of course, they may end up promoting development, particularly if they destroy political coalitions that have helped preserve inefficiencies.[23] In Egypt, today's mounting intercommunal tensions may eventually bolster economic freedoms by weakening some key elements of social control, including the state and the ideology of economic nationalism.

There is no easy answer, then, to the question of what Islamism implies for the future of the global economic order. Like other fundamentalisms, it is capable of doing economic harm, yet it can also serve as a hidden agent of economic advancement. Remember, too, that some of what passes as Islamic economics, like the establishment of Islamic enterprises, serves to alleviate potentially explosive social problems. How, then, should policymakers committed to a liberal social order respond to the economic activities of Islamists and to their calls for economic reform? What, if anything, should governments, funding agencies, research establishments, educational institutions, and the media be doing to meet the Islamist challenge to prevailing economic ideals, structures, practices, and relationships?

In the remainder of this paper, I shall offer a case for three classes of response. In the first place, one must expose the flaws and limitations of Islamic economics. Secondly, one needs to show that the economic prescriptions of Islamists have considerably less appeal than their leaders tend to claim. Finally, it is essential to devise creative solutions to the festering socioeconomic problems that have fueled the rise of Islamism. The last of these policies requires efforts at understanding the grievances, activities, and ideals of Islamists.

A FIRST RESPONSE TO ISLAMIC ECONOMICS: EXPOSE ITS FLAWS AND LIMITATIONS

The first response would involve disseminating information on the actual and potential effects of Islamic economics. I do not mean simply more scholarly research for the benefit of highbrow intellectuals. There is a need for writings at all levels, including works that will resonate with groups from which Islamist movements draw their rank and file. For all their inconsistencies and illusions, Islamist tracts have an emotional appeal that is often lacking in secular economic writings. They connect, therefore, with large numbers. The constituents of Islamic economics must be given exposure to the counterarguments. The lion's share would have to be shouldered by non-Islamists within countries where Islamists are making credible bids for power or are already in control. Outsiders can help out, but in the absence of local participation their efforts are likely to be perceived as evidence of cultural bias or foreign interference.

Because modern scholarship is essentially an outgrowth of European civilization, to criticize a non-Western movement is to risk being branded a tool of Western cultural arrogance. It would be wise, therefore, to integrate evaluations of Islamic economics into ordinary discourse on the merits of competing economic approaches. If criticisms of Islamic banking come to be confused with animosity toward Islam, matters resolvable through research and debate can turn into objects of cultural confrontation, possibly causing Islamist positions to harden. Another way to allay suspicions of cultural bias is to treat economically unsound policies issued in the name of non-Muslim faiths exactly as one is treating those advanced in the name of Islam. The American doctrine known as "Christian economics" harbors within it a protectionist, anti-market streak (Iannaccone 1993). If only to maintain the credibility of criticisms directed at Islamic economics, this movement should be treated no differently from the anti-market strands within Islamism. To be sure, one can strive for impartiality without managing to avoid the charge of bias. Fundamentalisms have a stake in perpetuating, if not heightening, the perception of deep divisions between themselves and some Godless, immoral, ignorant, foreign "other." Also, it is a common, and often effective, rhetorical tactic to deflect criticism by attributing dark motives to its bearers—a tactic that has been used successfully also by a wide array of secular leaders.

With respect to the specifics of what needs to be disseminated, efforts are needed to enhance general knowledge on the drawbacks of the nationalist and protectionist policies that have been adopted in the name of Islam, like those of Iran in the early years of Islamic rule. Greater awareness must be fostered of the ruses that Islamic banks employ to circumvent the prohibition of interest and of the ineffectiveness of the established zakat systems as instruments of poverty alleviation. It should become better and more widely understood that the distributional consequences of the Islamist-favored institutions are not necessarily equalizing. There should be greater familiarity with the unimpressive economic records of countries that have implemented Islamic economic reforms, like Iran and Pakistan. Finally, people who appreciate that Islamic economics is a reaction to a serious social crisis should also realize that, for all its lofty rhetoric, it fails to offer a comprehensive solution.

Much talk about Islamic economic behavior has been shallow rhetoric, for no one, from Maududi onward, has come close to defining it in practical terms. The Quran has not been of much help in this regard. It is not, after all, a treatise in political economy or a manual for economic governance. Of its thousands of verses, only about one-third of one percent provide economic directives.

And these verses offer enormous choice and flexibility to anyone who wants to use them to build an economic system. Accordingly, the juristic schools of Islam have differed with one another on numerous economic matters. Likewise, the great Muslim thinkers to whom contemporary Islamist writers commonly turn for authority and inspiration often held conflicting positions (Coulson 1969; Saleh 1986; Kuran 1989). The key implication that requires dissemination is that a wide variety of economic agendas can be given an Islamic justification; these include economically harmful ones. Economic restrictions imposed for the sake of cultural differentiation or confrontation—an enforced ban on interest, limitations on financial dealings with non-Muslims—may weaken a country economically, keep it on the periphery of the emerging global economy, and make it vulnerable to foreign domination.

In suggesting that one publicize the economically unsound elements of Islamist agendas, I am not proposing special treatment of religion in general or Islam in particular. On the contrary, I am proposing that economic challenges in religious garb be treated exactly like those in secular garb. To exempt Islamic economics from review and criticism would be to grant it a special privilege. When a labor union or trade association demands an anti-competitive regulation, there are ordinarily many challenges. Commentators point to the regulation's disadvantages and suggest alternative solutions to the problem that it would ostensibly solve. For a weightier example, when the young Soviet Union put its economy under central control and started to preach planning as a superior alternative to the market mechanism, the anti-market campaign met with swift intellectual opposition. Ludwig von Mises, Friedrich Hayek, and other scholars argued that central planning would prove unworkable. They showed that no government agency, however powerful, can disseminate information and coordinate individual actions as efficiently as markets operating under the rule of law. Although for many decades the anti-planning literature was widely dismissed as "reactionary," it managed to erode confidence in the Soviet economic system, thus helping to set the stage for the fall of communism.

My first policy suggestion, then, is to subject the economic proposals of the Islamists to the same critical examination that we bring to bear on any ordinary economic challenge. As of the mid-1990s, it bears mention, secular intellectual communities have paid scant attention to the economic content of Islamist agendas. Part of the reason lies in the fragmentation of the social sciences. Most economists consider religion outside their professional domain, while few area specialists and students of religion have had training in economics. Another reason why Islamic economics has received little attention is that its simpleminded prescriptions allow observers to dismiss all of its discourse as nonsense, even as its influence grows. Still another reason, at least in countries where Islamists have resorted to violence, is that thinkers with misgivings about the Islamist agenda are afraid to say so publicly. The assassinations of critics like Uğur Mumcu and the Egyptian thinker Faraj Foda have made many Muslim intellectuals reticent to speak honestly and openly about goals and activities identified with Islamism.[24]

SECOND RESPONSE: ESTABLISH THE LIMITS OF THE ISLAMIST APPEAL

A second measure to mitigate the economic damage of Islamism would be to counter the Islamists' efforts to portray themselves as spokespersons for heretofore quiescent majorities. When Islamists claim to represent the authentic voice of a society long silenced by secular pressures, they are often exaggerating.

Where Islamic banks are in operation alongside conventional banks, their share of Muslim deposits has remained under 20 percent; in some predominantly Muslim countries, it is as low as one percent (Moore 1990; Kazarian 1991, p. 149; Ağaoğlu 1994). Also, the customers of Islamic financial establishments have shown a remarkable readiness to withdraw their deposits at the first threat of insolvency.[25] By no means is it clear, therefore, that the goal of abolishing interest enjoys widespread acceptance. Nor is it clear that those who do accept the goal consider themselves bound by the Islamist injunction against interest. An Islamic economist might reply that "genuinely Islamic" banks would be more popular than the existing "nominally Islamic" banks that pay and charge interest in disguised form. Perhaps so, yet if none of today's hundreds of Islamic banks is "genuinely Islamic," it behooves us to ask why. One reason is that fewer Muslims consider interest immoral than the number who give public support to the anti-interest campaign. Another reason is that Muslims bothered by interest generally subordinate their qualms to objectives like financial security and simplicity.

It bears re-emphasis that many participants in the Islamic subeconomy are skillfully using Islamic symbolism as an instrument of economic self-advancement. Not every Egyptian who keeps money in an Islamic bank or operates an Islamic grocery store is a committed Islamist. Many Muslims who pass as Islamists know next to nothing about Islamic history, philosophy, and institutions; they have little interest in developing a distinctly Islamic social system; and they have made lifestyle adaptations under Western influences. Furthermore, Muslims who have taken the time to study Islam are not all under the spell of its angry, authoritarian, and inflexible side. Some are impressed instead by its gentle, tolerant, and pragmatic side; and they are ready to allow the co-existence of multiple interpretations of Islam, even to respect the rights of unbelievers (Lewis 1993, esp. chs. 8, 11).

İsmet Özel, Turkey's leading Islamist poet, readily acknowledges that few Turks are genuinely committed to the cause of creating an Islamic social order. He claims that only six percent of the Muslims in Turkey are true believers, though he considers this small share socially and politically pivotal (Meeker 1991, p. 215; Özel 1989, esp. pp. 149-162). Other leading Islamist writers take issue with the prevailing Islamist interpretation of Islam's economic mission. Suggesting that the established Islamic banks have little to do with Islam per se, they argue that some forms of interest are morally unassailable and that an effective ban on interest would hamper Muslim firms in the global marketplace.[26] The proponents of prohibiting interest are out of touch with reality, they say. An Islamist regime could never achieve widespread compliance with a ban, except by force.

Claims regarding the genuine popularity of Islamist prescriptions can be put to the test through elections conducted by secret ballot or surveys that give respondents anonymity. This is because social pressures that regulate openly conveyed preferences are absent from settings where people's choices remain unknown to others. To my knowledge, there have been no systematic surveys to determine the true popularity of Islamic economics. But certain predominantly Muslim countries have held secret-ballot elections that have given voters the option of supporting an Islamist party. In many of these elections Islamists have received far less support than the tenor of public discourse might have led one to expect. In Bangladesh, Islamists control several university campuses, largely because students with misgivings about Islamist causes are afraid to vocalize their views. Yet the Jama'at-i Islami of Bangladesh has never made a strong showing in national elections. In the 1991 election its support stood at 6 percent (Ahmed 1994, pp. 689-695). Pakistan provides a similar contrast. The Jama'at-i Islami of Pakistan never performs well in national elections. Its candidates routinely lose to rivals labelled un- or anti-Islamic. Benazir Bhutto, whom the Jama'at-i Islami has denounced from the day she entered politics, enjoys far more electoral support than most Islamist politicians.[27] In Turkey, one might observe, the leading Islamist party, the Welfare Party, came on top in the general elections of 1995. True, but it won with only one-fifth of the vote; four-fifths went to parties more or less committed to secular principles.

Social pressures do not necessarily work to the advantage of the Islamists. On the eve of the Algerian elections of June 1990 almost every observer expected the Islamic Salvation Front to lose decisively. It went on to win handily, thus suggesting that the fear of government reprisals had caused many of its sympathizers to conceal their inclinations (Roberts 1994, pp. 465-478). The point, then, is not that the publicly expressed preferences of Muslims always overstate their genuine sympathies toward Islamist causes. Rather, it is that one should never accept at face value the claim that Islamists form a cowed majority.

Certain segments of the secular intellectual community are predisposed to reject particular Islamist claims. For instance, most economists will readily reject the assertion that people brought up as Muslims will consider interest unjust. This is because economists are generally trained to believe that basic economic impulses and understandings exhibit no systematic differences across societies. Other segments of the intellectual community tend to be sympathetic to assertions of cross-national variation in values, goals, and attitudes. For instance, anthropologists and many area specialists are trained to notice, probe, and highlight differences, as opposed to similarities. Whatever their own values, many are prepared, therefore, to accept that Muslims are readier than non-Muslims to forgo the safety of fixed interest in favor of variable returns. Moreover, even if they find such a claim implausible, their professional norms may keep them from raising objections, lest they be accused of cultural bias. It so happens that interest has always played an integral role in Muslim economies (Khan 1929; Rodinson 1966/1973, ch. 3; Mandaville 1979, pp. 289-308). Islamic banking is a modern creation and nothing like it existed prior to the 1970s. Scholars who accept the notion that Islam mandates a distinct form of banking are unwitting accomplices, therefore, to an effort at myth making. In trying to respect an ostensible cultural difference, they are enhancing, usually unintentionally, the credibility of Islamists posing as defenders of an endangered local practice.

A related boost to the credibility of Islamists stems from the fact that the West has lost its once-booming cultural confidence. One manifestation of this loss is the ongoing uprising in the United States against the long-standing liberal curriculum in the humanities and the social sciences, now held responsible for deficiencies in the performance of certain groups (D'Souza 1991; Schlessinger 1991). Another is the intense self-criticism occasioned by the quincentennial of Columbus' landing in America, which stood in sharp contrast to the self-congratulatory atmosphere of a century earlier. Whatever their intrinsic merits, campaigns to denigrate Western culture as inherently racist, sexist, oppressive, and inegalitarian diminish the West's confidence in its own culture. Moreover, the resulting self-doubt makes it harder for secularist Muslims to defend Westernization. It also allows Islamists to point to Western malaise as evidence of the dangers of Westernization.

One should not seek to constrain Western self-criticism just because it yields side benefits to Islamism—or, for that matter, to other fundamentalisms. Nor should one compel Islamists to refrain from cultivating the best possible image for themselves—any more than one should make Coca-Cola refrain from manipulative advertising. However, one should avoid falling for, and basing policies on, inflated claims of Islamist support. This is not a trivial point. In certain countries incumbent non-Islamist regimes are making concession after concession to the Islamists, often on the grounds that Islamist demands command major support. To give an example, the Egyptian government is giving more and more time to religious programs on the state-controlled television network and seeking the approval of religious authorities in a host of policy domains. Such concessions are probably shortsighted. They serve to bolster the image of Islamist strength, make non-Islamists afraid to articulate their own beliefs, and embolden the Islamist opposition even further.

THIRD RESPONSE: LISTEN CAREFULLY

The responses proposed thus far are confrontational. One is intended to discredit the arguments on which Islamists base their challenges to prevailing socioeconomic structures. The other calls into question the popularity of these challenges. By contrast, the final suggested response is conciliatory. It is to listen carefully to the Islamist complaints about modernity, recognizing that some are motivated by tangible flaws of the social status quo. Secular regimes have every right to oppose misguided challenges to their policies. By the same token, they have an obligation to be receptive to criticism. The fact that Islamists are making some outrageous demands hardly means that their complaints are all unjustified and their suggestions uniformly worthless.

When Islamists complain about increasingly corrupt government, they are not pointing to a nonexistent problem. Government corruption is as old as government itself, yet its allocational and distributional consequences have grown with the state's expanding control over economic and social life. Nor are Islamists being unreasonable when they accuse incumbent regimes of wasting human resources. Many countries feature enormous unemployment and underemployment, partly because of excessive centralization and bureaucratic inefficiency. One need not be a militant anti-secularist to see that the basically secular regimes that have held power in Egypt, Turkey, Algeria, and Pakistan—to name just a few of the countries that have vociferous Islamist movements—have not been entirely successful at answering the frustrations of their swelling populations. Secular elites would respond that they, too, recognize the problems of corruption and unemployment, adding that they have proposed various reforms. Yet, the measures that many of them champion would probably compound the prevailing problems. To give an illustration, Mumcu tended to support widespread economic regulation. He never quite appreciated that Turkey's huge public sector creates enormous opportunities for favoritism, influence peddling, and graft, or that various bureaucratic controls inhibit job growth. His approach to solving Turkey's economic problems—tighter government controls, better planning—amounted to a more vigorous application of policies responsible for some of the country's economic disappointments.

While criticizing incumbent governments, Islamists have also been tackling certain social problems with considerable creativity. These efforts are not necessarily part of a grand plot to achieve power. Some Islamist services have emerged partly, if not mainly, to fill in gaps left by poorly functioning economies. As noted earlier, Islamic clinics, dormitories, clubs, camps, and schools serve various underprivileged and disaffected groups. That these services make a contribution to the wider Islamist cause does not diminish the benefits they bring to individuals who are not committed Islamists. Secular policymakers would do well, therefore, to explore how these services operate and why so many people are turning to Islamists for help.

Islamists do not necessarily offer viable solutions to problems such as unemployment and official corruption. In restricting economic freedoms, certain Islamist agendas would probably compound them. This is no reason, however, to dismiss all the grievances that underlie such agendas. On the contrary, the dangerous aspects of the agendas compound the urgency of finding ways to lessen the grievances that account for their popularity. Unless secular authorities reduce the excesses of government, limit the state's economic authority, and create an economic environment conducive to free exchange under the rule of law, illiberal fundamentalist agendas will enjoy some appeal. A century ago, when Engels denounced the work conditions in British factories, his criticisms prompted non-Marxists to institute reforms. This was doubtless a factor in the subsequent containment of communism. In any case, some Islamists are on the right track with respect to solving critical economic problems. Those who wish to restrict the economic reach of government and implement regulatory reforms are not just identifying serious sources of inefficiency and inertia; they are also helping to build a consensus for a shift in direction.

Adam Smith is widely remembered for *The Wealth of Nations* (1776/1937), his treatise on the connection between market freedoms and economic prosperity. Less well-known is his first major treatise, *The Theory of Moral Sentiments* (1759/1976), which explores, among other matters, how social and moral norms help ensure the social stability that allows human civilization to flourish. The lesser appeal of Smith's earlier treatise epitomizes the scant attention in secular discourse to the psychological, social, and economic functions of norms. Where Islamists, like other fundamentalists, take norms very seriously, many secular intellectuals and policymakers either dismiss them as unimportant or treat them as sources of inefficiency. The study of economic development is revealing. International differences in economic growth rates are commonly attributed to differences in resource endowments, trade policies, and social stability. Economists in the mainstream generally pay little attention to cultural variables like work habits, demand for education, attitudes toward innovation, and family structure.

Islamic economics promotes the view, we saw earlier, that behavioral norms contribute critically to economic success. They are right about the importance of norms,[28] even if they tend to overrate the social virtues of altruism and underrate those of selfishness. This is not to say that social definitions of right and wrong should stay fixed. Nor is it to deny the importance of laws, rules, and regulations. My point is that in vast reaches of human activity norms remain a cheap and effective way to discipline and coordinate individual actions.

If one requirement of a free economic order is that individuals enjoy broad economic freedoms, another is that they accept responsibility for their own circumstances and take primary responsibility for improving them. Certain Islamist inclinations go against the latter requirement, for instance, the tendency to blame foreigners for domestic economic ills—a tendency evident in Maududi's writings on the corrupting influence of Western culture. But not all Islamists are on the lookout for scapegoats. In South Asia the Tablighi Jama'at encourages self-improvement and self-reliance: its members are expected to decline gifts from others and to become more generous and more honest (Metcalf 1994, esp. pp. 710-717).

I remarked earlier that the observed support for Islamism should never be taken at face value, for some of it might be feigned. Islamist leaders would not necessarily object to this call for caution. They would quickly add, however, that one should be equally cautious with regard to the apparent support for secularist agendas and the observed opposition to public displays of religion. A prominent theme in the Turkish Islamist literature is that, since the Atatürk reforms of the 1920s, Turkish Muslims have had to keep their religious feelings private, at least in spheres officially divorced from religion. Here are a few revealing titles from Islamist presses: *The Postponed Islamic Life*, *Speaking in Difficult Times*, and *The Minister Who Recognized "Allah" After His Death* (Balcı and Balcı 1986-90; Özel 1984; Yılmaz 1990). The last title refers to Hasan Ali Yücel, Minister of Education from 1938 to 1946, a Westernizer who earned a reputation as a tough foe of traditional Islam. The book alleges that, contrary to his outward appearances, Yücel was a deeply religious person. After his retirement, it claims, he regretted having contributed to efforts to suppress Islam, as evidenced by his writing a long religious poem. Yücel did not publish the poem during his lifetime, though he instructed his heirs to publish it as soon as he died. The episode testifies, the book concludes, to Yücel's fear of being denounced as a closet-obscurantist. The veracity of the account is not a matter to be settled here. What matters is the book's central message, which is that Turks have tended to be less supportive of the secularist agenda in private than they have been in public. An implication of the message is that efforts to eject religion from domains such as politics and economics owe their apparent support to social pressures against overt dissenters.

That certain Islamists are prepared to intimidate, censor, and silence their opponents does not negate the fact that they, too, have had their expressive freedoms curtailed. When Islamists promote public displays of religiosity and cultivate religious symbols, this is partly, then, a reaction to past pressures that have made them lead insincere lives. The pressures could have come from official controls, established religious organizations, or simply the weight of public opinion. Whatever the cluster of sources, the pressures imply that the ongoing spread of Islamism stems from more than a rise in individual religiosity. A complementary factor is that religion is everywhere becoming more vocal, more open, and more demanding than it was, say, in the 1960s, a time when secularization was pushing religion progressively into the private realm.[29]

The notion that Islamists have felt and still feel expressively constrained is not one with which intellectuals of a secularist persuasion have much familiarity. Conscious as they are of the Islamist threat to their own freedoms—artistic, literary, sexual, intellectual, and political—they tend not to appreciate that Islamists feel expressively limited and threatened. The source of this ignorance is that secular intellectuals rarely bother to talk to Islamists, listen to their grievances, or read their writings. Were they to make an effort to understand the concerns of Islamists, they would learn that many Islamists have unmet spiritual needs. They would find that secular policies and pressures, like government controls on Islamic expression or the sanctions of secular public opinion, have left a huge void in some individuals. Such a realization would pave the way, perhaps, for fruitful communication between Islamists and their adversaries.

CONCLUSION

The central objective of this paper was to propose how policymakers committed to a free economic order should respond to the rise of Islamic economics and the Islamic subeconomy. Insofar as ongoing efforts to insert Islam into economic discourse and to confer a religious identity on formerly secular economic practices are bringing to the surface perspectives long submerged by secular pressures, they will advance individual liberties. The spirit of the free order calls for openness about thoughts, wants, and anxieties, including religious impulses. It offers the freedom to give one's actions whatever meaning one chooses, including some religious meaning. In and of itself, therefore, it is not a problem that certain social services are being defined as Islamic. Nor is it a problem that there now exist Islamic banks, Islamic grocery stores, and Islamic fashion shows. If Islamic economic enterprises give comfort to Muslims who consider Islam a precise way of life, they should be regarded as positive developments. The legitimacy of a social order is enhanced, not diminished, when a person who ascribes religious significance to the form of a loan is able to shun interest openly, proudly, and without apology.

Yet certain Islamist economic goals conflict with economic freedoms. Economic protectionism, which is pursued by some Islamists, would curtail individual rights to free exchange. Likewise, forcing all Muslims to shun interest would interfere with basic economic liberties. Such moves should be opposed vigorously, regardless of where they come from and irrespective of how they are justified. They should be rejected because they take away essential freedoms and also because they can be socially costly. Forcing all Muslims to shun interest could relegate Muslims to the sidelines of the international economy.

The promotion and enforcement of free economic order requires, therefore, drawing a distinction between, on the one hand, the liberty to live by Islamist economic rules and, on the other, the freedom to impose such rules on others. The latter clashes with the liberties of others, so it may be rejected. Thus, one may recognize the freedom to live by the rules of Islamic economics without recognizing a freedom to impose these rules on entire societies.

Islamists might never agree to make abiding by its tenets a matter of personal choice. But there is some room for optimism. The fact that they are making a primary issue out of allowing people to speak their minds, be themselves, and follow the calls of their conscience has an important implication. It means that Islamists may be, or may eventually become, amenable to granting reciprocal rights of self-expression to others. Perhaps, then, the focus of discourse on Islamic economics should shift to matters of openness and two-sided tolerance. A common ground could eventually emerge from an understanding that to force Islamic economic prescriptions on everyone is morally and practically equivalent to pressuring Islamists to keep their religious identities and views hidden.

Relative to Christianity, Islam has a good historical record with regard to religious tolerance. Historically, therefore, Muslims have felt less pressed than Christians to institute formal protections against religious tyranny. Contemporary conditions in many parts of the Islamic world are getting ripe, however, for a broad attempt at protecting freedom of thought and expression. With huge numbers of secularists and Islamists feeling expressively oppressed by one another, a stalemate between the two groups may fuel a mutual willingness to agree to a reciprocal exchange of basic liberties.[30] The liberties might include economic freedoms, including both the right to shun interest and the right to deal in it fearlessly.

ACKNOWLEDGMENTS

A version of this paper was presented to the Mont Pèlerin Society at its September 1997 convention held in Vienna. For many useful comments on earlier drafts, I am grateful to Mohammed Akacem, Ausaf Ali, Scott Appleby, Sohrab Behdad, Frederic Pryor, Emmanuel Sivan, and Etel Solingen. I also benefited from discussions with Tahseen Basheer, Cem Kozlu, Abdel Azim Ramadan, and Mursi Saad El-Din. These individuals hold diverse views on the matters addressed; none is responsible for my interpretations or policy recommendations.

NOTES

1. Here I generally use the last of these terms, mainly for simplicity. The first two are just as meaningful. As Al-Azm (1993, p. 97) points out, every Islamist group "is convinced that it is in the process not only of going back to the basics and fundamentals of Islam, but of reviving them as well, after a long period of, let us say, hibernation. They are reviving them as active beliefs and efficacious practices in the lives of people."

2. A militant Islamist captured by the Turkish police in early 1996 provided evidence implicating the Iranian government. Turkish-Iranian relations worsened dramatically following the revelations, and it has been reported that the Turkish government considered military reprisals. See *Hürriyet*, March 12, 1996, and later dates.

3. These ideals are defended and qualified by Buchanan (1975) and Hayek (1973-79).

4. See, for instance, Kuran (1986, pp. 152-154).

5. Few noneconomists worry about the financial feasibility or desirability of Islamic banking.

6. Influential contributions include Mannan (1970) and Chapra (1992).

7. For a survey of these and related developments, see Kuran (1993b).

8. The term refers to the practitioners of Islamic economics, who represent a small minority of all economists of the Muslim faith.

9. For related points, see Kuran (1996).

10. The timing of the first practical application of Islamic banking was influenced by the restructuring of the international oil market in the early 1970s. Much of the initial capital came from oil-rich Arab states seeking politically acceptable ways to recirculate their mushrooming petrodollars.

11. For several instructive case studies, see El-Ashker (1987, ch. 11). These studies provide information on the cosmetic aspects of Islamic establishments; they provide few insights into actual practices, suggesting that El-Ashker noticed nothing unusual worth reporting.

12. A similar argument is advanced by Baker (1990, ch. 8).

13. Based on a discussion with Mursi Saad El-Din, Cairo, August 2, 1993. For many complementary observations, see Sullivan (1994).

14. Such financing occurs largely through informal channels. See Hyman (1987, pp. 11, 19-20).

15. For the underlying logic, see Kuran (1995a, chs. 15-18).

16. Parts of this section draw on Kuran (1995b, pp. 167-169), where further details may be found.

17. For the underlying logic, see Rowley, Tollison, and Tullock (1988).

18. Such incentives arise wherever the legal system is weak. In Southern Italy, Gambetta (1993) demonstrates, they fueled the rise of the Mafia.

19. On the general phenomenon, see Elster (1983, ch. 3).

20. Information provided by Tahseen Basheer, in a Cairo interview, August 3, 1993.

21. For further examples of such divisions, see Kuran (1993b, 1995b).

22. For a comparative overview, see Kuran (1993a).

23. The underlying logic is developed by Olson (1982).

24. The logic that supports this observation is developed at length in Kuran (1995a).

25. For an example from Egypt, see Springborg (1989, pp. 45-61).
26. For instance, Bulaç (1988/1990, pp. 115-116) and Güngör (1981, pp. 11, 150, 241).
27. For some figures, see Mayer (1993, pp. 146-147, n. 31).
28. Wilson (1993) offers much supportive evidence.
29. See Casanova (1992, pp. 37-43). The religious revival of our times has also been attributed to religious deregulation. In this view, such deregulation is fueling an explosion of new religious forms and a huge increase in public religious expression, just as industrial deregulation stimulates production by allowing firms to satisfy a greater variety of needs. On the alternative view, see Chaves and Cann (1992, pp. 272-290). The two explanations are not mutually exclusive. Greater religious experimentation may be accompanied by greater religious openness.
30. Such a stalemate lies at the origins of today's religious tolerance in the West. Persecutions inflicted by Christians trying to impose their own orthodoxies on other Christians convinced astute Europeans and Americans that it would be mutually beneficial, first, to keep religion out of government affairs, and second, to grant individuals a broad array of formal expressive freedoms. See Dahl (1956).

REFERENCES

Ağaoğlu, E.A. 1994. "A Camel-wise Comparative Financial and Market Share Analysis of the Islamic Banks Currently Operating in Turkey." *METU Studies in Development* 21(4): 475-500.

Ahmad, M. 1991. "Islamic Fundamentalism in South Asia: The Jama'at-i-Islami and the Tablighi Jamaat of South Asia." Pp. 457-530 in *Fundamentalisms Observed*, edited by M.E. Marty and R.S. Appleby. Chicago: University of Chicago Press.

Ahmed, R. 1994. "Redefining Muslim Identity in South Asia: The Transformation of the Jama'at-i-Islami." Pp. 669-705 in *Accounting for Fundamentalisms: The Dynamic Character of Movements*, edited by M.E. Marty and R.S. Appleby. Chicago: University of Chicago Press.

Al-Azm, S.J. 1993. "Islamic Fundamentalism Reconsidered: A Critical Outline of Problems, Ideas, and Approaches." *South Asia Bulletin* 13(1-2): 93-121.

Amuzegar, J. 1992. "The Iranian Economy before and after the Revolution." *Middle East Journal* 46(Summer): 413-425.

Ayubi, N. 1991. *Political Islam: Religion and Politics in the Arab World*. London: Routledge.

Baker, R.W. 1990. *Sadat and After: Struggles for Egypt's Political Soul*. Cambridge, MA: Harvard University Press.

Bakhash, S. 1989. "What Khomeini Did." *New York Review of Books*, July 20, pp. 16-19.

Balcı, İ. and M. Balcı. 1986-1990. *Ertelenen İslami Hayat* [The Postponed Islamic Life], 4 vols. Istanbul: İklim Yayınları.

Behdad, S. 1989. "Property Rights in Contemporary Islamic Thought: A Critical Perspective." *Review of Social Economy* 47(Summer): 185-211.

________. 1994. "Production and Employment in Iran: Involution and De-Industrialization Thesis." Pp. 85-111 in *The Economy of Islamic Iran: Between State and Market*, edited by T. Coville. Tehran: Institut Français de Recherche en Iran.

Bhagwati, J. 1988. *Protectionism*. Cambridge, MA: MIT Press.

Buchanan, J.M. 1975. *The Limits of Liberty: Between Anarchy and Leviathan*. Chicago: University of Chicago Press.

Bulaç, A. 1988/1990. *İnsanın Özgürlük Arayışı* [The Human Quest for Freedom]. Istanbul: Endülüs Yayınları.

Casanova, J. 1992. "Private and Public Religions." *Social Research* 59(Spring): 37-43.

Chapra, M.U. 1992. *Islam and the Economic Challenge*. Leicester: Islamic Foundation.

Chaves, M. and D.E. Cann. 1992. "Regulation, Pluralism, and Religious Market Structure: Explaining Religion's Vitality." *Rationality and Society* 4(July): 272-290.

Chenery, H.B. 1979. *Structural Change and Development Policy*. New York: Oxford University Press.

Coulson, N.J. 1969. *Conflicts and Tensions in Islamic Jurisprudence*. Chicago: University of Chicago Press.

Dahl, R.A. 1956. *A Preface to Democratic Theory*. Chicago: University of Chicago Press.

D'Souza, D. 1991. *Illiberal Education: The Politics of Race and Sex on Campus*. New York: Free Press.

El-Ashker, A.A.-F. 1987. *The Islamic Business Enterprise*. London: Croom Helm.

Elster, J. 1983. *Sour Grapes: Studies in the Subversion of Rationality*. Cambridge: Cambridge University Press.

Etzioni, A. 1988. *The Moral Dimension: Toward a New Economics*. New York: Free Press.

Gambetta, D. 1993. *The Sicilian Mafia: The Business of Protection*. Cambridge, MA: Harvard University Press.

Güngör, E. 1981. *İslamın Bugünkü Meseleleri* [Contemporary Problems of Islam]. Istanbul: Ötüken Neşriyat.

Gürsoy-Tezcan, A. 1991. "Mosque or Health Centre? A Dispute in a *Gecekondu*." Pp. 84-101 in *Islam in Modern Turkey: Religion, Politics, and Literature in a Secular State*, edited by R. Tapper. London: I.B. Tauris.

Hayek, F.A. 1973-79. *Law, Legislation, and Liberty*, 3 vols. Chicago: University of Chicago Press.

Hyman, A. 1987. *Muslim Fundamentalism*. London: Institute for the Study of Conflict.

Iannaccone, L. 1993. "Heirs to the Protestant Ethic? The Economics of American Fundamentalists." Pp. 342-366 in *Fundamentalisms and the State: Remaking Polities, Economies, and Militance*, edited by M.E. Marty and R.S. Appleby. Chicago: University of Chicago Press.

Kazarian, E. 1991. *Islamic Banking in Egypt*. Lund: Lund Economic Studies.

Khan, M.S.A. 1929. "The Mohammadan Laws Against Usury and How They Are Evaded." *Journal of Comparative Legislation and International Law* 11: 233-244.

Kuran, T. 1986. "The Economic System in Contemporary Islamic Thought: Interpretation and Assessment." *International Journal of Middle East Studies* 18(May): 135-164.

________. 1989. "On the Notion of Economic Justice in Contemporary Islamic Thought." *International Journal of Middle East Studies* 21(May): 171-191.

________. 1993a. "Fundamentalisms and the Economy." Pp. 289-301 in *Fundamentalisms and the State: Remaking Polities, Economies, and Militance*, edited by M.E. Marty and R.S. Appleby. Chicago: University of Chicago Press.

________. 1993b. "The Economic Impact of Islamic Fundamentalism." Pp. 302-341 in *Fundamentalisms and the State: Remaking Polities, Economies, and Militance*, edited by M.E. Marty and R.S. Appleby. Chicago: University of Chicago Press.

________. 1995a. *Private Truths, Public Lies: The Social Consequences of Preference Falsification*. Cambridge, MA: Harvard University Press.

________. 1995b. "Islamic Economics and the Islamic Subeconomy." *Journal of Economic Perspectives* 9(Fall): 155-173.

________. 1996. "The Discontents of Islamic Economic Morality." *American Economic Review* 86(May): 438-442.

Lewis, B. 1982. *The Muslim Discovery of Europe*. New York: W.W. Norton.

________. 1993. *Islam and the West*. New York: Oxford University Press.

Mandaville, J.E. 1979. "Usurious Piety: The Cash Waqf Controversy in the Ottoman Empire." *International Journal of Middle East Studies* 10(August): 289-308.

Mannan, M.A. 1970. *Islamic Economics: Theory and Practice*. Lahore: Sh. Muhammad Ashraf.

Maududi, S.A.A. 1940/1985. *Let Us Be Muslims*. Leicester: Islamic Foundation.

Mayer, A.E. 1993. "The Fundamentalist Impact on Law, Politics, and Constitutions in Iran, Pakistan, and the Sudan." Pp. 110-151 in *Fundamentalisms and the State: Remaking Polities, Economies, and Militance*, edited by M.E. Marty and R.S. Appleby. Chicago: University of Chicago Press.

Meeker, M.E. 1991. "The New Muslim Intellectuals in the Republic of Turkey." Pp. 189-219 in *Islam in Modern Turkey: Religion, Politics, and Literature in a Secular State*, edited by R. Tapper. London: I.B. Tauris.

Metcalf, B.D. 1994. "'Remaking Ourselves': Islamic Self-Fashioning in a Global Movement of Spiritual Renewal." Pp. 706-725 in *Accounting for Fundamentalisms: The Dynamic Character of Movements*, edited by M.E. Marty and R.S. Appleby. Chicago: University of Chicago Press.

Mokyr, J. 1990. *The Lever of Riches: Technological Creativity and Economic Progress*. New York: Oxford University Press.

Moore, C.H. 1990. "Islamic Banks and Competitive Politics in the Arab World and Turkey." *Middle East Journal* 44(Spring): 234-255.

Mumcu, U. 1987. *Rabıta*. Istanbul: Tekin Yayınevi.

________. 1988. *Tarikat, Siyaset, Ticaret* [Tariqat, Politics, Commerce]. Istanbul: Tekin Yayınevi.

Murphy, K. 1992. "Islamic Extremists Declaring War on Egypt's Tourist Industry." *Los Angeles Times*, October 23, p. A6.

Naqvi, S.N.H. 1981. *Ethics and Economics: An Islamic Synthesis*. Leicester: Islamic Foundation.

Olson, M. 1982. *The Rise and Decline of Nations: Economic Growth, Stagflation, and Social Rigidities*. New Haven, CT: Yale University Press.

Özel, İ. 1984. *Zor Zamanda Konuşmak* [Speaking in Difficult Times]. Istanbul: Çıdam Yayınları.

———. 1989. *Sorulunca Söylenen* [Answers to Questions One Might Be Asked]. Istanbul: Çıdam Yayınları.

Papageorgiou, D., M. Michaely, and A. Choksi, eds. 1991. *Liberalizing Foreign Trade*, 7 vols. Cambridge, MA: Basil Blackwell.

Rahman, F. 1982. *Islam and Modernity: Transformation of an Intellectual Tradition*. Chicago: University of Chicago Press.

Rahnema, A. and F. Nomani. 1990. *The Secular Miracle: Religion, Politics, and Economic Policy in Iran*. London: Zed Books.

Rashwan, D. and E. Gad. 1993. "The Marriage of Militancy and Tradition." *Al-Ahram Weekly*, July 8-14, p. 3.

Reed, H.A. 1988. "Islam and Education in Turkey." *Turkish Studies Association Bulletin* 12: 2-4.

Roberts, H. 1994. "From Radical Mission to Equivocal Ambition: The Expansion and Manipulation of Algerian Islamism, 1979-1992." Pp. 428-489 in *Accounting for Fundamentalisms: The Dynamic Character of Movements*, edited by M.E. Marty and R.S. Appleby. Chicago: University of Chicago Press.

Rodinson, M. 1966/1973. *Islam and Capitalism*, trans. by B. Pearce. New York: Pantheon Books.

Rowley, C.K., R.D. Tollison, and G. Tullock, eds. 1988. *The Political Economy of Rent-Seeking*. Boston: Kluwer.

Saleh, N.A. 1986. *Unlawful Gain and Legitimate Profit in Islamic Law*. Cambridge: Cambridge University Press.

Saraçgil, B. 1993. "İslami Giyimde Moda ve Tüketim" [Islamic Clothing Fashions and Consumption Patterns]. *Yeni Zemin* (August): 74-76.

Schlessinger, A.M., Jr. 1991. *The Disuniting of America: Reflections on a Multicultural Society*. Knoxville, TN: Whittle Books.

Siddiqi, M.N. 1972. *The Economic Enterprise in Islam*. Lahore: Islamic Publications.

Smith, A. 1759/1976. *The Theory of Moral Sentiments*. Oxford: Clarendon Press.

———. 1776/1937. *The Wealth of Nations*. New York: Modern Library.

Springborg, R. 1989. *Mubarak's Egypt: Fragmentation of the Political Order*. Boulder, CO: Westview Press.

Sullivan, D.J. 1994. *Private Voluntary Organizations in Egypt: Islamic Development, Private Iniative, and State Control*. Gainesville: University Press of Florida.

Williamson, O.E. 1985. *The Economic Institutions of Capitalism: Firms, Markets, Relational Contracting*. New York: Free Press.

Wilson, J.Q. 1993. *The Moral Sense*. New York: Free Press.

Yılmaz, H. 1990. *Öldükten Sonra "Allah" Diyen Bakan* [The Minister Who Recognized "Allah" after His Death]. Istanbul: Timaş Yayınları.

ISLAMIC JURISPRUDENCE AND PUBLIC POLICY

Ali Rahnema

INTRODUCTION

The appreciation of the scope for public policy in Islamic law or its concern for such policies, requires a clear understanding of what constitutes Islamic law; the purview of the law in terms of the areas in which it intervenes in the everyday life of the people and the degree to which and the domains in which individual choice is allowed; the degree to which the letter and spirit of the law is amenable to the topics and issues that generally fall under the rubric of public policy and the extent to which the law or government policies may be changed to implement public policy. In this case, the ability of Islamic law to respond to new problems and therefore produce new laws and policies has to be addressed from a theoretical perspective.

This paper will attempt to demonstrate that a monolithic view of Islamic law even in theory and among Islamic jurists, let alone lay modernists, is a myth. It will also try to dispel the notion that Islamic law is infinitely rigid and static and will attempt to demonstrate that dynamic legal components coexist side-by-side with static and rigid principles. This paper illustrates that

International Review of Comparative Public Policy, Volume 9, pages 103-122.

ISBN: 0-7623-0268-2

the inherent elements in the law, in different schools of Islamic jurisprudence, lend themselves to the articulation and application of public policies. This paper will also demonstrate that certain topics or issues of public policy, in the broad sense of the concept, are eternally written within what could be considered as the *Shari'a*-based constitution of Islamic states. In these areas, based on immutable primary Islamic sources, a choice of public policies, in the sense of replacing the divinely-inspired policies with secular policies, does not exist. However, this paper will verify the proposition that in areas where new or modified laws, based on rational inferences or the tertiary sources of the Shari'a, are enacted, choices in public policy become a reality. Finally, this paper will discuss the policy tools that have been developed by Islamic jurists to circumvent inflexible, eternally fixed, and seemingly insurmountable legal principles, to demonstrate that where there has been a will on the part of Islamic jurists they have been almost always successful in bending the most inflexible laws to their designs. The above points, however, can only be addressed and discussed through a relatively detailed theoretical analysis of the content and composition of traditional Islamic law in its multifaceted aspects. The ongoing ideological and political debates within Islamic countries, reflecting different interpretations of Islamic law, will not be discussed in this paper nor will the legal practices of particular Islamic countries.

ISLAMIC JURISPRUDENCE AND THE ISLAMIC STATE

The Islamic state, by definition, is one that is based on, and committed to, the upholding and application of Islamic tenets in every aspect of life. Thus, all practical rules and injunctions have to be solidly rooted in the principles of Islamic law. The process of deduction and inference of practical rules from what may be considered as the constitution of Islam founded on the direct words of God and the sayings of the Prophet, requires a knowledge of the sources of Islamic law as well as the permissible methods of interpretation and deducing new laws. As will be shown, Islamic public policy either can be grounded in the original sources of Islamic law reflecting the concerns and problems of the guardians of the faith or can be interpreted and deduced from these sources, applicable to modern conditions and needs. To understand the theoretical and often practical constraints and opportunities of effectuating change in modern Islamic societies, which is the object of public policy, it is imperative to grasp the operation as well as the degree of flexibility of the legal foundation.

Far from being a neutral state of the laissez-faire type, the Islamic state is interventionist and absolutist. Its function is to manage and administer the Muslim community or the *umma* according to a divine code of conduct. This code is embodied in the Shari'a, which means the path to be followed. This is the path that Muslims believe to have been clearly indicated by God and one that leads to Him. Shari'a is, therefore, the sacred law of Islam and *fiqh* is the science of the sacred law. The legitimacy and authority of an Islamic state is, therefore, based on the application and observance of the Shari'a, which is believed to guarantee individual felicity in this world and salvation in the next. The principal lawgiver in Islam is God, to whom absolute sovereignty belongs. The notion of *towhid* or the singularity of God, on which the faith is founded, implies that He alone is the absolute sovereign of the world, His dominion includes all men, all those that He has created and chosen as His own vicegerents (57:5, 67:1).[1]

Based on this classical definition of divine sovereignty, individual human beings, irrespective of their numerical strength, cannot lay claim to sovereignty or to the power to legislate in areas where specific Quranic or Prophetic injunctions exist. Subsequently, at first instance it seems as if the Islamic state cannot be regulated and governed by any other law than that laid down by God and his Prophet. From this narrow perspective, the role of government in an Islamic state is restricted to the simple enforcement of the injunctions of God and the Prophet. In an Islamic society, therefore, a priori it seems as if there is little room for human-made laws. It is on this basis that many Islamic theoreticians and jurists have argued that democracy, as a political system in which the majority exercise their sovereignty through legislating change, is incompatible with the principles of Islam. Sayyid Abul Ala Mawdudi, a leading and influential Islamic revivalist, argues that Islam is "the very antithesis of secular Western democracy" (Mawdudi 1988, p. 159). In an Islamic society, the wisdom of people, the validity of their decisions, and their power to implement their aspirations is entirely subjected to the will of God as expressed in the Shari'a and interpreted by human beings.

THE SCOPE OF ISLAMIC LAW

Islamic law articulates and expresses a religion and a legal system. The issues elaborated and the problems addressed by it could be divided into two major areas. First, spiritual and religious obligations. These are affairs related primarily to the hereafter. Injunctions dealing with this aspect of Islamic law are categorized under the rubric of *ibada* "or laws concerning worship, religious duties, rites and ceremonies" (Klein 1985, p. 120). Injunctions concerning issues such as faith, prayers, fasting, *zakat*, pilgrimage, and holy war falls under the topic ibada. The clarification, explanation and elaboration of the rituals, rites and ceremonies that fall within the realm of ibada has by itself become a major area of research and scholarship in fiqh. Aside from the manner of performing religious duties, Islamic jurists have written at length on formalistic issues of preparing for such duties. For example, the complex varieties of rites of purification, ablution and washings in preparation for prayers have been the subject of innumerable treatises. Modern Islamic revivalists have often criticized the narrow focus of scholarship and research by Islamic jurists on issues dealing with rituals and ceremonies of ibada at the cost of those aspects of fiqh dealing with socioeconomic and political concerns. It is, however, understandable that the separation between state and religion induced Islamic jurists to preoccupy themselves with the spiritual rather than the worldly needs and affairs of their flock. The nascent resurgence of Islamic governments has led to a renewal of interest and scholarship in the worldly aspects of fiqh.

Aside from ibada, Islamic law is concerned with worldly and temporal affairs. Through legal injunctions it aims to regulate individual and social relations. The relevance and significance of this aspect of fiqh emerges once a state is committed to the general application of Islamic law. Countries in which Islamic tenets concerned with private religious affairs are upheld as the proof of adherence to Islam, yet the state has shed its commitment to upholding the worldly injunctions of Islamic law thereby separating religion from the affairs of the state, may not be considered as full-fledged Islamic societies.

The area of worldly or temporal affairs of fiqh is divided into three different legal categories. First, *mu'amelat* or laws pertaining to commercial, financial, and other economic contracts or agreements. Under this rubric, the rights and obligations of contracting parties are enumerated in transactions such as buying and selling, rent, hire, partnerships, gifts, loans, assignment or transfer of debts and usurpation (Mahmasani 1979, p. 20). Second, *munakehat* or laws pertaining to family affairs. These laws include injunctions on marriage and divorce, polygamy, various modes of dissolution of marriage, the period of retreat after divorce, dower, bequest, fosterage, and succession. Most of these injunctions fall under the category of family laws (Mahmasani 1979, p. 20; Doi 1984, p. 7). Third, *uqubat* referring to penal or criminal law in Islam. Since Islamic penal law is concerned with the punishment of crimes against religion or the public interest and particular individuals, it covers both torts as well as crimes (Doi 1984, p. 220). Uqubat deals with cases such as homicide, theft, adultery, homosexuality, defamation, highway robbery, use of intoxicating drinks, apostasy and "being corrupt on Earth," a crime punishable by death and usually associated with conspiring against or undermining the authority of the Islamic state.

The Shari'a not only enunciates the areas in which private and social life should be conducted according to divine law, but it also defines, classifies, and categorizes the injunctions of the Lawgiver in relation to Muslims. Human activities are classified on a scale of moral and ethical desirability from God's perspective. This scale of religious qualifications known as *al-ahkam al-khamsa*, or the "five qualifications," categorizes acts as obligatory duties (*wajib*), recommended (*mustahab*), permissible or left with a choice (*mubah*), reprobated (*makruh*), and forbidden or prohibited (*haram*). Experts on Islamic law (*fuqaha*, plural of *faqih*) have formulated the above categories, which are not specified in the Quran (Kamali 1991, p. 34). While the two extremes of wajib and haram are considered as legally binding duties, the three intermediate shades are essentially left to individual choice and discretion, though each carries a different weight in terms of moral desirability from the perspective of the Lawgiver.

Wajib is an obligatory act demanded of all Muslims. Performance of acts characterized as wajib leads to spiritual rewards in the hereafter and failure to act accordingly will result in punishments in this world and the hereafter (Kamali 1991, p. 324). Obligatory duties are sub-divided into individual duties and collective duties. Individual obligatory duties such as prayers, fasting, and *haj* (pilgrimage) are the responsibility of every individual and cannot be relegated to others. Collective obligatory duties such as *jihad* (holy war), *hisba* (promotion of good and prevention of evil), and building hospitals are, however, demanded of the community. This type of obligation is fulfilled if a sufficient number of community members perform the specified duties, although the remaining members of the community may refrain from carrying out the duties (Kamali 1991, p. 325; Schacht 1991, p. 121).

Performing the recommended acts, which are not binding, results in spiritual rewards, while not performing them does not have a punishment. Acts recommended by Islam include the creation of charitable endowments (*waqf*), attending the sick and giving alms to the poor (Kamali 1991, p. 327). Permissible (mubah) acts are those about which Muslims are left to make their own decisions. For example, based on Quranic injunctions, marriage is a permissible act (2:235). Interestingly, commitment of a sinful act in case of necessity (2:173) falls under this legal rubric as well (Kamali 1991, p. 327). Reprobated acts, as in the case of recommended acts, are not binding but constitute mere suggestions. The avoidance of these acts are recommended. Violation of such recommendations, however, will not result in punishment. An example of a reprobated act can be found in the saying of the Prophet that, "The most abominable of permissible things in the sight of God is divorce" (Kamali 1991, p. 332). Thus, divorce is a reprobated act that is allowed. Muslims, however, are strictly commanded to forego forbidden or haram practices. Haram is a binding command of the Lawgiver and its practice is punishable by law while its avoidance is rewarded. According to the Quran, Muslims are forbidden to eat dead carcasses, blood, and pork (5:3). They are also prohibited from engaging in usury, wine-drinking, and gambling (2:275; 5:90).

The five qualifications represent the width and scope of Islam's concern and preoccupation with individual and social acts. It reflects the degree to which it is a thoroughly interventionist religion. Even in cases where it does not engage in binding commandments it proclaims its morally rooted value judgments on innumerable issues. The pious believer or the ideal Muslim (*muttaqi*) would be required to regulate his life not only on the basis of the clear commands of the Shari'a, but also according to the preferences of the law enunciated in the gray areas of recommended, permissible, and reprobated acts. Islamic scholars readily admit that the detailed account of the five qualifications is also a reflection of the presumption that individuals are totally ignorant of "where the real good lies" and that they may enjoy things that are bad for them and take displeasure in things that are good for them (Muslehuddin 1986, pp. 64-65). For Muslims, the commands and preferences of Islamic law should constitute the sole correct and acceptable code of moral conduct, assuring them of felicity in this world and salvation in the hereafter.

THE SOURCES OF ISLAMIC LAW

During the rule of the Abbasids, which began in the eighth century, the process of systematizing Islamic law started in earnest. This movement found its expression in the legal debates that culminated in the emergence of the four major Sunni schools of Islamic jurisprudence: Hanafi, Maliki, Shafi'i, and Hanbali, all named after their founders. The detailed rules of Islamic law, or what is known as fiqh, is deduced from a legal source or base. The methodology of deriving laws from the source is referred to as *usul al-fiqh.* A law cannot be claimed Islamic and binding unless evidence and proof for it can be found in the accepted sources of the law. The establishment and identification of accepted sources of Islamic law is necessary before methods of deducing laws based on those sources could be addressed. The determination of legality or illegality, permissibility or impermissibility of an act requires the search for proofs or refutations in the valid sources of Islamic law. The sources on which fiqh is constructed and from which it obtains its legitimacy are called evidences or proofs (*adilla*). Religious injunctions and rules (*ahkam shar'iyya*) are, in the first instance based on religious proofs. Even though consensus on the general identification and categorization of the religious proofs exists among the different schools of jurisprudence, the differences and specific components of each category can lead to major differences. The development of different schools of jurisprudence gave rise to different rulings and laws rooted in differences of opinion on the specifics of what could and should be included and utilized as legitimate sources of Islamic jurisprudence and the proper and permissible method of deriving Islamic laws from those sources.

The religious proofs or the major sources of Islamic law refer to the four principal sources, namely the Quran, Sunna, consensus, and analogy. Jurists generally agree that justification of all four principal sources of Islamic law can be found in the verse, "O you believers! Obey God and obey the Messenger and those of you who are in charge of affairs. If you have a dispute concerning any matter, refer it to God and to the Messenger" (4:58-9). Obedience to God and His Messenger is a clear declaration that the Quran and the Sunna should be accepted as primary sources of Islamic law. Compliance with the decision of those "who are in charge of affairs" is argued to refer to the legitimacy of consensus, while referring disputes to God and to the Messenger is construed as a reference to analogy, where jurists are to extend existing laws in the primary sources by deducing new laws based on them (Kamali 1991, p. 10).

The four major Sunni schools of jurisprudence accept the validity of these four sources. The Shi'is uphold the first three sources but they replace analogy, as the fourth proof, with reason. The fourth source of Islamic law, be it analogy or reason, has lent itself to the employment of different rational methods of deducing laws and as such has constituted an important yet quasi-independent cluster of sources, on the basis of which laws have been promulgated by Islamic jurists. Concepts such as *istihsan*, or juristic preference on the basis of concern for equity and fairness, *maslaha mursala* and *istislah*, or unrestricted public interest and social benefit, became the essential rational basis of judgment. Today, Islamic injunctions can be traced to proofs of the Shari'a, ranging from those that are firmly rooted in the expression of the Lawgiver and are therefore divine, to those that are based on rational judgments only tempered by the spirit of Islam.

On the basis of the degree to which sources are divinely or rationally inspired and, therefore, the degree to which they are rigid or flexible, static or dynamic, we can categorize them into three major groups. First are the divine, direct or indirect, expressions of the Lawgiver of which the Quran and the Sunna are primary sources. The authority, jurisdiction, and acceptability of the Quran and the Sunna are independent of any rational defense of these sources. Since these sources are divine, they stand on their own and need no rational proof.

Second are sources that are not based on divine sources, yet they are deduced by varying degrees of reasoning and rational thought. Consensus and analogy fall under this category of secondary sources. Laws relying on consensus and analogy are usually in need of a basis or a document in the primary sources of the Quran or the Sunna. As such, the secondary sources are considered as dependent sources, usually incapable of standing on their own without a justification in the divine primary sources. Particular types of analogy, however, provide for a legally gray area, where judgments based on primary sources could gradually slip into decisions based purely on reason. The secondary sources could therefore be considered as part divine and part rational, both flexible and rigid.

Third are sources and methods of deducing Islamic laws that are essentially based on the independent reasoning of the jurist and what he believes to be the spirit rather than the letter of the law. Sources such as maslaha mursala, allow for Islamic law to be deduced and promulgated on the basis of the jurist's concept of equity and fairness or Islamic public interest. The inclusion of reason-based sources such as istihsan, which Imam Malik designated as nine-tenths human knowledge, introduces an important dynamic element into Islamic jurisprudence. These, rational, flexible, and dynamic sources fall under the tertiary sources of the Shari'a. Even though they are important enough to be categorized on their own, tertiary sources may be considered as a subdivision of analogy, because without the acceptance of analogical deductions on the basis of primary sources by the four Sunni schools, the application of reason as a method of inferring laws would not have been justifiable. The fact that tertiary sources of law display a rupture with the four principal religious proofs of the law is evident, yet the justification of such sources had to go through the prism of analogy. As far as the Shi'is are concerned, their fourth source, namely reason, opens the door to *ijtihad* or the process whereby every possible rational effort is made by a jurist to arrive at an appropriate ruling on the basis of the Shari'a.

THE QURAN

The most fundamental source of Islamic law is the Quran. The statement that Islamic laws, coordinating an Islamic society, are divinely pre-ordained finds its justification in the fact that such laws should be formulated on the basis of the letter or the spirit of Quran. The Quran, which literally means reading and recitation, was revealed to the Prophet by God, to direct men along the proper path. Quran is the very word of God revealed to the Prophet by the angel Gabriel in Mecca and then Medina. It constitutes a complete and unique text, the exact letter of which is the Lawgiver's design and will. As such, it specifies the moral, philosophical, social, political, and economic basis on which an Islamic society should be constructed. The verses revealed in Mecca, after which the Prophet began inviting the people to Islam, deal with theological issues such as the basis, requirements, rituals and practices of the faith, and its system of rewards and punishments in the hereafter. The verses revealed in Medina, during which the Prophet established the first Islamic state, deal essentially with legal, political, social, and economic rules and regulations necessary for governing the newly born state. During his lifetime, the Prophet did not compile the Quran into a single collection. The revelations recited by the Prophet were either memorized or written down disjointedly by his companions, since the Prophet did not know how to write (Mahmasani 1979, p. 128). The Quran, as available to us today, was compiled during the reign of the third Caliph, Uthman, who ordered the destruction of all previous copies and collections once the compilation was complete.

The Quran is believed to be unique and inimitable. Its revelation, over a period of approximately twenty-two years, to the prophet and his recitation of its contents, are considered a miracle. It is divided into 114 chapters (*suras*) of different lengths varying from very short ones, typical of the Meccan period, to very long ones, typical of the Medinian period. The Quran is further divided into some 6,200 verses (*ayat*), the exact figure of which, due to different methods of calculation, is disputed by different schools. The verses in the Quran cover three broad areas: the "science of speculative theology," "ethical principles," and "rules of human conduct" (Qadri 1986, p. 186). Islamic jurisprudence is directly concerned with the third category of verses, revealed in Medina and destined to transform the behavior of individuals and thus reform society according to divine injunctions (Qadri 1986, p. 186). However, less than one-tenth of the verses relate to law and jurisprudence. Most of the verses deal with matters of faith and morality (Kamali 1991, p. 19) The number of verses dealing with practical problems and legal matters is estimated at 350, while the number of those dealing with commercial transactions is about 70, economic matters, 10, family affairs, 10, and those addressing criminal matters is no more than 30 (Kamali 1991, p. 20).

Quranic texts and, subsequently, legal injunctions may be divided into two broad categories. First, injunctions that are obvious and definitive (*qat'i*). The meaning of such texts are specific and clear. The reader is able to grasp the content and message of these injunctions immediately. Definitive texts cannot be subjected to interpretation. The rulings of the Quran on essentials of the faith, such as fasting and prayers, are categorized as definitive and incumbent upon all Muslims (Kamali 1991, p. 21). Second, those injunctions the meaning of which is not specific or immediately apparent. These veiled or speculative (*zanni*) verses are open to interpretation and ijtihad (Kamali 1991, p. 21). Based on different interpretations of the same speculative Quranic verse, jurists have arrived at as many as seven or eight different juristic conclusions (Kamali 1991, p. 21). Quranic texts, as well as those of the Sunna, lend themselves to another categorization based on how specific (*khass*) or definitive a text is in contrast to how general (*amm*) or speculative it may be. The specific (khass) text is definitive, while the general (amm) text is speculative and subject to interpretation and ijtihad (Kamali 1991, p. 24).

The Quranic proclamation that "God has permitted sales but prohibited usury" (2:275) is argued to be a general (amm) ruling. The reference to the permissibility of sales, renders any kind of sale lawful. The Sunna, however, outlaws a variety of sales such as that of an unborn animal, unripened fruits on trees, or unproduced goods. In the case of sales, it could be argued that the general or speculative text of the Quran was made specific and definitive by the Sunna (Kamali 1991, p. 24). The specific texts of the Quran such as the punishment of eighty lashes for false accusation, is an example of a definitive ruling. The texts of the Quran on *shura* or consultation (3:159, 42:38) are typically general rulings in which no clear indications are provided on the process and rules of consultation. As the text lends itself to speculation, modernist Muslims have read into it a clear argument for democracy in the sociopolitical and economic realm and have derived from it political systems varying from representative parliamentary democracy to revolutionary workers' councils.

Mahmasani maintains that the task of interpreting the verses of the Quran has led to different and conflicting opinions and judgments. He opines that the most authentic interpretation could be found among the Prophet's companions, who were closest to him and were therefore knowledgeable about the why, where and when of each revealed verse (Mahmasani 1979, p. 128). Mahmasani's position might lead one to expect that the interpretations of the four Rightly Guided Caliphs, Abu Bakr, Omar, Uthman, and Ali, would evince a great degree of homogeneity and similarity since their authors were all witnesses to the circumstances of the revelations and learned the inner meanings of the verses directly from the Porophet. Individual interpretations of Quranic texts by the four Caliphs, however, indicates marked differences.

On the issue of wealth and its distribution, the position of the third Caliph, Uthman, differed from that of the other three. Uthman is said to have been in favor of the individual pursuit of wealth, while the other three are said to have been egalitarian and even biased against the wealthy who conspicuously flaunted their riches. During the last days of his life, Omar is reported to have said that "[i]f I continue to be in a position of power and responsibility, I will take away the excess wealth of the rich and divide it among the poor" (Va'ezzadeh-Khorasani 1990, p. 21). Informed that Sa'd ibn Abi Waqqas, a prominent Islamic figure and warrior, had built a palace to seclude himself from the public, Omar ordered the palace to be burnt down (Ibn Taymiya 1983, p. 25). In contrast, Uthman is said to have encouraged individual enrichment without due consideration for the condition of the poor, and to have given free reign to those who accumulated wealth (Va'ezzadeh-Khorasani 1990, p. 25).

Va'ezzadeh-Khorasani (1990, p. 23) argues that such conflicting interpretations and positions could result from reference to different and even contradictory verses in the Quran. He concludes that Omar probably based his position on a verse in the Quran which states that the wealthy will be punished on the Day of Judgment with the molten gold and silver they accumulated (9:35), while Uthman's interpretation is probably based on another, in which people are invited and encouraged to wear and display their adornments and beautiful clothing (7:31) .

The fact that different verses in the Quran could be called upon to support different, and at times opposing, opinions on a single issue was apparent even to the Rightfully Guided Caliphs. In the directive that Imam Ali, the fourth Caliph, gave to his envoy who was to confront and negotiate with the Khawarij, who had rebelled against their Caliph, he wrote: "Do not argue with them by the Quran, because the Quran has many faces. You would say your own and they would say their own; but agree with them by the Sunna, because they cannot find escape from it" (Ali 1984, p. 567).

THE SUNNA

Other than enunciating God's manifest and direct revelations in the form of the Quran, the Prophet was also inspired by God and pronounced his inspirations in his own words. Such internal revelations (*wahy batin*) differ from manifest revelations not in their source, which is God, but the fact that internal revelations are inspirations while the Quran is His direct word. The sayings and utterances of the Prophet, known as *ahadith* (plural for *hadith*) which is a significant component of his Tradition or Sunna, is, therefore, based on internal revelations (Kamali 1991, p. 15). The word Sunna has many literal meanings, among them, the clear path, custom, tradition and the way of life (Qadri 1986, p. 189). The everyday words and utterances of the Prophet, in addition to his acts and his tacit consent or acknowledgment of statements, constitute the Tradition or Sunna of the Prophet. Hadith is the report of sayings, acts, or acknowledgments by the Prophet. The Sunna is "what was practiced" and the Hadith is "the record of what was practiced" (Qadri 1986, p. 190). It is suggested that at the end of the first century, the Sunna of the Prophet became a part of Islamic legal theory at the hands of jurists in Iraq (Kamali 1991, p. 45). The Tradition of the Prophet, in its entirety, is regarded by all Islamic schools of law as the most important source of Islamic jurisprudence, second only to the Quran. The Sunna is related to the Quran in three main ways. Its rules may merely confirm and reiterate the Quran, it may explain, clarify, or specify its rulings, or it may enunciate new rulings on which the Quran had been silent (Kamali 1991, pp. 60-62). For example, the punishment of *rajm*, or death by stoning for adultery when committed by a married man, has its origin in the Sunna. The Sunna according to the Twelver Shi'i jurists, also includes the words deeds and acknowledgments of the twelve infallible Shi'i Imams. The Twelver Shi'i's broadening of the authoritative sources capable of exploring and explaining the hidden meanings of the Quran can lead to a proliferation of positions and paths.

During his lifetime, the Prophet is said to have discouraged his companions from compiling his sayings in a written form; however, he instructed them to disseminate them orally (Mahmasani 1979, p. 133). Encouraging the faithful to follow the Quran and his own tradition as the path to salvation, the Prophet is reported to have said; "I leave with you two guides; if you follow them faithfully, you will never go astray: the Quran and my practice" (Qadri 1986, p. 191). While the Prophet was alive, the authenticity of his sayings could be verified. After his death, however, the absence of a rigorously documented collection of his sayings and acts gave rise to legal problems as well as to some confusion over proper Islamic conduct in the community. Words and acts fabricated by individual opportunists for private, racial, tribal or political interest, and unreliable sayings attributed to the Prophet, were in circulation and used to support individual positions.

Jurists and experts of Hadith are unanimous on the existence of extensive forgeries in the Hadith literature.

Even though the Sunna is accepted as a primary source of Islamic law by all Islamic schools of jurisprudence, not all existing Hadiths are accepted as sources of law by all these schools. The Shi'i school accepts and uses those Hadiths the reporting of which can be traced to the infallible Shi'i Imams. They call these Hadiths, *akhbar* or reports (Mahmasani 1979, p. 66).

Abu Hanifa (d. 767), the founder of the Hanafi school and a contemporary of some of the Prophet's Companions, is reported to have been extremely selective in accepting the authenticity and validity of Sunna. It is suggested that he accepted only a few Traditions as genuine (Qadri 1986, p. 95). The number of Hadiths, the authenticity of which he accepted and subsequently used as legal evidence, is put at only seventeen (Mashkour 1989, p. 170). Hanifa, who resorted to reasoning in his Islamic rulings, is known as the founder of the speculative tendency of *ahl-e ra'y*, or "the people of opinion." Jurists of this tendency believed in rational interpretation or independent opinion as a source and a means of arriving at Islamic rulings.

Malik ibn Anas (d. 796), founder of the Maliki school, compiled, accepted, and relied on many Hadiths which he deemed authentic, even though they were solitary and reported by only one narrator (Mahmasani 1979, p. 47). Even though he was a renowned expert on Hadith, he was not a Traditionist. His emphasis on Hadiths for legal deductions was no greater than any of the other three sources of the law. Considering Medina as an exceptional city that reflected the customs and practices of the Prophet, Malik attached great importance to the traditions of the city. In his selection of Hadiths it is said that Malik rejected those which reported an act or a custom that went against the established tradition and practice of the city of Medina (Qadri 1968, p. 114).

Shafi'i (d. 820) was the founder of the Shafi'i school. He is generally considered as the person who articulated the methodology of fiqh. It is said that it was Shafi'i who spelled out the basic outline of four principal sources of the law, which was subsequently accepted by Islamic jurists (Kamali 1991, p. 5). Although both the Quran and the Sunna are considered as primary sources, the primacy of the Quran over the Sunna is widely accepted by most Islamic jurists. However, for Shafi'i, the Sunna constitutes a legal source as important as the Quran. Shafi'i is reported to have maintained that both the Quran and the Sunna "are from God though the reason for their revelations differ" (Qadri 1986, p. 128). In cases of a difference between the injunctions of the Quran and those of the Sunna, Shafi'i argues that the Quran cannot abrogate a Tradition; that has to be done by another Tradition (Qadri 1986, p. 128).

For Ahmad Hanbal (d. 857), the founder of the Hanbali school, research into the Sunna provides the most solid approach to the achievement of a sound legal solution. His verdicts, therefore, were primarily based on the Tradition of the Prophet. Hanbal accepted and collected more than forty thousand Hadiths in his book *Masnad al-Imam Ahmad.* The Imam is said to have accepted and relied on discontinuous, solitary, and weak Hadiths (Qadri 1986, p. 141). If the Tradition of the Prophet did not provide solutions for the problems that had been brought to his attention, Hanbal sought the legal opinion of the Companions of the Prophet. He is reported to have said that, "Anyone who rejects the Traditions of the Prophet brings himself to the verge of death" (Qadri 1986, p. 139). Hanbal is known as the real founder of the Traditionist tendency of *ahl-e Hadith* or "the people of Hadith." Jurists of this tendency place heavy emphasis on the Hadith and reject any and all application of human reason as a source of law. The different schools of Islamic jurisprudence span the extremes of total rejection of human reason to quasi-reliance on human reason as a source of law. Shi'i jurists are also divided between those who belong to the people of Hadith, called the *akhbariyoun* or the people of khabar (another name for Hadith), and the people of independent judgment or *ra'y*, called the *usuliyoun.*

IJMA' OR CONSENSUS

Consensus or *ijma'* among Muslim jurists on a particular legal issue, during a specific period of time, constitutes the third source of Islamic law. As long as the Prophet was alive, there was no reason to rely on anything but the Quran and the Prophet. With the death of the Prophet, the medium through which divine laws could be enunciated was no longer available and individuals, as God's vicegerents, had to apply their reason to deduce legal opinion to deal with new circumstances. An opinion that would become law as a result of consensus is the product of human reasoning and as such ijma' is a rational proof of Islamic law, based on the letter and spirit of the primary sources. The principle of ijma', or general agreement among the companions, was successfully applied for the first time in the election of Abu Bakr as the first Caliph (Muslehuddin 1986, p. 146). The event, however, challenged by the Shi'is, brought the crucial issue of the exact meaning of consensus to the foreground. It was argued that certain Companions such as Sa'd ibn Ubad, never conceded the election of Abu Bakr, thus invalidating the claim to a meaningful consensus (al-Rafi'i 1987, p. 61).

Accepting consensus as a legal proof, different Islamic schools of jurisprudence offer different interpretations of who is eligible and qualified to participate in the process of reaching an agreement, and of how consensus is to be reached. The Shi'i school defines consensus as the agreement of Shi'i jurists or *ulema* on an issue. The Hanafi school maintains that consensus can be reached by the public agreement of the jurists or through the silence of other jurists once the legal opinion or judgment of one Islamic jurist has become public knowledge (Mahmasani 1979, p. 145). According to the Hanafis and the majority view among the ulema, a consensus is valid only through the participation of the *mujtahid*s or experts in Islamic law. This view excludes the lay opinion. A minority view represented by Shafi'i and al-Amidi suggests that consensus did not imply the agreement of a select few, but that of the "public" or "the entire community" (Shafi'i 1987, pp. 286-287; Kamali 1991, p. 173). According to this view, which expands the scope and modern practical application of consensus, ijma' includes the agreement of both the laymen and the mujtahids.

The Maliki school, preoccupied with the sanctity of the city of Medina, maintains that consensus is only valid if it is reached by the residents of this city. The three other schools of Sunni jurisprudence all reject the rigidity of the Maliki position. The Shafi'i school also rejects the Hanafi method of ascertaining the validity of a judgment through the silence of other jurists (Mahmasani 1979, p. 145). Shafi'i is further reported to have accepted consensus as a proof only when it applied to "the principles of duties established through a definitive text or through well established traditions" (al-Rafi'i 1987, p. 60). Whereas, according to Malik, consensus was limited to the citizens of Medina, Hanbal limited the validity of consensus only to the practice of the Prophet's Companions (Mahmasani 1979, p. 144). Certain Hanbali jurists further limit the application of consensus by reducing it to the agreement of the Four Rightly Guided Caliphs (Doi 1984, p. 68). The original Hanbali's disdain for the application of reason led them to place such rigid conditions that would effectively foreclose the possibility of legislation through consensus by generations after the Companions.

In the history of Islamic jurisprudence, deducing new laws on the basis of consensus was, however, limited to the very short period between the death of the Prophet and the "closing of the door of ijtihad." Since consensus was the application of independent thought, with the closure of the "door of ijtihad" in the tenth century, the dynamic capacity of Islamic law came to an end. All problems after the tenth century had to be resolved with laws promulgated until the tenth century. Modernist Islamic scholars and jurists who argue that the door of ijtihad is open, hold that consensus provides the mechanism for the modernization and updating of Islamic law according to evolving circumstances and changing conditions. Politically, consensus is argued to be the justification for a democratic state in which government and legislation is based on "public opinion." Consensus can certainly act as a tool for establishing public policy priorities and objectives in Islamic societies. The extent to which consensus can be considered as a dynamic, innovating and change-inducing factor depends on its definition and according to the different schools of jurisprudence, which themselves are subject to changing opinions.

QIYAS OR ANALOGY

Analogical deduction or *qiyas* constitutes the fourth proof or source of Islamic legislation for the Sunni schools of jurisprudence. The Shi'is, however, reject the validity of analogy and replace it with *aql* or human reason. Analogy is a process of legal deduction according to which the jurist, confronted with an unprecedented case, bases his arguments on the logic used in the text of the Quran or the Sunna and applies the same logic to the case at hand. The authority of analogy cannot be upheld unless reason, applied by jurists, is also accepted as a source of law. Legislation on the basis of analogical deduction is, however, generally limited to matters of social and financial affairs.

Mahmasani's explanation of the process of qiyas indicates how Islamic law may be applied to making public policy. He maintains that "[t]he general principle is based on the understanding that every injunction guarantees a beneficial and welfare satisfying objective. The underlying principles of laws are none other than these welfare satisfying conditions. Therefore, if we can deduce the cause of an injunction in the primary sources, we can apply analogical deduction to cases with similar causes and pronounce a valid injunction" (Mahmasani 1979, p. 146).

Two types of analogical deductions emerge on the basis of this definition. First, there are the analogical deductions based on clear and apparent examples in the primary sources. For example, Imam Ali's inference that the penalty for wine-drinking, not stipulated in the Quran or the Sunna, should be equivalent to the punishment for slander and false accusation specified in the Quran, since an intoxicated person will tell lies which is similar to false accusation. These types of analogical deductions are called "strict analogies" (*qiyas tam*). Second, there are analogical deductions based on causes that are not clear and apparent in the primary sources. In these, the jurist relies primarily on his own reason and pronounces an injunction on the basis of the implicit meaning of a primary text. These are called "sound analogies" (*ta'wil*) (Muslehuddin 1986, pp. 129-130). The application of human reason, rooted in, and conditioned by, the primary sources, thus provides a potent and perpetual instrument for problem solving and adaptation in the form of consensus and analogical deduction. Injunctions based on these secondary sources, therefore, represent a synthesis between the divine revelations and the words and acts of the Prophet, which could be considered as fixed and static, and human reason, which is flexible and dynamic.

The proponents of qiyas and ijtihad have argued that emerging problems, especially those related to financial transactions, are numerous and one cannot find relevant injunctions for each and every particular event or problem in the primary sources. Shahristani, a renowned Islamic jurist has argued that "since clear and directly pertinent injunctions in the primary sources are limited and events are unlimited and the unlimited cannot be confined to the limited, many problems have remained without appropriate Islamic solutions. This situation necessitates the application and practice of ijtihad and qiyas by jurists" (Howzeh 1988). In his search for suitable responses to emerging problems, Abu Hanifa provided greater scope for human reason and discretion. He took the final step of passing judgments not on the basis of the explicit meaning of texts in primary sources (sound analogy), but on what the jurist believed to be in the "spirit" of, and "beneficial" to, the general objectives of Islam. This type of ruling is usually believed to be based on "public interest" and the welfare of the Islamic community.

The closing of the door of ijtihad, among the Sunni schools of jurisprudence, signaled an end to the capacity of the law to develop itself with changing times. The institution of reflection and exercise of rational thought based on the divine sources or the spirit of Islamic law was arrested and gave way to the uninspiring custom of accepting one of the established schools and imitating its precepts. Thus the law became introspective, backward looking and pedantic, losing touch with social development and change, throughout the ages. Jurists became concerned with formalistic and casuistic issues the use of which for society was extremely limited. Instead of dealing with major and pressing socioeconomic and political problems of their epoch, jurists limited their independent judgments to secondary and petty issues of minor significance. For example, they ruled that "melted butter into which a mouse had fallen and drowned, could not be used as fuel for lamps because the air would be thus polluted by the impurity of the flesh of a dead animal" or that "it was not permissible to ride a camel which had drunk wine because of possible contact with the forbidden substance through the sweat of the camel" (Coulson 1990, p. 81). Conservatism in the application of independent judgment among the jurists led to the gradual decline of the Arab civilization (Mahmasani 1979, p. 30).

TERTIARY SOURCES

Once the application of human reason in legislation was implicitly endorsed by both Sunni and Shi'i jurists, the door was opened to greater speculation and interpretation on deducing laws from the primary sources. This tendency allowed for an increased reliance on the general spirit of Islam rather than the particular letter of the law. Based on their individual assessment of the spirit of Shari'a, jurists singled out criteria or standards, they believed to be the quintessential attribute and object of the law, on which legal judgments could be based. The sources of the law were expanded from the unalterable letter of the law to include the highly speculative and therefore maneuverable domain of the spirit of the law. Those single standards, which became accepted as sources of Islamic law, constitute the elements of the tertiary sources. Imams of different schools, to the extent that they came to accept the use of such standards, identified different quintessential attributes of the spirit of Islam. Each, however, justified their newly articulated yardstick on the grounds that it furthered the public welfare of Muslims and facilitated their everyday life. Tertiary sources or standards provide Islamic law with a considerable degree of freedom, leverage, and flexibility to deal with modern problems and issues.

The basis on which Abu Hanifa pronounced his Islamic rulings is called *istihsan*, which literally means "to approve, or to deem something preferable" (Kamali 1991, p. 246). Based on the notion that the spirit of Islam is governed by equity and fairness, istihsan or juristic preference is employed by the jurist to avoid unfairness, which may result from the strict application of the letter of existing laws. The jurist employing istihsan, applies his independent reasoning and exercises his personal opinion to serve the public interest and uphold justice. According to the Hanafi jurist al-Sarakhsi, since the avoidance of hardship constitutes a cardinal principle of religion, istihsan departs from analogical deduction "in favor of a ruling which dispels hardship and brings about ease to the people" (Kamali 1991, p. 247).

Malik, identified the cardinal principle of religion as that of satisfying the "public good or interest." Islamic jurists agree that the objective of the Shari'a is to protect and secure the "five necessities" (*zaruriyat khamese*) of life, religion, reason, progeny, and property (Siddiqi 1988, p. 258). To protect these necessities, laws and injunctions would have to be based on the principle of securing public interest or benefit (*maslaha*). The Maliki school proceeded to replace strict analogy with the wisdom of the jurist in his pursuit of securing public interest. The tertiary source of law developed by Malik is called *al-maslaha al-mursala*, which has been translated as unrestricted public interest, social benefit, public interest, and public weal. This concept is synonymous with istislah. The jurist wishing to minimize or prevent hardship, while assuring the protection of the "five necessities," would rule on the basis of maslaha mursala.

Shafi'i, a firm believer in "strict analogy," rejected Abu Hanifa's istihsan and Malik's maslaha mursala on the grounds that the discretionary power of jurists, based on independent reasoning and unrestrained by clear evidence in the primary sources, was tantamount to secular legislation interfering with divine law which was eternal (Mahmasani 1979, p. 158). Despite Shafi'i's official and seemingly inflexible opposition to independent reasoning, the Shafi'i jurist, al-Amidi, has suggested that Shafi'i did resort to istihsan when necessary (Kamali 1991, p. 259). Based on public interest and employing *istidlal* (argumentation), Shafi'i is said to have deduced laws on the basis of a variant of istihsan (Doi 1984, p. 8).

Hanbal, whose school of jurisprudence is known as the most averse to analogical deduction, let alone reasoning, is also said to have used independent judgment. On issues involving transactions, where no direct evidence was found in the primary sources, he too relied on securing public interest. He used Malik's concept of maslaha mursala, but he called it *istislah* (Doi 1984, p. 81). Prominent Hanbali jurists, such as Ibn Taymiya and Ibn Qayyim al-Jawzi, later broke with the opposition of their master toward reasoning. They engaged in and justified the application of ijtihad in order to improve the welfare of the community (Qadri 1986, p. 143).

Finally, overcoming the pressing needs and necessities of the Muslim community is a basis for deducing legal injunctions, hence the Islamic maxim that "necessity permits prohibited things." This is an ad hoc criteria that allows jurists to confront emergencies. It is clearly considered as a temporary measure, suspending the primary ordinances of the faith. Referring to the five necessities, Ghazali's statement that "all prohibited things become permissible in case of necessity," opens the door to a multitude of innovation once this cardinal standard is invoked. Even the drinking of wine as a medicine can become permissible (Muslehuddin 1986, p. 182).

CIRCUMVENTIVE TOOLS

From the very early days of Islamic society, the incongruity and disparity between the specific and clear injunctions of the Shari'a and the commercial needs and practices of the community gave rise to the articulation of a set of practices known as legal tricks or devices. These were legal devices that were employed to formally legalize an act that had been clearly categorized as forbidden by the Shari'a. In such cases, the letter of the law was manipulated to obtain ends essentially opposed to the spirit and intent of the Shari'a. The fact that the acts or practices employed for circumventing the intent of the law are permissible by the Shari'a, provides legal tricks with a semblance of legitimacy. The religious admissibility of legal tricks, which deceitfully transforms an act forbidden by the Lawgiver into a permissible one, has long been a source of debate and controversy among Muslim scholars and jurists.

The most prevalent cases in which legal tricks have been employed are in circumventing the clear and immutable prohibitions of *riba* or usury and *gharar* or any form of speculative risk in contracts. Through legal sale transactions, parties could effectuate an illegal usurious transaction, without the explicit payment of interest. Numerous types and kinds of legal tricks have become prevalent practice in the Muslim world (Qadri 1968, pp. 266-267; Coulson 1990, p. 139).

Abu Hanifa and Shafi'i are said to have allowed the use of religious tricks (Qadri 1968, p. 334). The justifications used to support their positions is said to range from invoking dire financial necessity on the part of the borrower, and therefore the use of the maxim that "necessity permits prohibited things," to the argument that "God alone knows the motive of man," therefore, the motives of individuals are left to God, while only the formal or external deeds are judged (Qadri 1968, p. 335). Jurists in the Hanafi, Shafi'i and Shi'i schools are said to be lenient toward the acceptance of legal tricks (Mahmasani 1358, p. 212). Maliki and Hanbali jurists are on the other hand firmly opposed to legal tricks. Malikis argue that the motive of the contract is essential and determines the legality or the illegality of it. If the motive of the parties is illegal then the act is illegal.

Like many other issues in Islamic law, the position of scholars from different schools of jurisprudence differs on the degree of permissibility of legal tricks. The fact, however, that a clear and firm injunction such as that on usury can be circumvented and stood on its head, not only by individual Muslims but even by states that rule in the name of Islam, demonstrates that Islam lends itself to a large degree of interpretation. Within certain limits, very much like western law, legislators, be they Islamic jurists or laymen, as the ultimate interpreters of the law, have the power and could if they wish, fashion the law in just about any manner they choose. They could concentrate the power of the state in the hands of one jurisconsult, in the case of Khomeini, and allow him to go as far as claiming the right to suspend the basic principles of the faith, such as prayers and pilgrimage, for reasons of state. They could also minimize the role of religion in the socioeconomic and cultural affairs of individuals as is the case in countries such as Pakistan and Indonesia.

The spirit and letter of Islam is similar to that of all laws in the sense that it wishes to protect the public interest and promote public welfare while facilitating the everyday life of the people. Public interest and welfare is defined within an Islamic paradigm. If the application of Islamic law in certain countries seems contradictory to these objectives, it is obviously due to fundamental differences in the perception of what constitutes social interest, welfare, and ease of life between Islamic and non-Islamic paradigms. Just as any state that allows for government intervention to effectuate change, Islamic law, which constitutes the basis of an Islamic state, is amenable to the concept of public policies. The particulars of such policies would obviously have to be measured according to Islamic standards set by the Islamic state adopting them. The fact that the letter of the law represents socioeconomic and cultural conditions of some fourteen centuries ago has enabled those who wish to establish a modern Islamic state to interpret the law and update it according to the changing times. The updating of Islamic law has generally been associated with greater reliance on and use of the spirit of the law rather than its letter. Despite a cerebral resistance against modernization of Islamic law by the official custodians of Islam, the attainment of political power when in opposition and its maintenance once in power has forced a greater degree of adaptation, accommodation, and modification upon Islamic leaders. The point, however, is how far can the guardians of Islamism allow the modernization of Islamic law before it looses its Divine component and becomes thoroughly secular.

NOTE

1. All references to the Quran are based on the translation and commentary by Abdullah Yusuf Ali (1989).

REFERENCES

Ali, A.Y. 1989. *The Holy Quran; Text, Translation and Commentaries*. Brentwood, MD: Amana.

[Ali ibn Abi Talib]. 1984. *Nahjul Balagha: Sermon, Letters and Sayings of Imam Ali.* Rome: European Islamic Cultural Center.

Coulson, N.J. 1990. *A History of Islamic Law*. Edinburgh: University Press.

Doi, A.R.I. 1984. *Shari'ah: The Islamic Law*. London: Ta Ha Publishers.

Howzeh. 1988. (Summer).

Ibn Taymiya. 1983. *Public Duties in Islam*. Leicester: The Islamic Foundation.

Kamali, M.H. 1991. *Principles of Islamic Jurisprudence*. Cambridge: Islamic Texts Society.

Klein, F.A. 1985. *The Religion of Islam*. London Curzon Press.

Mahmasani, S. 1979. *Falsafe-ye Qanoongozari dar Eslam*. Tehran: Amir Kabir.

Mashkour, M.J. 1989. *Dictionary of Islamic Sects*. Mashhad: Astan Qods Razavi.

Mawdudi, A.A. 1988. "Political Theory of Islam." In *Islam–Its Meaning and Message*, edited by K. Ahmad. Leicester: The Islamic Foundation.

Muslehuddin, M. 1986. *Philosophy of Islamic Law and the Orientalists*. New Delhi: Taj Printers.

Qadri, A.A. 1986. *Islamic Jurisprudence in the Modern World*. New Delhi: Taj Company.

Rafi'i, M. 1987. *Islamuna*. Middlesex: The Muhammadi Research Trust.

Schacht, J. 1991. *An Introduction to Islamic Law*. Oxford: Clarendon Press.

Shafi'i. 1987. *Risala*. Cambridge: The Islamic Text Society.

Siddiqi, M.N. 1988. "The Guarantee of a Minimum Level of Living in an Islamic State." In *Distributive Justice and Need Fulfillment in an Islamic Economy*, edited by M. Iqbal. Leicester: The Islamic Foundation.

Va'ezzadeh Khorasani, M. 1990. *Majmou'eh-ye Farsi Avalin Majma' Barrasiha-ye Eqtesad-e Eslami*. Mashhad: Astan Qods Razavi.

ISLAM AND HUMAN RIGHTS POLICY

Ann Elizabeth Mayer

INTRODUCTION

For some decades various institutions and ideologues have proposed that there should be a distinctive Islamic approach to human rights questions. With the founding of the Islamic Republic of Iran in 1979 and the surge in Islamic fundamentalism, how Islam should affect human rights policy ceased to be a theoretical question and became of practical political importance, prompting a burgeoning of the relevant literature (Afkhami 1995; Afshari 1994; Chehabi 1990; Dwyer 1991; Mayer 1995; Mernissi 1992; An-Na'im 1990; Waltz 1995). As the following selective review of some relevant initiatives will indicate, exactly how Islam should relate to human rights policy remains a matter of controversy.

BACKGROUND

The contemporary Arabic term for human rights, *huquq al-insan*, as well as the counterparts *hoquq-e ensani* and *hoquq-e bashar* in Persian, have only recently come into use. The ideas of the right of the individual Muslim, *haqq*

International Review of Comparative Public Policy, Volume 9, pages 123-148.

ISBN: 0-7623-0268-2

al-'abd, as distinguished from the right of God, *haqq allah*, were familiar to Islamic jurists, but the right to protection for private property, which could be invoked against incursions by the state, was the only right recognized by them that actually corresponds to a modern right (Coulson 1957). Islamic legal culture was more geared toward defining the duties that it was incumbent upon the believer to perform than toward setting forth the rights of the individual vis-à-vis the ruler. According to the traditional perspective, rulers had obligations to rule according to Shari'a law and to treat their subjects justly; their subjects were to obey them unless ordered to do something sinful. Legal and political instititions that could place effective curbs on rulers' despotism and injustice were not adopted till the nineteenth century. Before then, the only way to combat tyranny and oppression was through rebellion. In contrast, human rights theory presumes that governments will rest on modern norms of constitutionalism that encompass popular sovereignty, meaning that governments are responsive to the popular will.

Contentious issues in Islamic thought since the nineteenth century have been the degree to which Western constitutionalism, schemes of effective restraints upon the exercise of government power, and democratic freedoms are compatible with adherence to Islamic law. By the end of the twentieth century, the focus had shifted to questions about whether international human rights could be accommodated in an Islamic system. The problem has been that there is no consensus on what Islam entails. The Islamic tradition, being rich and multifaceted, contains many ideas that anticipated modern rights, as well as others that are hard to reconcile with rights. Instead of saying that Islam per se dictates contemporary Muslims' positions on human rights, it is more accurate to say that Muslims' attitudes toward the merits of international human rights tended to shape their perceptions of the implications of the Islamic tradition for rights. In sum, there seems to be no monolithic Islam that bears on rights policy, only a series of "islams" that accord with Muslims' varying interests and philosophies as these bear on rights.

THEORETICAL PROBLEMS

There are thorny theoretical problems facing Muslims who want to develop public policy on human rights using Islamic standards. Some of these relate to how to define Islamic positions on issues that are already covered by international law; others relate to finding methods for extending Islamic principles to cover domains that were not treated in Islamic jurisprudence in the past, like the relationship between the modern nation-state and the individual.

Adding to the complexity of the task, as Muslims attempt to elaborate Islamic principles on human rights, they face existing human rights law in the form of numerous conventions and declarations that have been produced under the auspices of the United Nations since World War II. Human rights are part of the system of modern international law, which is supported by the consensus of nations. All Muslim states are United Nations members and are also governed by the system of international law. This is acknowledged in the Charter of the Organization of the Islamic Conference (OIC), the major international organization linking Muslim countries. OIC members reaffirm "their commitment to the UN Charter and fundamental Human Rights, the purposes and principles of which provide the basis for fruitful co-operation [sic] amongst all people" (Moinuddin 1987, p. 186). Thus, any Islamic human rights project should take into account the existing international system of human rights, as failure to do so can only lead to conflicts and confusion.

In August 1990 a meeting of OIC foreign ministers endorsed the the OIC's newly-minted Cairo Declaration on Human Rights in Islam, thereby putting forward a set of supposedly Islamic rights without having first dealt with the theoretical problems that this involved. The rights protections in the Cairo Declaration were feebler than those afforded by their international counterparts (Mayer 1994, pp. 327-350). Therefore, the OIC initiative suggested a weakening of Muslim countries' commitments to follow international law. Various Cairo Declaration rules were also impossible to reconcile with OIC members' conflicting obligations in terms of their ratifications of international human rights conventions (Mayer 1994, pp. 348-350). Since a number of Muslim countries had ratified many conventions, the resulting conflicts were in some cases extensive.

In a resolution on the declaration issued by the Twenty First Islamic Conference of Foreign Ministers in Karachi on April 25-29, 1993, the ministers reaffirmed the OIC commitment to "the principles contained in the Charter of the OIC as well as the 'Cairo Declaration on Human Rights in Islam' as general guidelines and the Charter of the United Nations."[1] By failing to stipulate that OIC members were bound by international human rights conventions, where modern human rights norms are comprehensively set forth, the ministers seemed to imply that they were only recognizing the much smaller number of rights afforded in the documents mentioned. Having failed to reaffirm any of the conventions, the OIC resolution casually acknowledged the existence of various international and regional human rights instruments to which OIC members were parties, without sorting out the relationship between these and the Cairo Declaration or explaining how the OIC could expect its members to adhere simultaneously to different sets of incompatible standards. One was left to wonder: Was the Cairo Declaration meant to supersede existing obligations under international law? Of course, another possibility was that the whole OIC initiative on human rights was so slipshod and so devoid of input from the many Muslims who were expert in international human rights law that none of those involved had even bothered to examine the problems that endorsing conflicting OIC resolutions and declarations would entail.

The Cairo Declaration was presented to the World Conference on Human Rights that was held in Vienna in 1993 as the OIC contribution to the conference—as if it represented the views of Muslim countries. The declaration thus presupposed a consensus on Islamic rights that was belied by political realities. Not only did the principles in the declaration conflict with Muslim nations' treaty commitments, they also clashed with many rights provisions in the constitutions and domestic laws of member countries, which in turn differed from each other.

The problems were not merely ones of the fit between OIC principles and relevant domestic and international rights commitments. The issuance of the Cairo Declaration also embroiled the OIC in difficult, unresolved issues of Islamic law. Islamic jurists had always envisaged Islamic law as the ultimate, definitive criterion of legality. They did not develop a jurisprudence that contemplated today's reality, where Muslims may be simultaneously subject to two different, competing systems of legality, one Islamic and one a set of universal secular principles grounded on the consensus of nation-states. They also did not envisage the possibility that individual Muslims might opt to be guided by secular norms on rights issues, so that the binding force of Islamic law could be challenged by Muslims appealing to a system of international human rights law based on entirely different premises. This is no longer a mere hypothetical possibility, because international human rights principles have won a considerable following among Muslims, as valuable correctives to the ills afflicting Muslim countries, where human rights are routinely and often egregiously violated by undemocratic regimes. As rights-consciousness has expanded in Muslim countries and a number of important human rights NGOs have emerged, Muslim supporters of human rights have treated the international standards as normative, even when they conflicted with aspects of Islamic jurisprudence, like the discriminatory features of Islamic personal status law. That is, Islamic human rights principles were being put forward by the OIC and other institutions at a time when, at the grass roots level, secular rights concepts were influencing Muslims' attitudes and expectations.

Governments of Muslim countries perceived the growing influence of human rights ideas as a threat, which encouraged them to concoct new sets of Islamic rules on human rights, in which Islamic criteria would be employed to override and circumscribe human rights so as to render them toothless. Conservative Muslims had reason to applaud Islamic human rights schemes that would use the authority of religion to delegitimize secular versions of rights in order to maintain old hierarchies and forestall any expansion of freedoms. Being reactive in nature, the Islamic rights schemes were naturally more intelligible in terms of the politics of human rights than in terms of Islamic theology and law. That they are fraught with theoretical deficiencies should therefore come as no suprise.

What would be the basis for the authority of Islamic rights schemes promulgated by governments of modern nation-states in terms of Islamic jurisprudence? A document like the Cairo Declaration was produced via a process unlike the processes by which Islamic rules have been traditionally derived from the sources. Instead, the process mimicked the way that public international law is made, in that representatives of nation states acted as the decision makers. That is, provisions were adopted in the Cairo Declaration simply because they had won the assent of nation-states. However, nation-states have no authority to declare Islamic law. The fact that Muslim states endorse principles may give these principles some legal cachet among their fellow states, but agreements by governments of nation-states signify nothing in terms of the Islamic theory of sources (Mayer 1990a). The OIC seems to have elected to dismiss the importance of nailing down the Islamic premises behind the declaration, neglecting even to try to explain how a "declaration" by Muslim nation-states could fit in the scheme of Islamic jurisprudence, where such state-sponsored "declarations" were unknown.

The problem of the dubious Islamic authority of the Cairo Declaration was aggravated because of the nature of the regimes involved. The states in the OIC system had, for the most part, no claim to have any Islamic character, and the pretensions of certain states like Saudi Arabia and Iran to have Islamic governments were contested by many of their own citizens as well as by other supposedly Islamic states. One would be hard-pressed to offer persuasive Islamic rationales explaining why Muslims should consider themselves religiously bound by principles agreed to by regimes as disparate as Qadhdhafi's eccentric, vaguely Maoist military dictatorship; Iran's Shi'i theocracy; Saudi Arabia's absolute monarchy with its Wahhabi orientation; Saddam Hussain's totalitarian Baathist system; the elected government of secular Turkey; or Sudan's Sunni fundamentalist military clique. Bitter enemies like Iran and Saudi Arabia had heaped scorn on each other's official versions of Islam. What credentials did these states have either individually or in common for dictating to Muslims what Islamic law should be—whether on human rights or on other subjects?

Another theoretical problem compromising the credibility of the Cairo Declaration was that the Muslim states that had created that document had, for the most part, strikingly poor human rights records; countries like Indonesia, Iran, Iraq, Libya, Saudi Arabia, and Syria counted among the most flagrant rights violators in the world. When countries notorious for disregarding human rights exerted themselves on behalf of winning the acceptance of the Cairo Declaration in lieu of international law, it did nothing to promote the authority of that document. Indeed, with such states backing it, it was natural to assume that the declaration would serve the policies of these regimes and strengthen the hand of governments at the expense of the rights and freedoms of their oppressed and restive citizens. In these circumstances it was no surprise to see that supposedly Islamic criteria were employed in the Cairo Declaration exclusively as a mechanism for curtailing the rights and freedoms afforded to Muslims under international law.

A list of representative provisions can illustrate the character of the declaration, which has more problematic provisions than valuable ones. Critical is the provision in Article 24 providing that all the rights and freedoms stipulated in the declaration are subject to the Shari'a, subordinating all rights and freedoms to vague Islamic derogations. In lieu of provision for equal rights for all persons regardless of sex or religion, Article 1 states that "all men are equal in terms of basic human dignity and basic obligations and responsibilities [not rights], without any discrimination on the grounds of race, colour, language, sex, religious belief, political affiliation, social status or other considerations." Article 6 provides that "woman is equal to man in human dignity," omitting to provide for equality in rights and imposing on the husband the responsibility for the support and welfare of the family, thereby accommodating discriminatory laws in the area of personal status. In contrast, in a constructive step, Article 13 provides that men and women are entitled to fair wages "without discrimination." Article 5 provides that on the right to marry there should be "no restrictions stemming from race, color or nationality," not prohibiting restrictions based on religion, thereby accommodating the bans on Muslim women marrying outside the faith and on Muslim men marrying others than Muslims, Christians, or Jews. International human rights are religiously neutral, but Article 10 reveals the strong Islamic bias of the OIC scheme, stating that Islam is the religion of unspoiled nature and prohibiting "any form of compulsion on man or to exploit his poverty or ignorance in order to convert him to another religion or to atheism." Article 9 calls for the state to ensure the means to acquire education "so as to enable man to be acquainted with the religion of Islam," which could support mandatory instruction in Islam, even for non-Muslims. Article 2 prohibits taking away life except for a Shari'a-prescribed reason, and Article 19 states that there should be no crime or punishment except as provided for in the Shari'a. These articles permit the death penalty to be imposed for crimes such as adultery or apostasy from Islam and would accommodate punishments like crucifixion, stoning to death, amputation, and floggings, which conflict with international law prohibiting cruel, degrading, and inhuman punishments. Obviously, these articles are meant to allow countries like Iran, Saudi Arabia, and Sudan to continue to apply penalties taken from the Shari'a that are banned by modern standards of humane criminal justice. Given the serious deficiencies of criminal procedural rules in the Islamic tradition, the provision in Article 19 assuring a defendant "a fair trial in which he shall be given all the guarantees of defence" combined with the assurance in Article 25 that the Shari'a "is the only source of reference or the explanation or clarification of any of the article of this Declaration," leaves open the possibility that few, if any, of the safeguards of modern criminal procedure are included. Article 20 prohibits torture, albeit in an equivocal formulation (Mayer 1995, pp. 342-343). Article 18 stipulates a right to privacy in the conduct of private affairs, in the home, in the family, and regarding property and relationships, which could be used constructively to ensure a zone of privacy. Article 15 sets forth "rights of ownership" to "property acquired in a legitimate way," which is guaranteed against confiscation "except for a necessity dictated by law." Article 22 provides that freedom to express opinions is allowed in such a manner as would not be contrary to the principles of the Shari'a, thereby significantly qualifying this freedom, since how Shari'a qualifications would affect contemporary issues of freedom of expression is left open to speculation. Article 14 prohibits *riba*, which can mean usury or interest, a very peculiar precept to include in a scheme of human rights. Reflecting the Third World setting in which Muslim nations elaborate their positions on rights, Article 11 prohibits colonialism and states that "peoples suffering from colonialism have the full right to freedom and self-determination." Notable by their absence are provisions calling for the observance of democratic principles in political systems, protecting freedom of religion, freedom of association, freedom of the press, and guarantees of equality in rights and equal protection of the law.

Islamic human rights schemes like the Cairo Declaration recognize only a small range of rights compared with the vast number established in international conventions, but the rights that they do provide are not original but are borrowed from international law and then qualified or distorted. These appropriations of ideas taken from non-Islamic sources are not acknowledged or explained, but the language used is sufficient to betray their Western origins. After selectively appropriating certain substantive rights from the international models, the authors of Islamic human rights schemes then impose on the borrowed rights Islamic qualifications or derogation clauses that subordinate the rights or freedoms involved to "Islamic" limitations. There is no established theoretical basis for saying that foreign rights concepts can be Islamized by virtue of superimposing Islamic conditions on them. Moreover, requiring such Islamic conditions is an exercise in vacuity so long as specific definitions of what these conditions mean are not forthcoming. Leaving open how Islamic criteria should apply to limit human rights norms that have been borrowed from a non-Islamic source, international law, is obviously unsatisfactory from the standpoint of Islamic jurisprudence, and in terms of their practical consequences, the vague Islamic qualifications prove nefarious. When Article 24 of the Cairo Declaration says "All the rights and freedoms stipulated in this Declaration are subject to the Islamic Shari'a," it leaves the governments involved free to interpret the limits that Islamic law imposes as broadly as they please. Since many modern civil and political rights are designed to protect the rights of the individual against the state, allowing the state complete discretion to define the scope of rights and freedoms can render rights illusory.

Employing Islamic criteria to qualify and potentially nullify rights clashes with international law, according to which human rights law should reign supreme, overriding all conflicting domestic law, whether religious or secular. International law allows certain specified curbs to be placed on rights and freedoms but does not allow any religious law to override them (Mayer 1995, pp. 733-776). This is another area where clashes with international law have been created but not acknowledged.

With Islam being conceived of as a limitation on rights, the result of using Islamic human rights schemes to supplant international law is naturally that Muslims (and potentially also non-Muslims subject to the jurisdiction of Muslim states) wind up with inferior rights. No theoretical basis for assigning Muslims to an inferior standard of human rights has been offered, and candid discussion of the negative consequences for Muslims that these Islamic rights entail has been avoided by the governments and institutions that promulgated them. Instead, Muslim countries like Saudi Arabia that sponsored the OIC declaration have emphasized other supposed advantages to having Islamic rights, trying to portray Islam's human rights, because of their divine character, as offering stronger, uniquely efficacious protections for rights (Mayer 1994, p. 375).

No person or organization with competence and credibility in the area of human rights has endorsed the Islamic versions of human rights, recognizing that they have been inspired by policies of trying to supplant international human rights and strengthening the hands of governments at the expense of rights and freedoms. The retrograde character of the Cairo Declaration was roundly denounced by the International Commission of Jurists (ICJ) when it was submitted in 1992 to the U.N. Human Rights Commission, on which occasion Iran and Iraq both pressed for its acceptance (Vichniac 1992). Adama Dieng, the Senegalese Secretary General of the ICJ condemned the Declaration for challenging the intercultural consensus on the international human rights standards, introducing discrimination against non-Muslims and women, deliberately restricting certain fundamental rights and liberties, and ratifying under the cover of Shari'a the legitimacy of practices like corporal punishment (Vichniac 1992).

Around the globe, the strongest advocates of international human rights have tended to be outside governments, and Muslim societies are no exception. Independent Muslims and human rights NGOs in Muslim countries have not been persuaded to adopt the Islamic human rights principles being purveyed by countries like Saudi Arabia and Iran. Instead, they have continued to work courageously under highly adverse conditions for the advancement of human rights according to the international standards. Many Muslim feminists have been particularly vigorous in demanding adherence to international norms, which guarantee women full equality. In other words, the "Islamic" rubric pasted on the human rights, afforded in schemes like the Cairo Declaration, has not made the emasculated rights principles that they contain more palatable to knowledgeable Muslims who seek to forward the cause of human rights.

All this is not to say that clashes between Islam and human rights are necessarily inevitable. Islam can be developed in an altogether more constructive way that would enhance human rights, as the work of Abdullahi An-Na'im demonstrates (An-Na'im 1990). The inability of Islamic thought to date to offer more positive contributions to human rights does not imply that the Islamic heritage is deficient or that human rights cannot be successfully developed within an Islamic framework. There is ample material in the legacy of Islamic civilization for building a satisfactory system of protecting rights, but building this system is impeded by entrenched forces of reaction and a hostile climate for intellectual endeavor that threatens innovative and creative thinkers in Muslim societies. Not surprisingly, in a climate of fear and repression, Islamic thought has tended to be elaborated in ways that ratify existing hierarchies and legitimize repression, avoiding dealing with the actual rights problems facing Muslim societies. That is, the weakness of the theoretical literature in this area ties in with the political circumstances in which it is elaborated. Like the Muslims who live in the repressive environments of today's Muslim countries, Islam itself is a prisoner of politics. Reactionary forces are suspicious that every intellectual breakthrough will lead to a political threat and exploit appeals to Islamic authority to anathematize Muslims who want to experiment with new ideas that might transform Islam into an instrument of liberation. Influential clerics and fundamentalist leaders have attacked Muslims who dare to call for human rights and tolerance or who propose reforming and revitalizing Islam to accommodate rights and freedoms. Thus, Islam has become associated with the condemnations of the ideas and even threats to the lives of Muslims like Fazlur Rahman, Mohammed Arkoun, Toujan al-Faisal, Sadiq al-'Azm, Mehdi Bazargan, Fatima Mernissi, Abdullahi An-Na'im, Ayatollah Taleqani, Said al-Ashmawi, Nasr Hamid Abu Zayd, Taslima Nasreen, Nawal El-Saadawi and many others. Mahmud Muhammad Taha, a daring Sudanese reformist thinker who was strongly supportive of human rights, was excoriated by defenders of Islamic orthodoxy and was finally executed as an apostate in 1985 by the Numeiri dictatorship, which was then pursuing a fundamentalist agenda (An-Na'im 1986). This record means that Islam has become associated with rights violations when it is actually the political systems involved that are really to blame for Islam being used in this manner.

Since those with vested interests in maintaining the status quo are deeply threatened by revolutionized Islam, dynamic Muslim intellectuals who want to challenge old ways of thinking about religion are generally compelled to speak only obliquely or to go abroad to carry out their work. However, their working in exile means that they are cut off from their natural constituencies and this also attenuates the impact of their ideas in their own societies. One winds up with a situation where important figures are obliged to seek sanctuary in the West and to write in Western languages about issues that are vital for the regeneration of Islam and for the modernization of Islamic thought, leaving young and inquiring Muslims to learn from teachers who will inculcate in them the false notions that Islam demands acquiescence in policies antithetical to rights and that international rights embody alien values.

At the theoretical level, some of the most vigorous critiques of the inherited corpus of Islamic thought as it pertains to human rights issues have come from feminists, who have spoken out despite the grave risks that calling for the equality of women presents. One of the most salient features of Islamic human rights schemes is that they deny women the right to equality and non-discriminatory treatment that is guaranteed to them under international law, and provide Islamic rationalizations for the legal discrimination facing Muslim women under customary and traditional mores. Muslim feminists have challenged the view that Islam mandates an inferior status for women, arguing that discriminatory elements of Islamic law are not set forth in the text of the Quran or supported by the example of the Prophet, but were invented by male interpreters. Their view is that the Islamic sources, when properly understood, support the equality of the sexes.[2] Proponents of Islamic schemes of human rights that deny equality to women have to struggle to sustain retrograde interpretations of Islamic requirements in the face of feminist critiques. Islamic feminism is turning out to be a potent force for Islamic renewal and one of the most important factors contributing to the integration of human rights within a reformed Islam.

The reactions to contemporary Islamic feminism have varied. Depending on their outlook, Muslims may welcome proof that Islam can be reconceived as a progressive force that supports women's advancement and participation in society. Other Muslims may denounce Muslim women who question what Islamic jurists and clerics or fundamentalist leaders have decreed about women's status. Impeding the ability of women to carry out the mission of Islamic reform is the fact that they have long been excluded from Islamic institutions and are not recognized as members of the ulema or *fuqaha*. There is no woman Sheikh of al-Azhar, no woman mufti, no woman Ayatollah. Thus, women lack some of the institutional credentials and prestigious titles that men wield when speaking on behalf of Islam and that lend authority to the positions that men articulate. The rise of militant Islamic fundamentalism, which has shown hostility to women's equality and has called for women to remain subordinated to male authority, menaces Islamic feminism, and since the Islamic revolution in Iran, many accomplished Iranian feminists have felt obliged to flee abroad to seek refuge from repression and fundamentalist violence. The loss occasioned to the cause of reform has been considerable.

ISLAM AND RIGHTS POLICIES IN APPLICATION

Human Rights Policy in the Islamic Republic of Iran

The rights policies of Iran have turned out to be as problematic as those of the OIC and are perhaps even more convoluted. In the 1960s and 1970s under the rule of the Shah, Iranians took a leadership role in the United Nations in promoting the development of international human rights law, and Iran ratified the International Covenant on Civil and Political Rights (ICCPR). However, human rights were not respected in practice under the Shah. The 1979 Revolution awakened hopes that a new era of respect for human rights was dawning. Unfortunately, human rights policy under the Islamic Republic offered changes in style and rhetoric rather than changes in substance. Persecutions and executions of political opponents, routine recourse to cruelty and brutal torture, censorship, unfair trials, and other abuses continued, to which were added aggravated forms of discrimination affecting women and religious minorities.

Since the revolution, Iran has manifested ambivalence about whether it is bound by international human rights, and it has struggled to define a tenable "Islamic" position vis-à-vis human rights. As an example of Iran's ambivalence, one could cite the fact that the Islamic Republic did not choose to terminate its status as party to the ICCPR, even though it regularly violated ICCPR principles and often issued statements at odds with its principles. In imprudent remarks that have been cited by critics of Iran's human rights policy, Iran's U.N. Ambassador, Sa'id Raja'i Khorasani, proclaimed that, because of its Islamic values, the Islamic Republic would have no qualms about violating human rights. According to the paraphrased record of his speech, he announced that:

> conventions, declarations and resolutions or decisions of international organizations, which were contrary to Islam, had no validity in the Islamic Republic of Iran ... The Universal Declaration of Human Rights, which represented secular understanding of the Judaeo-Christian tradition, could not be implemented by Muslims and did not accord with the system of values recognized by the Islamic Republic of Iran; his country would therefore not hesitate to violate its provisions, since it had to choose between violating the divine law of the country and violating secular conventions (Mayer 1994, p. 316).

Thus, Raja'i Khorasani effectively asserted that, being guided by Islamic precepts, Iran was no longer bound by international human rights standards. The values of the latter system derived, according to him, from an alien, Judeo-Christian tradition, as distinguished from the Islamic tradition. He thereby implied that Muslims, because of their religion, could only claim a standard of rights set by Islamic law, which required violations of international standards. This was tantamount to stipulating that Muslims could only claim rights inferior to those enjoyed in the West. However, along with his repudiation of human rights, Raja'i Khorasani offered the inconsistent assertion that secular and non-Muslim states who could not live up to the divine standards of Islam should at least meet the minimum requirements established by international organizations (Mayer 1994, p. 316). By this he implied that international rights protections, being based on secular law, should somehow rank below the standards afforded by Islamic law. Iran's Islam, he seemed to want to say, did protect human rights, but according values that, due to their divine origin, were superior to those based on mere secular authority.

Obviously, Raja'i Khorasani's alternating presentations of "Islam" as the reason for the regime's denials of human rights and "Islam" as the guarantor of a higher standard of rights than what was afforded by international law were contradictory. One assumes that the contradictions came about because the regime relied on Islam as a legitimating device in two different ways. On the one hand, Iran needed a rubric like "Islam" to justify its rights violations. On the other hand, the regime's sole base of legitimacy was Islam, and the more this "Islam" became associated in people's minds with human rights violations, the more unpopular and less credible this "Islam" would become, making it less useful as a means of enhancing legitimacy. (A similar predicament has led to similar confusion in the remarks of official Saudi spokesmen, who have alternately appealed to Islam to justify their government's resistance to calls for democratization and rights and extolled Islamic rights principles as affording perfect protection for rights.) The proposition that Islam mandated an inferior standard of rights protection for Muslims, which flowed from Raja'i Khorasani's stance, was not something that the regime wanted to broadcast as such. In maintaining that Muslims could not have the same rights as non-Muslims, Raja'i Khorasani was essentially reviving the argument of Europeans in the era of nineteenth-century European Imperialism. Western legal scholars of the period had held that international law protected the rights of European citizens when in Muslim territory and did not protect Muslim subjects of Middle Eastern sovereigns. If Muslims were mistreated in ways that would violate international law if the rights of Europeans were at stake, it was of no concern to international law as the Imperialists understood it (Mayer 1990b, pp. 133-158). Although any critical review of Iran's human rights practice would reveal that Iran has repeatedly used "Islam" as the reason for deviating from international standards, spokespersons for the Islamic Republic have, for the most part, preferred to take the route of hypocrisy, seeking to obscure as far as possible the degree to which Iran's Islamic rights entailed curtailing internationally-protected human rights. In so doing, they have revealed that they actually believe that international human rights are the binding standards and that Islamic rationalizations do not suffice to justify violating these.

There are reasons for saying that Iran's post-revolutionary human rights record should be seen as a reflection of political choices, not a reflection of Islamic requirements. Many of Iran's most egregious rights violations—like torture and the brutal suppression of political dissent—had nothing to do with Islamic law. Actions by the regime had already revealed the politically contingent nature of Iran's official "Islam" well before January 1988, when Ayatollah Khomeini made his famous pronouncement that the Islamic state was entrusted with absolute power that freed it to do as it chose—even if this meant violating fundamental pillars of the Islamic religion like the pilgrimage (Mayer 1993, p. 193). His pronouncement meant that the government was not beholden to Islamic strictures and that Islam could be ignored whenever this was needed to serve the political interests of the regime.

The provisions of the 1979 Iranian Constitution deserve to be examined in any discussion of Islam and human rights policy. They consist of Western principles of constitutionalism, upon which "Islamic" ideas have been grafted. Far from offering rights principles conceptualized de novo on an Islamic basis, Iran borrowed its human rights concepts from Western constitutions and from modern international law. Iran's theory that these can be somehow indigenized by imposing "Islamic" limits simply replicates an element in other Islamic human rights schemes.

Islamic qualifications are repeatedly stressed in Iran's constitution, but not in a coherent way. Although the constitution is replete with Western ideas and principles, Article 4 pretends that this is not the case, providing that all laws, including the constitution itself, must be based on Islamic criteria, and in the Preamble, it is stated that the constitution is based on Islamic principles. This should mean that the rights principles, being part of a constitution based on Islamic criteria, should be Islamic. However, Article 20 stipulates that human rights are to be subordinated to Islamic criteria, which seems to indicate that the rights principles are not deemed to be Islamic—whereas the criteria that override rights are deemed to be Islamic. That same article expressly provides that all citizens enjoy human rights [*hoquq-e ensani*] according to Islamic standards [*mavazin-e eslam*]. Article 21 provides that women's rights are guaranteed in accordance with Islamic criteria. Muslims are enjoined to honor the human rights of religious minorities and to treat them according to the dictates of virtue and Islamic justice in Article 14.

It is significant that separate Islamic qualifications are specifically imposed on the rights of women and non-Muslims, without specific Islamic qualifications being imposed on the rights of male Muslims. This is tantamount to signalling that a hierarchy is being established in which Islamic criteria are being employed to ensure that male Muslims rank at the top and in which Islamic principles became the pretexts for depriving women and non-Muslims of human rights. Here one sees a direct correlation between the use of Islamic criteria to restrict human rights and the regime's policies, which discriminate on the basis of sex and religion in ways that violate principles of international law.

The intention to resort to Islamic law at the expense of rights can be detected in some provisions that do not specifically invoke Islamic law. Article 19 of the Iranian Constitution provides: "The people of Iran, regardless of their ethnic, family and tribal origins shall enjoy equal rights. Color, race, language and the like shall not be a cause for privilege." This formulation avoids altogether the question of whether equality can be denied on the basis of sex or religion, even though in any system based on Islamic law one would expect that the issues of equality of women and non-Muslims would be among the most sensitive. The discrimination against both women and religious minorities in the wake of the revolution indicates that the omission of these two categories was not accidental. One contrasts the equivocation in Iran's equality provisions with the clear statement in the UDHR, which in Article 1 provides that "[a]ll humans are born free and equal in dignity and rights," and in Article 2 provides that:

> [e]veryone is entitled to all the rights and freedoms set forth in this Declaration, without distinction of any kind, such as race, colour, sex, language, religion, political or other opinion, national or social origin, property, birth or other status.

Paradoxically, Iran's constitution, in which Islamic criteria are used to subordinate women, includes in Article 20 an unqualified guarantee of equal protection of the law [*qanun*], the secular law, for all citizens, both men and women—without mention of any Islamic qualifications. The inclusion of a secular equal protection clause in the Iranian Constitution is perfectly emblematic of its awkward blend of Islamic rules and incongruous ideas taken from Western constitutions and international human rights law. Furthermore, the Iranian Constitution also guarantees the equality of men and women in Article 3.14 before the secular law without imposing any express Islamic qualifications. However, referring back to other provisions like Articles 4 and 20, one is reminded that "Islamic criteria" ultimately override all other provisions in the constitution, making the guarantees of equality under the secular law nugatory.

Article 10 does not mention women, but it calls for the family to be safeguarded according to Islamic rights and morality [*hoquq va akhlaq-e eslami*]. What "Islamic rights and morality" entailed was revealed immediately after the revolution when the regime moved to abrogate the important reforms advancing women's status that had been made in the Iranian Family Protection Act of 1967, as amended in 1975. Women were pressured to abandon their jobs and to return to the home, where they were expected to devote themselves to maternal and housewifely duties. Many women were removed from desirable professional and governmental positions, and their educational opportunities were sharply curtailed. Furthermore, the age of marriage was lowered from eighteen to thirteen years for girls and temporary marriage, an institution condemned by Iranian feminists, was promoted. After the revolution, there were many setbacks for the progress toward equality that Iranian women had made in the 1960s and 1970s.

In Article 13, the Iranian Constitution recognizes only three religious minorities—Zoroastrians, Jews, and Christians. These three are given the right, within the limits of the secular law, to perform their own religious rites and to follow their religious law in personal matters. That is, the constitution does not expressly cite the Islamic rules on *dhimmi* status to assign these minorities to the inferior position that they traditionally occupied in Iran, even though these rules are behind the singling out of minorities for special treatment. Similarly, Article 144 provides that the army is to be for people who adhere to Islamic ideology and who are faithful to the goals of the Islamic Revolution, thereby revealing that non-Muslims will be excluded from the military, just as dhimmis traditionally were. Again, Islamic rules on dhimmi status that underlie this discriminatory treatment are not specifically adverted to, apparently because the regime is uncomfortable with admitting that it discriminates against Iranian citizens based on their religous affiliation.

Article 14, after guaranteeing the rights of non-Muslims, proceeds to warn that this guarantee protects only "those who do not become involved in conspiracies and activities which are anti-Islamic or are against the Islamic Republic of Iran." Since what activities might qualify as "anti-Islamic" is left open, this provision provides sweeping grounds for curbing the associational rights of members of religious minorities. This is in addition to a general provision in Article 26 that enables the government to curb the activities of groups, including "minority religious associations," if they are contrary to the principles of Islam or the Islamic Republic. Taken together, Articles 14 and 26 suggest that non-Muslims are presumed to be more likely to oppose Iran's Islamic government than Muslims are and that their associational activities are presumed to have anti-Islamic tendencies. Of course, since Iran's official Islamic ideology legitimizes discrimination against non-Muslims, opposition on the part of non-Muslims would be natural.

The Islamic Republic has accumulated a record of exceptional religious intolerance. Iran persecutes persons who are deemed to have abandoned Islam or, in the case of Baha'is, persons who are deemed to be descended from persons who converted from Islam, who may be compelled to choose between death and returning to Islam. The historically intense antagonism toward Baha'ism in Iran accounts for why Article 13 does not recognize Baha'ism as a religion and correlates with the harsh persecution inflicted on the Baha'i minority, which has earned the Islamic Republic repeated condemnations by the international community and by independent human rights organizations.[3] The persecutions and executions for apostasy of Baha'is and other Iranians who converted from Islam were not mere aberrations tied to the initial revolutionary upheaval; such persecutions continued into the 1990s. Attacks on Iranians for their religious beliefs have been accompanied by official attempts to hide the religious motives behind them, which reveals Iran's lack of confidence about whether a policy of criminalizing religious belief can be justified in today's world. Iran's mistreatment of non-Muslims directly clashes with the principle of freedom of religion, an unqualified freedom in international law, which comprises the freedom to change religion.[4] With the exception of Ayatollah Khomeini's notorious *fatwa* condemning Salman Rushdie, the Iranian regime has sought to hide its denials of religious freedom and to disguise its policies of prosecuting converts from Islam, non-Muslims, and Muslim religious dissidents as prosecutions for other capital offenses, such as spying or treason (Mayer 1995, pp. 178-180).

Post-revolutionary Iran has also shown considerable ambivalence about whether Islam suffices to justify treating women in ways that violate international norms. Iran wants to be perceived as a revolutionary Islamic society, and in this regard its retrograde policies affecting women alienate many observers and detract from the positive impression it wants to make. Its veiling rules have prompted international rebukes. However, since its supposedly "Islamic" standards for women's dress have become symbols of the regime's policy of reinstating authentic Islamic values, the regime cannot back away from these, no matter how reactionary they may appear.

Iran has resorted to stern sermons on Islamic morality, stringent criminal laws that include flogging penalties for non-compliance with Islamic dress rules, and aggressive policing to try to cowe women into submission. Despite years of vigorous propaganda and harsh repression, non-conformity with the much-resented requirements that women should wear either the *chador* or other approved versions of Islamic dress has continued to manifest itself. In 1996, the regime felt obliged to enact a law imposing fines and prison penalties for violations of Islamic dress rules.

Women's resistance to official dress requirements has been acutely embarrassing. The more obvious the need becomes to rely on police measures to ensure compliance with what are officially described as "cultural" norms, the more Iran's pattern of repression of women begins to resemble a regime of apartheid, in which the state establishes and maintains the domination of one group and the systematic oppression of another. This suggests parallels with Saudi Arabia and its reactionary policies affecting women, which can only further damage Iran's image.

Iran has been placed on the defensive when forced to explain to the international community why it needs to resort to force to compel conformity with what are supposed to be indigenous cultural norms. In attempts to discredit manifestations of women's dissent from the regime's policies, Iran, like Saudi Arabia, has tried to disassociate non-conforming women from the local culture, charging that their protests are products of Western conspiracies or that the women involved are "Westernized." Embarrassed by negative publicity about its treatment of women, the Islamic Republic has felt compelled to lie, denying that its Islamic dress requirements were being forcibly imposed on resisting Iranian women. In February 1993, responding to a critical U.N. report on Iran's human rights record, which included negative evaluations of its discrimination against women, Iran, in its defense, asserted that its dress requirements and personal status laws affecting women were part of religion and, as such, were willingly accepted by Iranians. Of course, the aim was to convince the U.N. that there was no state policy of discriminating against women. Iran's response posited the existence of uniform Islamic cultural values that were shared by all Iranians and voluntarily adhered to:

> Considering the fact that the majority of the Iranian people are Muslim, the holy Islamic codes form the basic guidelines for the laws...[Women] freely accept regulations regarding marriage and the limitation of the rights and duties of both men and women, based on such Islamic criteria...The legal philosophy of the observation [sic] of "Hijab" for women and men in an Islamic society is fully credible and within the context of conventional international laws. Therefore, since more than 95 per cent of people in Iran are Muslim and thereby follow the laws prevailing upon [sic] an Islamic society, all follow and support the appropriate laws and regulations on the rights of society...Although some political groups believe that women should observe "Hijab" entirely and have criticized the Government for not responding to this matter, until now there have been no confrontations with females who do not observe the "Hijab" properly (Commission on Human Rights 1993).

Several points are noteworthy in this statement. Faced with charges that punishing women for non-compliance with Islamic dress requirements violated international rights norms, the regime felt obliged to deny its actual domestic policy of aggressive policing of women's dress. Instead of asserting that its "Islamic values" justified punishing women for not veiling, the regime argued that its Islamic dress rules applied to men, as well as women, even though the record demonstrates that the regime's preoccupation has been with forcing women, not men, to comply with Islamic dress rules. Furthermore, despite the fact that it claimed that its dress rules were based on Islamic law, by the year 1993 the regime feared that the argument that Iran's distinctive "Islamic" policies on rights justified breaching international law would not hold water. The regime's public relations strategy dictated presenting Islamic dress rules as congruent with international norms.

In Autumn 1994 Ayatollah Khamene'i insisted that, contrary to Western propaganda saying that women's rights were trampled on in Iran, women's rights had actually been strengthened.[5] In the same period, the Islamic Republic decided to pose as the defender of human rights, going on the offensive to charge that it was another country, France, whose dress requirements were violating women's rights. In November 1994 Ayatollah Yazdi, the head of Iran's judiciary and human rights commission, denounced France for expelling Muslim girls from public schools if they insisted on wearing the veil. As an Ayatollah, he might have been expected to attack France for cultural insensitivity or lack of respect for Islamic values. Instead, Yazdi invoked international human rights law, charging that France was violating human rights principles in the U.N. Charter, which, he said, emphasized the right to religious freedom.[6] Of course, in seeking to pose as a defender of international human rights where Muslim women were concerned and in resting his case on the inviolability of the norm of freedom of religion, Yazdi had moved onto treacherous ground, indeed. No principle of human rights law had been more egregiously violated by the Islamic Republic than the norm of freedom of religion, making Iran's sudden protest on behalf of this norm sound strange, indeed. Yazdi's protest provoked awkward questions. If Muslim women in France had the right to dress in accordance with their consciences and their personal religious convictions, regardless of whether this violated the long-established French policy of laicité, then why should not Muslim women in Iran likewise enjoy the same right? Why did Muslim women lose their freedom to dress in accordance with their own religious beliefs once they were on the territory of the Islamic Republic?

This could not be an issue according to Ayatollah Khamene'i, who supported the new line that Islamic dress was a cultural tradition: "But our women want to protect their hijab. They love chador. Chador is our national costume. Chador is an Iranian tradition before being an Islamic tradition. It belongs to our people. It is our national dress."[7] Thus, to avoid being charged with violating human rights and discriminating against women, the *faqih*, the linchpin of the Islamicity of the regime, chose to present wearing the chador as a popular choice mandated by Iranian tradition, not one dictated by the commands of Islamic law—a remarkable change. In inconsistently appealing to the supremacy of Islamic criteria on rights and trying to reconcile official policies with international law, the regime was entangling itself in a web of contradictions that precluded developing a coherent policy line on human rights.

The regime did not succeed in convincing the Iranian people that its policies of denying freedoms were acceptable. Despite exposure to intermittently harsh repression, Mehdi Bazargan and his associates in the Freedom Movement remained stalwart in their demands for democratic freedoms and human rights. After Bazargan's death in 1995 and after fifteen years under an intolerant and repressive theocracy, the spirit of resistance was not quelled. In 1995, for example, at great risk to themselves, eighty-six Iranian activists published and signed a statement of protest against clerical misrule, condemning the regime for preventing the Iranian people from governing themselves and participating in determining their destiny. They also proclaimed their indignation over being denied even those rights provided in Iran's own Islamic Constitution. In this protest, they touched on the root of the problem, which was that the government's actual policy on rights matters was one of ignoring any rights principles, regardless of whether they were secular or nominally "Islamic." The arrest and detention without trial of the prominent writer and intellectual Ali Akbar Saidi-Sirjani in March 1994 seems to have been prompted by his criticisms of the government for denying Iranians rights stipulated in their constitution and his assertions that Iran's pre-Islamic tradition supported respect for individual rights and fighting tyranny, which, in context, seemed subversive to the clerical hierarchy governing the country. When he died in custody in suspicious circumstances in November 1994, the assumption was widely made that torture had caused his demise.

Rushdie's death warrant looks different now that people who were once strong supporters of the Islamic Revolution, like the philosophy professor Abdolkarim Sorush, who has come around to questioning the value of clerical government, are treated in similar fashion. In 1995, Foreign Minister Ali Akbar Velayati encouraged the assassination of Sorush by claiming that his expressions of opinion were harming the people's national and ideological foundations and were antagonistic to the nation and comparing him to Ahmad Kasravi, the secular intellectual who was assassinated in the forties.[8] In a similar vein, Ayatollah Khamene'i attacked the people in the universities who were criticizing the theocratic system, *nezam-e ruhaniyat*, condemning this as sedition. He warned those that were undermining the clergy in Iran that if "ulterior motives" or "ill-intent" were involved, the Islamic system would slap these people hard in the face.[9]

Meanwhile, while indulging in such repression, the regime has flailed about in attempting to respond to U.N. and foreign criticisms, attacking the motives of its accusers and insisting that the charges that it violates human rights are lies, slanders, politically motivated, and based on misinformation and misunderstandings. Iran was obviously disturbed by the March 1995 U.N. Human Rights Commission condemnation of its performance. Many of the assertions of rights violations related to Iran's policy of following retrograde clerical interpretations of Islam, such as the discrimination against non-Muslim minorities, the Rushdie death sentence, the forcible imposition of Islamic dress and the sexual segregation of women, and the amputations of the fingers of thieves. Yazdi tried to insinuate that the report was an attack on Islam:

> The new issue is this: who says that human rights devised by the West should be applied to the whole world? Who has said such a thing? Some nations have their own cultures, some nations have their own religions, you cannot describe as human rights violations the issues which concern their religion and culture. You cannot impose the human rights as you have translated and defined them...Islamic human rights differ from the Declaration of Human Rights. Islam has its own rules and regulations...The punishment ordained by God's law and mentioned in the Koran cannot be ignored. Human Rights must be Islamic human rights.[10]

Faced with the predicament of how to respond to U.N. critics of its human rights record, Yazdi felt compelled to go back to Raja'i Khorasani's earlier idea that Islam required Iran's human rights violations, so that the regime could not be blamed for these. Here, however, the defense was offered for internal consumption in an effort to discredit external critics of Iran's rights record by portraying Iran's critics as being insensitive to Iran's Islamic obligations.

Human Rights Policy in Saudi Arabia

Saudi Arabia's policy on rights has been to deny human rights across the board while posing as a strong proponent of supposedly Islamic human rights. Even this supremely reactionary regime found it impossible to ignore the clamor for increased rights. It felt compelled to respond to the mounting popular disaffection of the 1990s, when anger at the the Saudi monarchy was aggravated by factors like the faltering economy, mounting exasperation over the corruption and immorality of the royal family, the regime's refusal to democratize and its stifling of dissent, and the arbitrary administration of justice. The Saudi alliance with the United States against Iraq in the Gulf War intensified the restiveness of many Saudis. A secular petition was submitted to King Fahd by liberal Saudis in 1990 and another submitted by conservative ulema in 1991, both demanding reforms, the adoption of principles of constitutionalism, and respect for certain rights (Middle East Watch 1992, pp. 6-7, 59-62). In the early 1990s disaffected Shi'is launched campaigns to embarrass the regime over its bad human rights record in general and its mistreatment of Shi'is in particular. The restlessness of Saudi women briefly surfaced in 1990 when a group of professional women mounted an historic driving protest in Riyadh. Before doing so, they sent a petition to the governor of Riyadh insisting that their demand to drive was consonant with Islamic principles (Doumato 1995, p. 139). The protestors were punished for daring to challenge Saudi custom, and that same custom was belatedly sacralized by a fatwa claiming that women driving violated Islamic precepts. Another significant protest against Islam being deployed to justify Saudi autocracy occurred in May 1993, when Saudi Arabia's first independent human rights organization, the Committee for the Defense of Legitimate Rights (CDLR), was launched by conservative Sunnis who were far from accepting international rights standards but characterized their opposition to Saudi misrule as a struggle for human rights.

Confronted with these challenges, the regime undertook initiatives like promoting the Cairo Declaration and made some cosmetic concessions to popular demands for constitutionalism and human rights, such as the issuance of the 1992 Basic Law, which was composed in secret by a committee led by Prince Nayif, the Minister of the Interior. Although meant to convey the impression that certain principles of constitutionalism had been adopted, not surprisingly, the Basic Law confirmed the right of the ruler to do as he pleased.

Article 7 provides that the Quran and Sunna reign supreme over the Basic Law and all other laws of the State. That is, according to this article, the Basic Law is not the supreme law of the land, but is subordinated to the Islamic sources. However, in a noteworthy omission, how the sources are to be interpreted and by whom is never precisely defined in the Basic Law. In these circumstances, what Islamic law mandated would ultimately reflect the regime's preferences, so that Saudi policy and Islamic policy would be deemed identical.

The Basic Law allows for none of the civil and political rights associated with democratic governance. All power is firmly situated in the hands of the monarch. Article 5 states unequivocally that the system of government in the Kingdom of Saudi Arabia is monarchy—that is, monarchy pure and simple, as opposed to a constitutional monarchy. Although there is mention of a consultative assembly and a judicial branch, careful appraisal of the rest of the Basic Law reveals that the system contains no checks whatsoever on the King's power. The Saudi Basic Law has no function other than placing a constitutional facade before an absolute monarchy. Article 6 includes a provision that citizens [*al-muwatinun*] are to pay allegiance to the King in accordance with the Quran and Sunna, in submission and obedience. This is a call for the kind of obedience that was in the past exacted by traditional amirs and sultans from their subjects and rests on a philosophy fundamentallly at odds with modern constitutionalism.

Article 26 provides that the state protects human rights in accordance with the Islamic Shari'a, thereby imposing vague Islamic limitations on rights that cancel out the rights involved. Only the rudiments of a few civil and political rights provided are set forth (Mayer 1994, pp. 348-349, 363-364). One looks in vain for a guarantee of freedom of religion, one of the most basic principles of modern human rights. This gap correlates with the Saudi disregard for the principle of freedom of religion, which has had particularly harsh consequences for the Shi'i minority (Amnesty International 1993a, 1993b, p. 6).

Heedless of the inconsistency involved, after the promulgation of the Basic Law, Prince Saud al-Faisal presented the very different set of rights provisions in the Cairo Declaration at the 1993 World Human Rights Conference, as if they were definitive. The Saudis' affirmance of two disparate and inconsistent sets of "Islamic" standards on rights provides some indication of the lack of intellectual seriousness with which the Saudi regime approached human rights issues. Since the government had no intention of adhering to any rights, the details of the rights provisions were not important, the only concern being establishing "Islamic" bases for deviating from international law norms. One assumes that the Saudi rulers hoped that their campaigning for the Cairo Declaration would serve to associate the regime with the cause of human rights and improve its tarnished image, while committing itself to nothing. This was possible, because, although it contained more rights than the Saudi Basic Law, the wording of the Cairo

Declaration was intentionally kept vague and evasive, as the foregoing discussion of its Islamic qualifications on rights has indicated.

The contradictions in Saudi positions on rights were manifold. Article 1 of the Cairo Declaration at least pays women the compliment of recognizing that their status needs to be addressed, even though it fails to guarantee them equality in rights. The Saudi Basic Law, in contrast, completely ignores the existence of women. The Cairo Declaration in Article 20 prohibits torture, but the Basic Law, in contrast, offers no prohibition of torture. Since the prohibition of torture is of vital importance in the modern human rights system, the inconsistency between the two documents on this point is significant. Many other provisions in the Cairo Declaration have no counterparts in the Basic Law, like the Article 19 affirmation that all individuals are equal before the law without distinction between ruler and ruled; the Article 9 right to education for all; the Article 11 ban on slavery; the Article 13 right to work for fair wages that do not discriminate between men and women; and the Article 23 right to participate directly or indirectly in the adminstration of the country's affairs.

At the 1993 Vienna Conference Prince Saud al-Faisal asserted that the Cairo Declaration was a document that embodied "an expression of the will of over one billion people," a highly dubious claim, since the OIC and its member nations had made no effort to ascertain the views of Muslims on rights questions before issuing the declaration. He spoke as if Islam had its own efficacious scheme of rights, asserting that Islamic law was "a comprehensive system for universal human rights," meant not as moral exhortations but as "legislative orders." However, although endorsing the Cairo Declaration, Prince Saud al-Faisal did not call for rejecting the idea of universality of rights, claiming that he accepted that the principles and objectives of human rights were universal but that in their application it was necessary to show consideration "for the diversity of societies, taking into account their various historical, cultural, and religious backgrounds and legal systems."[11] That is, he seemed eager to downplay the conflicts with international human rights in the Cairo Declaration and to try to persuade the international community that in supporting this OIC initiative Saudi Arabia was not breaking with the international consensus on rights. Despite their efforts, the Saudis were not successful in getting the Vienna conference to endorse the Cairo Declaration as a legitimate alternative to the rights set forth in international law.

As in the case of Iran, the official adoption of Islamic versions of human rights by Saudi Arabia in no way mitigated the rights abuses occurring in the country, which continued as they had been before Saudi Arabia had endorsed the Cairo Declaration or promulgated its Basic Law. By its practice, Saudi Arabia showed that, like Iran, it was as indifferent to rights that were supposedly based on Islamic principles as it was to rights established in international law.

Not surprisingly, the watered-down "Islamic" human rights in the Basic Law or the Cairo Declaration that the Saudi regime was promoting did not persuade the long-suffering populace that meaningful improvements in the standard of rights protections had been made.

Like the members of Iran's Freedom Movement, the rights activists in the CDLR found that there was no tolerance in their country for persons who demanded a higher standard of human rights protections and called for adherence to the rule of law. The communications by the CDLR, which had to move its main operations offshore to London due to the harsh persecution of its members, managed to reach disaffected Saudis via new means of communication like faxes. After its leader was forced into exile, the CDLR indicated that it aimed "to elucidate and assert the concept of human rights in Islam and eliminate the distortions created by the Saudi regime's erroneous practices and severe abuses and massive violations of human rights" and that it would adhere to the Islamic sources and Islamic ethics in its work (CDLR 1995, p. xiv). That is, it challenged the Saudi government's monopoly over how Islam pertained to rights. Some measure of the threat that the regime perceived in such independent human rights advocacy carried out under an Islamic rubric is indicated by the crackdown on relatives and associates of CDLR members in 1994-95. In the course of efforts to discredit the CDLR and prosecutions of its supporters, the CDLR was denounced for straying beyond the pale of Islam, and the council of senior religious scholars denounced the CDLR as heretical. Thus, human rights advocacy that was not under the control of the regime and that claimed an Islamic basis for challenging the legitimacy of Saudi repression was seen as such a serious threat that it became elevated to the status of a capital crime.

CONCLUSION

Muslim nations, which have generally poor records of compliance with human rights law, have great difficulty reconciling their human rights policies with the governing international human rights standards. Against this background, various "Islamic" versions of human rights have been promoted that actually borrow extensively from Western rights models and employ Islam simply to weaken or nullify rights protections, converting Islam into a rationale for denying human rights that are guaranteed under international law. These Islamic schemes of human rights have been developed opportunistically without first laying down sound jurisprudential foundations for constructing rights on an Islamic basis and without dealing with the theoretical problems that are entailed by introducing non-conforming Islamic schemes of rights into the modern system of international law. Despite promulgating Islamic rights, in various ways regimes have continued to pay tribute to the authority and popularity enjoyed by international human rights in Muslim milieus, including trying to disguise the degree to which their policies and practices actually deviate from international norms. Meanwhile, Muslims' appreciation of the value of human rights seems to be growing, and the impetus for fusing elements of the Islamic heritage with international human rights law is likewise spreading. Unless Muslim countries democratize, this important task is likely to be accomplished by Muslims who are working in opposition to state policies on human rights.

NOTES

Note: Relevant articles can be found in journals such as *Human Rights Quarterly* and *Index on Censorship*, and information is regularly compiled, analyzed, and published by independent human rights groups such as Amnesty International, The Lawyers Committee for Human Rights, and various parts of Human Rights Watch, like Africa Watch, Asia Watch, and Middle East Watch. Furthermore, despite the exposure to reprisals and severe censorship, some independent human rights groups in Muslim countries also manage to publish reports.

1. See the OIC Resolution No. 41/21-P. on Coordination Among Member States in the Field of Human Rights.
2. A variety of sources and articles relevant to this question can be found in a recent conference volume (Afkhami 1995). The works of Fatima Mernissi are particularly interesting in this regard (Mernissi 1991).
3. See, for example, Final Report on the situation of human rights in the Islamic Republic of Iran by the Special Representative of the Commission on Human Rights, Mr. Reynaldo Galindo Pohl, pursuant to Commission resolution 1992/67 of 4 March 1992, U.N. Doc. E/CN.4/1993/41 28 January 1993, 54-55, and the sections on Iran in the annual reports of Amnesty International and Middle East Watch.
4. Article 18 of the Universal Declaration of Human Rights provides: "Everyone has the right to freedom of thought, conscience and religion; this right includes freedom to change his religion or belief, and freedom, either alone or in community with others and in public or private to manifest his religion or belief in teaching, practice, worship and observance." In the International Covenant on Civil and Political Rights, Article 18.1 begins: "Everyone shall have the right to freedom of thought, conscience and religion."
5. BBC SWB ME/2126/MED, October 14, 1994.
6. Live broadcast by Ayatollah Yazdi on March 10, 1995, Reuter Textline Monitoring Service, March 13. 1995, available in LEXIS, Nexis Library, ALLWLD File.
7. BBC SWB, ME/2126/MED, October 14, 1994.
8. BBC SWB EE/D2510/ME, January 15, 1996.
9. BBC SWB EE/D2484/ME, December 12, 1995.
10. Reuter Textline, BBC Monitoring Service, March 13, 1995, available in LEXIS, Nexis Library, ALLWLD File.
11. Islam Guarantees Human Rights, Says Saud, *Riyadh Daily*, June 17, 1993, available in LEXIS, Nexis Library, Saudi File.

REFERENCES

Afkhami, M., ed. 1995. *Faith and Freedom: Women's Human Rights in the Muslim World.* London: I.B. Tauris.

Afshari, R. 1994. "An Essay on Islamic Cultural Relativism in the Discourse on Human Rights." *Human Rights Quarterly* 16: 235-76.

Amnesty International. 1993a. *Saudi Arabia: Religious Intolerance: The Arrest, Detention and Torture of Christian Worshippers and Shi'i Muslims.*

________. 1993b. *Saudi Arabia: An Upsurge in Public Executions.*

Caesar, J. 1991. "Big Saudi Brother." *Christian Science Monitor*, January 4, p. 18.

CDLR (The Committee for the Defence of Legitimate Rights). 1995. *Year Book '94-95*. London: CDLR.

Chehabi, H.B. 1990. *Iranian Politics and Religious Modernism: The Liberation Movement in Iran under the Shah and Khomeini.* Ithaca, NY: Cornell University Press.

Commission on Human Rights. 1993. Economic and Social Council, 49th Sess., Agenda Item 12, at 7. E/CN.4/1993/41/Add. 1: 7-8.

Coulson, N. 1957. "The State and the Individual in Islamic Law." *International and Comparative Law Quarterly* 6: 49-60.

Doumato, E. 1995. "The Ambiguity of Shari'a and the Politics of 'Rights' in Saudi Arabia." Pp. 135-160 in *Faith and Freedom: Women's Human Rights in the Muslim World,* edited by M. Afkhami. London: I.B. Tauris.

Dwyer, K. 1991. *Arab Voices: The Human Rights Debate in the Middle East.* Berkeley: University of California Press.

Mayer, A.E.. 1990a. "The Shari'ah: A Methodology or a Body of Substantive Rules?" Pp. 177-198 in *Islamic Law and Jurisprudence*, edited by N. Heer. Seattle: University of Washington Press.

________. 1990b. "Current Muslim Thinking on Human Rights." Pp. 133-158 in *Human Rights in Africa. Cross-Cultural Perspectives*, edited by F. Deng and A. An-Na'im. Washington, DC: The Brookings Institution.

________. 1993. "The Fundamentalist Impact on Law, Politics, and Constitutions in Iran, Pakistan, and the Sudan." Pp. 110-151 in *Fundamentalisms and the State: Remaking Polities, Economies, and Militance*, edited by M. Marty and R.S. Appleby. Chicago: University of Chicago Press.

________. 1994. "Universal versus Islamic Human Rights: A Clash of Cultures or a Clash with a Construct?" *Michigan Journal of International Law* 15(Winter): 303-404.

________. 1995. *Islam and Human Rights: Tradition and Politics*, 2nd ed. Boulder, CO: Westview.

Mernissi, F. 1991. *The Veil and the Male Elite: A Feminist Interpretation of Womens' Rights in Islam.* Reading, MA: Addison-Wesley.

________. 1992. *Islam and Democracy. Fear of the Modern World.* Reading, MA: Addison-Wesley.

Middle East Watch. 1992. *Empty Reforms: Saudi Arabia's New Basic Laws* (May).

Moinuddin, H. 1987. *The Charter of the Islamic Conference.* Oxford: Oxford University Press.

An-Na'im, A. 1990. *Toward an Islamic Reformation: Civil Liberties, Human Rights and International Law.* Syracuse, NY: Syracuse University Press.

________. 1986. "The Islamic Law of Apostasy and Its Modern Applicability: A Case from the Sudan." *Religion* 16: 197-224.

Nazir, K. 1990. "Women's Demo was a Stupid Act, says Naif." *Middle East News Network*, November 16, 1990, available in LEXIS, Nexis Library, ALLWLD File.

Vichniac, I. 1992. "La commission internationale de juristes denonce un projet de 'declaration des droits de l'homme en Islam.'" *Le Monde* February 13, p. 6.

Waltz, S.E. 1995. *Human Rights and Reform: Changing the Face of North African Politics*. Berkeley: University of California Press.

WOMEN, SEXUALITY, AND SOCIAL POLICY IN ISLAMIC CULTURES

Haideh Moghissi

I would like to say to the young men formed in our Muslim civilization that it is highly improbable that they can value liberty—by which I mean, relating to another person as an act of free will, whether it be in bed, in erotic play, or in political debates in party cells or parliament—if they are not conscious of the political import of the hatred and degradation of women in this culture.

—Sabbah (1988, p. 5)

INTRODUCTION

The Islamic concept of sexuality is male-centered and hierarchical. The explicit and implicit view of woman as the object of male gaze, desire and derision, as all-body, whom God created to provide men companionship, sexual pleasure, and to produce children, excludes equality of rights between the sexes. Woman's expression of her desires and the pursuit of her interests contradicts the interests of man and challenges man's God-given rights over women. Underpinning the sexual and moral beliefs and practices in Islamic societies is the conception of women as weak in moral judgment and deficient in

International Review of Comparative Public Policy, Volume 9, pages 149-168.

ISBN: 0-7623-0268-2

cognitive capacity, yet, sexually forceful and irresistibly seductive. The susceptibility of women to corruption, in this view, explains the obsession with sexual purity in Islamic cultures and justifies surveillance of women by family, community, and state.

Yet, it is not the unchanging character of Islamic conception of sexuality and its essential difference from other traditional religions in this area that explain the plight of women under Islamic rule. Indeed, Islam may not be any more obsessed with human sexuality and female sexual purity than other religions. What makes Islam different is its insistence on upholding religious instructions governing the believers' sexual behavior and moral conduct as the rules of social conduct enforced by the law. This means that medieval Islamic scriptual statements and religious commands have immediate impact on the everyday life of women (and men) in Islamic societies. Islamic instructions and views, regarding the physical, psychological, and sexual differences between the sexes, favoring male authority over women, interpreted as divine laws, are translated into the articles of family law and civil legislation. Religious gendered sexual ethics and moral standards regulate the status of the citizens and determine women's access to social and economic resources. This is the case, to varying degrees, for example, in Iran, Sudan, and Afghanistan, where the state is controlled by Islamic fundamentalists; in Saudi Arabia, where state power and religious authority have always been concentrated in the hands of the same power elite; and it is the case in a large number of other Islamic societies where the state has returned to the Shari'a under the pressure of new Islamic political movements.

The central argument in this paper is that the combined impact of sociohistorical economic and political retardation of Islamic societies, along with its articulation of indigenous customs and patriarchal cultural values, conspire to sustain the authority of misogynist religious commands over the lives of women. The non-separation of religion and politics, the unclear divide in Islamic societies between law and policymakers and religious-spiritual leaders, have smothered the civil society. In the majority of countries of the Islamic world, civil society—where the formation and contestation of progressive counter-hegemonic forces takes place—has not existed except in rare historical moments. This area has been closed to women, religious minorities, and progressive secular forces. Religious leaders and political rulers—except when their interests clash—equally benefit from the convergence of politics and religion and from a stifled civil society. By giving a divine character to political commands, they try to protect themselves against progressive radical challenges.

One expects that it should go without saying that this sociopolitical dynamism is not conducive to change in favor of women. Any meaningful change in the treatment of women in Islamic societies has to start by the recognition of women as autonomous full citizens, which includes legal equality for women in family law and other civil legislation. But if the principles of the Shari'a are to be maintained, women cannot be treated any better. Women cannot enjoy equality before the law and in law. The Shari'a distinguishes between the rights of human beings on the basis of sex (and religion). Shari'a is not compatible with the principles of equality of human beings. Change in Islamic societies is multi-dimensional, but it has to start, perhaps, with the rule of law, state accountability, and separation of state from religion.

FEMALE SEXUALITY AS THE SYMBOL OF COMMUNAL DIGNITY

In almost all human societies, cultural beliefs surrounding female sexuality, perceived as the site of motherhood, wifehood, fertility, purity, modesty, and selflessness, construct beliefs and values pertaining to differing feminine and masculine roles, and gender relations. However, female sexuality (or the control and the protection of it) finds a more complex political meaning in Islamic societies. Seen as the symbol of Islamic social order and cultural continuity, restraining and disciplining female sexuality seems to draw the boundaries between "Muslim culture" and non-Muslim Other.

In Islamic societies, a woman's body generates fascination and pleasure. It is exploited for procreation, and as a symbol of communal dignity. It is manipulated and its activities are codified. It is covered and confined. It is disciplined for defiance and is mutilated in anticipation of trespassing—all this often sanctioned legally and, particularly, culturally. The female body is the site of struggle between the proponents and opponents of modernity and is used as a playing card between imperial and anti-imperial political forces. Sexuality, the site of love, desire, sexual fulfillment, and physical procreation, in Islamic societies is, at the same time, for women, the site of shame, confinement, anxiety, compulsion. "With the first drop of her menstrual blood, every Muslim girl becomes a temple of her family's honor" (Minai 1981, p. 100). Managed independent of her desire and will, sexuality for women becomes the legal possession of the Islamic community, *umma*, and, by extension, the state. Laws pertaining to marriage and divorce speak clearly of the inability of women to enjoy full legal status. The marriage contract and the termination of it, divorce, are negotiated between state and the male citizens (i.e., father in the case of marriage, and husband in the case of divorce). Young virgin women, according to the Islamic Shari'a, need the permission of their fathers or guardians to enter a marriage contract; fathers can legally marry off their underage daughters for a set price, *mahr*; and a man can end the marriage contract without the consent or even the knowledge of his wife. The diverging interpretations of the Quranic rulings and various legal traditions and reforms launched in Islamic societies in the area of personal status, have done little to remove women's legal powerlessness in marriage and divorce.

The rise of Islamic fundamentalism[1] and the increasing political influence of Islamists, through intimidation and terror or popular support, have, in fact, removed the little control that women had achieved over their lives in different Islamic societies. The legal authority of men, and the public community via the state, over the female body sometimes finds astonishing expressions. For example, in some provinces of post-revolutionary Iran, the influential clergy would go so far as to arbitrarily announce the annullment of marriages of militant communists, who were considered as *kafar* (apostate), regardless of the couples' wishes. Egypt saw similar bizarre forms of asserting the authority of the Islamic community over female sexuality when, in 1993, a group of Islamist lawyers filed a suit to divorce a Cairo professor from his wife, against the couple's wishes and without their knowledge, on the grounds that he was a *kafar* (Colla 1993).[2] The provisions of the Shari'a incorporated in the civil and criminal codes establishes, for the state and for individual men, control over women's bodies and, at the same time, stipulates women's lack of control over their own bodies, sexual expression, and procreative power. This is illustrated, for example, by *hijab*, provisions regarding *zina* (extramarital sex), "honor" killing, and by women's limited rights to custody in cases of divorce.[3]

The question is, however, why is female sexuality so tightly linked with communal honor and politics in Islamic cultures? What are the sociohistorical structures and processes that legitimize the regulation of female sexual and moral conduct by Islamic states? According to one argument, the blame must be laid at the door of European colonialism. Perhaps the resistance of Islamic societies to change women's familial status is the reaction of a culture that has been shamelessly stereotyped and inferiorized—like none other—for its treatment of women. Hence, from this point of view, it was colonialism that made the "Muslim woman" and her rights central to its imperial policy in the Middle East. Another approach, making references to specific verses of the Quran and the Shari'a, conceptualizes subjugation of women in terms of Islamic beliefs, divine in character, static, and unchangeable.

Much historical evidence lends credence to the first claim. The "Muslim woman," her sexuality, sexual power and sexual enslavement have, historically, been the subject of fascination, and fantasies of the European male viewer and the cause of cultural anxiety and rage for the Muslim man. In the eyes of the Western viewer, the "Muslim woman" was "far more Oriental than the man." Her "oriental mystery never failed to charm, her resources never exhausted."[4] Unfettered by a reality that did not fit this imagery, the Western male viewer constructed a plausible image of Muslim woman, the enslaved source of sexual pleasure and sexual possession, through "studio fantasies," using prostitutes as models (Graham-Brown 1988; Alloula 1986).

In this colonial imagery, the domesticated, subjugated, unenlightened Other as opposed to the liberated, independent and enlightened Western Self legitimized colonial power relations. The "Muslim woman" was to be exploited by the Western man, but protected from her own ignorance and enslavement by the Muslim man. The Muslim women's role and status became a stick with which the West could beat the East (Malti-Douglas 1991, p. 3).

The "civilizing" and "liberating" colonial policies in the area of women's rights further aggravated the Muslim male. French colonial rule in Algeria, for example, by using prostitution as a means of social coercion, targeted families unwilling to collaborate with the new rulers and gave nationalism and the decolonization movement a distinctively gendered character—invoking cultural and religious resistance (Lazreg 1995, p. 55). Hiding women from the gaze of the Western viewer, and guarding women's bodies and their minds from changes produced by foreign intervention, symbolized protection of Islamic identity, communal dignity, and social and cultural continuity.

Despite its persuasiveness, however, this argument has several problems. After all, colonial rule in other parts of the world also involved the invention of stereotypical images for the native population, the brutal abuse of female sexuality, and sexual objectification and exploitation of native women. Everywhere women and the conquered territories "were concurrently exploited as part of the bounty due to the conquering Europeans." For example, the myth about the "New World" was that there were "young and beautiful women, who everywhere were naked, in most places accessible, and presumably complacent" (Atkinson 1992, pp. 5-9). However, in the Americas we cannot see the stubborn resistance to change the legal status of women after the fall of colonialism that we observe in Islamic societies.

Furthermore, the patterns of colonization by European powers were different in different parts of the Islamic world. Yet, Islamic societies, with some exceptions over many centuries, disclose greater similarities than differences in their rigidity in the treatment of women, particularly in the area of women's legal rights and personal status. This is illustrated, for example, by making provisions of the Shari'a the basis of women's legal rights in personal status and legislation pertaining to marriage and divorce, inheritance, and custody—all of which define and confine women's participation in social and political activities (Mayer 1995, pp. 79-122; Anderson 1976, pp. 100-146). There were substantial differences between the lengthy occupation of Algeria by the French and the "civilizing" and "Frenchifying" policies of the colonizer—specifically targeting Muslim women—and the relatively short presence of the British in Egypt, where the power of local rulers and Ottoman legal traditions, remained virtually intact.[5] Still, in both cases women's status and rights, to a varying degree, remained within the confines of the Shari'a.

Likewise, the totalizing and universalizing approach that looks only to the Quranic injunctions and Shari'a laws to explain the surveillance of women in Islamic societies is inadequate. For it disregards the basic fact that Islam, like any other religion or ideology, has a contingent nature and is the product of its articulation with indigenous cultures and societies. The spatia-temporal existence of Islam points to heterogeneity of "Islamic culture." The idea of Islam as a kind of meta-culture obscures the reality that, as Aziz Al-Azmeh has noted, there are as many Islams as the conditions that sustain them—as many "Islamic cultures" as different geographical, social conditions, size of wealth and educational levels can produce (Al-Azmeh 1993, pp. 6-8). The similarities among Islamic societies in the application of the principles of Shari'a, therefore, should not cloud over significant differences between various interpretations of the Quran and Shari'a, in different time frames, and in different settings, and the political context that determine the extent of their observance. For instance, polygamy, taken as Islam's engendering signifier, is prohibited in some countries of the Muslim world, such as Turkey and Tunisia; the temporary marriage, *mut'a*, a practice limited only to Shi'i Shari'a, allowing a man to "marry" as many women as he wishes for a set time and price, is strictly prohibited among the Shi'i Ismailis of East Africa[6]. The Constitution of Shi'i Imami Ismailis, in fact, provides that a marriage "may be solemnized between two members of the community only if neither party has a spouse living at the time of marriage" (Anderson 1976, pp. 110-111). Likewise, there are differences among Shi'i, Maliki, Hanbali, and Hanafii schools of law on such issues as compulsory marriage and child marriage, that is, a guardian's right to contract a marriage on behalf of his minor ward.

Indeed, the relative variance in a religious and political tradition, stretching from Indonesia and Malaysia to Morocco, suggests that Islamic traditions and values could be accommodating and moldable in proportion to the strength of local customs and cultural practices and to the processes of social and economic development. The Nigerian Islam, which represents an accommodation between Shari'a injunctions and pre-Islamic African customs and mores, is different from the secularized Islam practices in former Soviet Azerbaijan where Islam represents a cultural-ethnic identity rather than a way of life. The impact of 70 years of Soviet legislation and secular social policies on Azerbaijan is that nearly all adult women work at salaried jobs, people eat pork and drink alcohol, and there are few "Muslims" who are well informed about Islam (Dragadze 1994, pp. 156-162). Both Nigerian and Azarbaijani Islams have significant differences with Khomeinist-Islam in Iran, where state legislation, invoking Shari'a interpretations, invents "Islamic" rules to further restrict women's physical mobility and participation in public life. The existence of different "Islams" practiced in the Muslim world means that the local customs and practices, *urf*, have made effective contributions in the stipulation of personal status legislation within territorial borders. Hence, Islam cannot be taken as the sole signifier of the situation of women in Islamic societies.

Even under Islamic rule, class and wealth, to a great extent, define women's life options and gender experience.

Nonetheless, colonial or homegrown, externally imposed or locally generated, compelled by the Quranic injunctions and Shari'a rulings or the erratic interpretations of local ulema (jurists), "Muslim woman," her sexuality and her moral conduct has remained a central preoccupation of Muslim men over many centuries. This preoccupation has been translated into structures, policies, legal practices, and personal status codes that determine women's life options and the extent of women's participation in public life. In other words, despite the capacity of Islam to adapt and to change face and force under the influence of various social, political, and economic stimuli, the systematic, vigorous, and often violent, opposition to change is a grim reality in many Islamic societies.

This reality raises several questions as to what cultural and religious suppositions and what complex sociocultural and political circumstances sustain the formidable influence and authority of Islamic laws, and traditions perceived as Islamic, over women's lives? What are the psychological impulses and mental frames in Islamic cultures that tolerate, or even licence, the violent disciplining by men of female sexual and moral conduct? These questions, in the end, must merge with the problematic of the rise of Islamic fundamentalist movements in contemporary Middle Eastern and North African societies. Islamic fundamentalists, by embracing female body as the symbolic representation of communal dignity—in their opposition to the West—and by drawing only on the Quran and orthodox texts to explain, as divine, the historically-developed subjugation of women in Islamic societies, recycle the whole colonial conception of Islam and women's right, as static, unchanging, and unchangeable order. By manipulating the female body as a playing card in oppositional politics, Islamists, in fact, embrace, however unsought and uncomfortable, the views of Western colonizers. This itself raises questions about the history of Islam and women's rights as well as introducing concerns about the possibility of future change in Islamic cultures.

ISLAMIC CONCEPT OF SEXUALITY

Although different interpretations and traditions within Islam makes it hard to generalize, it can be said that Islam is a sex-affirming cultural and religious tradition. The Islamic attitude toward sexual pleasure may, indeed, be a major dividing line between Islam and other traditional religions. Strictly interpreted, both Judaism and Christianity approve of sexual relations between men and women, but only in marriage and only for procreation. Historically, both Judaism (which was not ascetic and encouraged marriage and procreation) and Christianity (which celebrates celibacy and virginity—because marriage is a worldly obligation that might interfere with devotion to God), condemned any kind of sex that did not lead to reproduction. In early Christianity, sexual pleasure was a sin if children were not the object (Tahhahill 1989, p. 153). Christians even promoted the idea of sexless marriage—that is, even the sexual union of a couple that was legally and religiously sanctioned should not be "infected" by lust (Bullough 1976).

Islam, by contrast, disapproves of celibacy. The Prophet is believed to have considered marriage for a Muslim as half of his religion. According to the Maliki and Hanafi schools, marriage is obligatory for men. Marriage "shields him from promiscuity, adultery, fornication, homosexuality," and is commendable "even for a person who has strong will to control his sexual desire, who has no wish to have children, and who feels that marriage will keep him away from his devotion to Allah" (Doi 1989, pp. 33-34). Islam opposes celibacy and celebrates sexual pleasure as a legitimate right of the believer. Sex in itself is regarded as a sacred function within the domestic field. For this reason, the Prophet paid special attention to the relations between a man and his wife. The promises made to the believer of the "good life" awaiting him in Paradise, a place in which sexual indulgence with "eternally young," "fair," and "wide eyed" women seems to be man's only activity, can, perhaps, reveal what constituted ultimate happiness for the Muslim believer (Sabbah 1988, pp. 91-97). Eternally lasting physical pleasure and unrestricted access to the female body, as the source of physical pleasure, would be delivered to the believing man in Paradise, as a reward for his piety, good deeds, and self-control in life.

While approving of sexual pleasure, the Islamic orthodox view develops, at the same time, a justification for sexual hierarchy, with women as sexual objects at the service of men. The Quran makes men "the managers of the affairs of women," requiring righteous women to be "obedient, guarding the secret for Gods guarding," advising women to "cast down their eyes, guard their private parts and reveal not their adornment...save to their husbands."[7] It assigns men the task of admonishing women they fear may be rebellious.[8] The sure outcome of this palpable sexual hierarchy, incorporated into family laws in Islamic societies, is that women's very existence is serving men, sexually and emotionally. Women are a "tillage" for the male believer. He can come unto his "tilth as he wishes and forward his souls."[9] If a wife refuses her husband's sexual demands, she is to be punished.

Along the same line, decoding Islamic Paradise, Fatna Sabbah suggests that the paradisal female model, the *huri*, represents the ideal female and, at the same time, the ideal society for the Muslim believer. The *huri* "is created to be consumed as a sexual partner, her value comes from her physical beauty, which God gives as a gift to the believer." She is passive and is stripped of human dimension. "She has been created for one sole destiny: to be consumed by the male believer" (Sabbah 1988, pp. 96-97). Given the fact that, as will shortly be discussed, religious instructions in Islamic societies are at the same time state legislation, this concept of sexuality has specific legal consequences for women.

The Shi'i jurisdiction in Iran, for example, lays down the rights and obligations of women, based on the view that it is the woman's religious duty to submit to all sexual demands of her husband. Ayatollah Khomeini clarifies this point beyond any doubt: "A woman who has been contracted permanently, must not leave the house without the husband's permission and must surrender herself for any pleasure that he wants and must not prevent him from having intercourse with her without a religious excuse" (Khomeini 1984, p. 318). He further makes the ruling that: "If the wife does not obey her husband in those actions mentioned in [the] previous problem, she is a sinner and has no right to food and clothing and shelter..."[10] In the Sunni tradition also a married woman has the obligation to be faithful and obedient to her husband and to admit him to sexual intercourse [when he so desires] (Chaudhry 1991, p. 34).

In the end, Islam's hierarchical concept of sexuality coupled with gender discrimination, in the realm of sexual love and desire, means that even though—unlike other traditional religions—Islam recognizes sexual desires and the need for their fulfillment for both sexes, it makes it harder in practice for women to achieve sexual satisfaction. Female sexual desires and a women's right to sexual fulfillment through the institution of marriage is recognized and legally sanctioned in Islamic tradition. A married woman has the right to sexual fulfillment and the husband is required to respect that right by not forgoing intercourse with his wife for more than four months. Yet, the Muslim man, it is presumed, have sexual desires that cannot be satisfied with only one woman at a time. Therefore, he is allowed to marry four women and hire as many temporary "wives" as he can afford through the institution of Mut'a—as is still the case in Shi'i tradition. This special concession made to men, in effect, nullifies women's rights to sexual pleasure recognized in Islam. Since Islam strongly discourages solitary sex and severely punishes homosexuality, marriage is the only permissible framework for women to seek sexual pleasure. In polygamous unions, women's rights to sexual pleasure is confined to one fourth of a man.

This apparent contradiction in Islamic views of female sexuality may be the root cause of Islam's obsessive concern with the rights and wrongs of female sexual conduct. No doubt, in a tradition that looks with disfavor on "illegitimate" children, the importance of ensuring physical paternity and an unconfused bloodline is an important reason for the surveillance of women's moral conduct. Yet, it is Muslim men's explicit and implicit fear of female sexuality and her seductive power that explains the seclusion and surveillance of women in Islamic cultures (Mernissi 1985, pp. 30-31). For, if female sexuality is seen as active, but a woman's movement for fulfillment of her desires are forcefully restrained, sexual transgression may appear to be inevitable.

Moreover, since a man may not be expected to have much control over his sexual desires, men are understood to be victims of female seductive power.

The fear of the intensity of female sexual desire and her seductive power, indeed, viewing woman-as-body and an exclusive physical entity, is explicit both in Islamic legal discourse and particularly in erotic discourse (Sabbah 1988, p. 50). Men's fear that they will be unable to live up to the intensity of women's sexual energy is the source of their anxiety. First, it makes men preoccupied with sexual performance, with prolonging intercourse and with searching for a sexual strategy that can meet female expectations in bed. Hence, sexual relations constitute a continuing crisis for the believer, because they divert attention from God (Sabbah 1988, p. 90). Second, women's seductive power is a threat to Muslim social order. Woman, in Imam Ali's words, is wholly evil; and the worst thing about her is that she is a necessary evil (Bouhdiba 1985, p. 117). She is necessary because, as expressed in the Quran, her womb is a field that can cultivate the seeds of man. The prominent interpreters of the Quran, like Imam Ghazali, saw women's power as the most destructive element in the Muslim social order.

It should go without saying that the fear of women's seductive power is not unique to Islam. In Jewish scriptual tradition, women seem to have had an aggressive and insatiable sexual drive to the extent that men were only the willing victims of women's enticement. Likewise, Christian priests blame women, whom they saw at fault, for causing men's sexual arousal. In fact, St. Paul believed that women's seductive powers were so great that they caused even Angels to sin (Bullough 1976, pp. 76, 178-179). The grip of these statements over sexual morality in Western civilization is well known. Nonetheless, in the West the standards of sexual morality and immorality in private, which does not harm or cause offense to other persons, are not normally enforced by law. For example, a sexual relationship outside marriage is not a legal offense unless it is aggravated by circumstances such as lack of consent, rape, or with an underage person. Islamic law, by contrast, holds that any sexual relationship is a crime unless it is between husband and wife (Coulson 1969, pp. 77-78). This distinction causes Islamic morality to have a more paralyzing impact on the lives of women (and men). Religious moral instruction and standards constitute the basis of women's legal rights and personal status. The observance of Islamic moral standards is enforced by law.

Jewish and Christian clergy were not statesmen. They, of course, manipulated the state to enhance their own power and influence, as well as the influence of their faith. Social, economic, and political developments in European societies—including establishment of modern legal system, recognition of the rights of the individual vis-à-vis the state, universal suffrage and equality before the law, in different historical periods in various European countries—diminished the arbitrary power of the Christian church and the possibility for it to manipulate the state and the courts to enforce its vision of moral and immoral conduct on the citizenry. With the separation of church from state, the rule of law in the West was substituted for the rule of the divine. This opened the way for the formation of a relatively strong civil society—a crucial development toward accountability of the state. These developments have been instrumental in the struggle of women in European societies for gender equity and justice.

In the case of Judaism, the experience of Jewishness for women, one would think, has been, and still is, different depending on which part of the world they live in, as well as on the class status of members of the Jewish community. In the state of Israel, the Orthodox movements have historically used the state both to gain more resources for their institutions and to impose as many religious practices as possible on Israeli society (Yuval-Davis 1992, p. 206). Women's legal rights in matters of marriage and divorce have been shaped in the process of cooperation and compromises made between the nationalistic (and militaristic) goals of the state and the patriarchal values of the Jewish Orthodoxy—vehemently defended by the Orthodox coalitions in the parliament. The articulation of Jewish patriarchal values (which regard women not as individual citizens with distinct interests and needs, but primarily as the guardians of Jewish home—wives—and producers of future generation of Jews—mothers—with the "security" needs of the state of Israel has resulted in compromising the secular principles at the expense of women. The enactment of the 1953 Marriage and Divorce Law (ending civil marriage and divorce by the recognition of the authority of Rabbinical Courts in these matters) is a case in point (Landau 1993, pp. 127-132; Jorgensen 1994, pp. 285-287). However, the state never assumed the responsibility of overseeing women's moral conduct. Observing the "right Jewish way," which involves following rigid religious instructions in every sphere of life, to a large extent, has remained a personal choice that individuals make in order to be part of a specific Orthodox and Ultra-Orthodox group.

Islam, however, is a religion as well as a legal and political tradition. It also embodies sexual moral and ethical principles that are strictly enforced. The spiritual leaders in Islamic tradition, have been, at the same time, the legislators and administrators of the affairs of the Muslim community. Statements and instructions of Islamic clergy relating to sexual morality and women's personal conduct are not found only in orthodox Islamic texts or practices of Muslim fanatics. They are principles which are enforced through criminal legal system. Sexual offenses in Islamic cultures must be understood in this context.

The fear of female sexuality and concerns over women's moral conducts are not, therefore, limited to orthodox views of bygone generations. Ayatollah Motahhari's writings in pre-revolutionary Iran, for example, demonstrated his deep concerns over the presence of unveiled women and its "disastrous" effect on the Iranian youth. Women, he thought "were turning the youth, who should be the symbol of strength, will-power and productivity, into weak, pleasure-seeking and lustful oglers" (Motahhari n.d., pp. 91-93). Fearing the tempting power of female sexuality, Motahhari argued in favor of making the veil mandatory in the institutions of higher education. The appearance of unveiled women in educational institutions, offices and factories, Motahhari thought, is sexually provocative and diverts men's attention from productive activities. Motahhari was not convinced that even hijab could curb women's sexual and seductive power, for he suggested that female and male students in universities should be separated in classrooms by a curtain. The suggestion was put into practice in post-revolutionary Iran.

It should go without saying that Motahhari's ideas are typical of arguments by Muslim conservatives in the Islamic world. Opposing the "Western" ideals of human rights, Abul Ala Maududi, for example, considers women's chastity—to be preserved through keeping her housebound and in *Purdah*—as one of the "basic principles of human rights in the Islamic world" (Mayer 1995, pp. 100-101). He instructed that "[T]he women should wrap themselves up well in their sheets, and should draw and let down a part of the sheet in front of the face." Another Islamic jurist in Pakistan, Justice Aftab Hussain, argues that purdah keeps both sexes chaste and they avoid gazing at one another (Chaudhry 1991, pp. 107-108). It should be emphasized, however, that while all Islamic conservatives advocate hijab or purdah for women, they do not necessarily agree on the idea of domesticity for women. Moreover, despite the influence of the rhetoric of domesticity, Islamic movements and states, in practice, draw women to public life, albeit within all-female networks and essentially in the service of sexual segregation in public life (Gruenbaum 1992; Moghissi 1995; Tessler and Jesse 1996).

The Islamic idea that women have pernicious seductive power, which endangers Muslim social order and translated into laws and legal practices, affect the believer and non-believer, Muslim and non-Muslims alike. Thus, the Islamists, whether as the functionaries of the state or members of Islamic umma, take it upon themselves to guard the "moral purity" of women in their societies. Indeed, the Islamic principle of *Amr-i bi Mar'ouf va Nahy-i az Monkar* (ordering good and preventing evil) rules out any recognition and respect for the concept of individual right to choice and equal protection under the law. In this context, the clutches of the Guardians of Islamic Shari'a are felt in every aspect of the lives of women (and men). For instance, in Saudi Arabia women are barred from driving. In Sudan, the middle-class professional women, seen as the "same as prostitutes," are harassed and questioned by the regime, as well as by self-appointed moral guards, about their presence in public and their relationships to the men in their company (Hale 1996, p. 192). In Iran, the Chief of Police in Tehran rules that women are not "to smile at strange men, because a woman's smile might arouse satanic lust" (*Iran Times*, 17 June 1994). These practices take their legitimacy from the idea that female sexuality has to be confined, tamed, and controlled for the good of the community. The state's control over women's moral conduct, therefore, goes so far as to policing women's movements, looks, and smiles in public.

The ideological construction of female sexuality as the symbolic representation of Muslim identity and its centrality in the Fundamentalists' "cultural purification" schemes means that women lose much more than men do in the process of the Islamization of the already Muslim societies. The legal sanctioning of various forms of gender violence is the most immediate consequence for women of Islamization schemes. The lowering of the legal marriage age to nine for girls, in post-revolutionary Iran, promotes sexual abuse of girls within the institution of marriage. The unprecedented increase in the number of violent crimes committed against women under the pretext of defending "family honor," alarming even to the Iranian officials, [11] is also the inevitable outcome of the introduction of the Islamic Penal Code in the 1980s. The right of men to kill their wives, sisters, and mothers for "immoral" behavior already existed in the Iranian Penal Code in pre-revolutionary Iran. The enactment of the Law of Retribution and Punishment, *qisas*, by the Islamic Republic, however, sets a price on human life, with the lives of women having lesser value than that of men. The blood money, *diya*, (the sum of compensation payable to the family of the victim), for a murdered woman is, according to the law, half that of a murdered man.[12] It would go without saying that the victims of this cruel medieval legal tradition are almost entirely poor women.

Revitalization of ancient (and often bizarre) legal practices, such as stoning of women accused of zina, reduces women's legal status, promotes hostility toward women, and de-sensitizes society toward violence against women. The provisions of the 1979 Hudud Ordinance (Islamic criminal punishments), and the Law of Evidence, introduced by Zia-ul-Haq in Pakistan, for example, specifically affected women. The part of the Ordinance that affected women most seriously was zina, which encompasses both extramarital sex and rape. A man and a woman accused of zina were sentenced to death by stoning or a hundred lashes. The same provision also was applied to the case of rape. To determine whether the case before court was zina or rape (that is whether or not the women consented to sex), the court required either the confession of the rapist or the presence of four morally "trustworthy" Muslim men. By prescribing that there must be four Muslim male adults as eyewitness, the law excludes the testimony of women altogether, even in the case of their own rape. Moreover, by blurring the line between zina and rape, and given the fact that even in modern legal systems, establishing the guilt of a rapist is difficult, Zia's 1979 law ended up discriminating against women and helping rapists get off the hook (Mumtaz and Shaheed 1987, pp. 100-106; Haq 1996, pp. 158-175). Hence, among the women who received public lashes, were fined and imprisoned at the time, was a blind domestic servant, Safia, who had survived multiple rapes by her employer. The legislators of this cruel ruling leave no room for doubt that in their view women are to blame for sexual crimes committed against them.

Similarly, the move to "Islamize" the already Islamic Sudan, starting with Numeiri's declaration of Sudan as an Islamic Republic and its strict adherence to Shari'a in 1983, particularly affected women. With the passage of the Islamic Penal Code, for example, the Numeiri regime dropped the law against infibulation that had been enacted under British rule and had remained on the books until 1983. The Islamization policy was further intensified under the fundamentalist military junta after Numeiri's downfall in 1989. Among the first "Islamic" acts of the "revolutionary" government was curtailing women's freedom of movement and the imposition of Islamic dress code, hijab, which followed the suspension of the constitution outlawing all political parties, unions, and professional associations, and the imposition of Islamic Shari'a on non-Muslims (Hale 1996, pp. 88-91, 111).

Islamic practices, such as the execution of prostitutes and the forcible re-imposition of hijab on women and the harassment, arrest, and flogging of those seen as improperly dressed in Sudan, Iran, Algeria, and Afghanistan, take their rationale from the Islamic concept of female sexuality as the main cause of social corruption. Whether hijab is a Quranic injunction, aiming to protect the privacy of women from the public gaze, or was meant to put a barrier between two men (the Prophet and his companion-visitor) which was reduced, over time, to "a scrap of cloth that men have imposed on women" (Mernissi 1991, pp. 94-97), its imposition punishes women for their assumed seductive power. The pretext is "protecting" the Muslim community from corruption. There are apologetic interpretations of the veil that present it as "just any other piece of clothing" worn "to beautify the wearer much the same way western women wear make-up" (Hoodfar 1993). The veil can also be misrepresented as the source of empowerment to women or as "an anti-consumerist claim" by women for their right to modesty (El Guindi 1996). Such mystifications, however, do not change the fact that the veil signifies objectification of women. It represents chastity and modesty, as defined by men, and it symbolizes male authority and control over the female body. The fact that some women in the Islamic world, under specific social and cultural conditions and as a result of Islamic socialization processes, "choose" the veil, and that a group of Muslim female diaspora in the West adopt the veil, as a symbol of Muslim identity—essentially in protest to racist structures, relations, and attitudes in the host countries—does not change the meaning of the veil as an instrument of sexual control and domination. The reality is that the veil is still forced on millions of Muslim women through terror or police coercion. Present day Algeria and Iran, where women may lose their lives for defying the hijab code, are cases in point.

WOMEN'S STRUGGLE FOR EQUALITY AND SELF-DETERMINATION

The absence of the concepts of equal rights and equal protection under the law for women in the Islamic system of justice, and the compromises increasingly made by the so-called secular states in Algeria, Egypt, Tunisia, and Bangladesh, make one wonder about the future prospect for women. Indeed, one may be curious as to the meaning of secularism in an Islamic-Middle Eastern context. In Algeria, the state, under both Houari Boumedienne and Chadli Ben-Jadid, tried to appease the Islamic Salvation Front (FIS) by playing the women's card. That is, in 1981, further restrictions were put on women's social rights, legal status, and personal freedom through such measures as placing a temporary restriction on a women's ability to travel abroad alone and by invoking Islamic Shari'a in the new Family Law in 1984 (Cheriet 1992). The same policy was adopted by the Egyptian state in the 1980s. Through the enactment of a new law in 1985, the 1979 Personal Status Law, which, to a certain extent had reduced male privileges in marriage, divorce, and child custody was repealed (Hatem 1994; Badran 1993). The Iranian Civil Code, enacted in the mid-1930s, was also based on the Shari'a, particularly on issues pertaining to family status. Yet, the clerical opposition to Reza Shah and his son Mohammad Reza Pahlavi demanded the application of a more rigid interpretation of the Shi'i Shari'a. To the clerics, the return to Islam on issues related to women's legal status was essential for regaining cultural and political autonomy (Moghissi 1994, pp. 44-46).[13]

With the rise of Islamic fundamentalist movements, whose leaders prescribe an Islamic solution for all social, economic, and political crises, the misogynist medieval Islamic traditions and presuppositions are injected into the civil society by means of new legislation. Under the banner of asserting Muslim identity against the infidel West, personal and social relations are further Islamized. The focal point in Islamization schemes, and the most apparent signifier of a return to "true Islam," is the "cultural purification" of the society. The purification of the female body and soul from the infection of Western influence, and what is perceived as Western values, is central to this utopian scheme. Political repression and heavy handed treatment of the secular opposition, on the one hand, and making concessions to fundamentalists in the areas of women's legal status and personal freedoms, on the other, have been the pattern followed by the state in almost all Middle Eastern countries. Ironically, these policies, added to the West's—particularly the United States—growing economic and political domination in the region and its commitment to an unconditional support of the State of Israel, become a breeding ground for Islamic fundamentalists.

The growth in the recruiting power of Islamic fundamentalist movements may make the prospects of change even gloomier for women, at least in the near future. Paradoxically, Islamization policies have added fuel to anti-oppression struggles in the region. Despite extensive campaign by Islamic fundamentalists to present the idea of gender equity as un-Islamic, feminism as colonialist in its goals and Western feminists as women's main enemy, the most promising signs of resistance come from women themselves. From *mashreq* to *maghreb*, women in Islamic societies are launching the most effective resistance against both Islamic fundamentalists and the repressive regimes in their societies. Women seem to have found a new energy and conviction to fight patriarchy in its many forms (Hale 1996; Moghissi 1995; Amin and Hossain 1995; Chaumeil 1995; Cummins 1995; Hatem 1993; Gruenbaum 1992). Women are coming to the sobering awareness that our participation in anti-colonial and nationalist movements failed us completely. In many cases, women removed a senile colonial patriarchy only to be faced by a vigorously native, nationalist patriarchy, harder to challenge precisely because of its authentically indigenous character. Middle Eastern women are at the forefront of the struggle for democratization of the Islamic societies.

It is precisely because of the rising voices of discontent that the secular Middle Eastern women are under fire from both Islamist and hypocritical rulers in their societies. Jahanara Imam, murdered by Jama'at-i Islami in Bangladesh, Khalida Messaoudi and Nabila Djahanine, assassinated by the FIS sympathizers in Algeria, and Homa Darabi, who set herself on fire in protest of the suppression of women's human rights in Iran, are but a few names that come to mind. Many thousands of women locked in the prisons of Middle East countries, and many other hundreds of thousands have put their lives on the line by challenging the traditional patriarchal social order, and question the rights of state and religious leaders to use and abuse women in their own political games.

In the past, discriminatory treatment of women and strict gendered legal and moral codes of conduct also existed in Western societies. The separation of state from religion, economic development, and political democracy in non-Muslim societies of Europe planted the seeds for gender equity. Women's full citizenship status was achieved after secularization of legal and political life, when social and economic forces of modernity opened the way for the counter-hegemonic forces and voices within the civil society. Consequently, legal and formal discrimination against women were removed from many areas of social life. These are the initial steps, in the right direction, that have to be taken in Islamic societies. Many times in the course of Middle Eastern history the intellectual and popular struggles moved in this direction, only to be pushed back by the reactionary forces of Islamism and imperialism. The disastrous balance sheet of Islamic rulers in, say, Iran, Sudan, and Afghanistan, one would hope sooner or later, should bring home the reality for the masses in Islamic societies, that the solutions to their social, economic and political crises are worldly and secular. Social justice, political democracy and economic development, desperately needed in the Middle East and other Islamic societies are not what the Islamic rulers or leaders of fundamentalist political movements have in mind or are capable of delivering to their societies. Democratic change in Islamic societies is irresistible and inevitable. Women would benefit more than men from such a change.

NOTES

1. I use the term fundamentalism to refer to new Islamic political movements within the Middle East. I share Sahgal and Yuval-Davis's definition of the term and their view that the popular usage of the term which obscures the fact that fundamentalism is not peculiar to Islam, aims at abusive labelling of Muslims. They define fundamentalism in the following way: [P]olitical movements which have a religious imperative and seek in various ways, in widely differing circumstances, to harness modern state and media powers to the service of their gospel. This gospel is presented as the only valid form of religion. It can rely heavily on sacred religious texts, but it can also be more experiential and linked to specific charismatic leadership. Fundamentalism can align itself with different political trends in different countries and manifest itself in many forms. It can appear as a form of orthodoxy—a maintenance of "traditional values"—or as a revivalist radical phenomenon, dismissing impure and corrupt forms of religion to "return to original sources" (Sahgal and Yuval-Davis 1992, p. 4).

2. In August 1996, the Court of Cassation, Egypt's highest court, upheld the ruling of a lower court that Professor Nasr Hamed Abu Zeid, must divorce his wife because he has denounced Islam (*Guardian Weekly* 1996).

3. In Shi'i school of Shari'a the right of the divorced mother to the custody of her male child is until he is two years old and of her female child until she is seven years old. The right of mother to custody varies in Sunni shools of Shari'a, from seven years for male child and nine years for female child in Hanafi school, to until puberty for boys and till marriage for girls in Maliki (Anderson 1976, pp. 142-143). What is important to note, however, is that in all Shari'a schools the father maintains the legal guardianship (*velayat*) of the children.

4. Douglas Sladen, quoted in Mabro (1996, pp. 61, 67).

5. On the conditions of women and their struggle for liberation in nineteenth-century Egypt see Badran (1995).

6. On the institution of Mut'a see Haeri (1989).

7. Sura XXXIV:30, *The Quran Interpreted,* translated by Arberry (1964).

8. Sura IV:35 (Women).

9. Sura II:223 (The Cow).

10. It is worthwhile to note that Shi'i jurists go into great details in describing all kind of sexual activities in which men might get involved, from incest, sodomy to sex with animals. None of these sins is unforgivable in Islamic tradition.In each case, men are required to pay a compensation money (jezieh) to wash their sin.

11. The increase in the number of homicide cases, committed against young women in southern province of Khuzitan alarmed even the Chief Justice, Ayatollah Yazdi, who instructed the judges not to acquit the murderer without proper investigation (*Zan-e Rooz* 1993).

12. The murder of women for protecting "family honor" is prevalent all over the Middle East. A report by the Al-Fanar-Palistinian Feminist Organization estimates the number of murders or attempted murders to be within the range of 20 to 40 cases a year in Palestinian society (Al-Fanar Report 1995). See also Abu-Odeh (1996).

13. For more information on Family Law in Iran and women's legal rights and status in the family during Reza Shah and Muhammad Reza Pahlavi see Aghajanian (1986), Bagley (1971), Momeni (1972), Bamdad (1979), and Sanasarian (1983).

REFERENCES

Abu-Odeh, L. 1996. "Crimes of Honour and the Construction of Gender in Arab Societies." Pp. 141-194 in *Feminism and Islam: Legal and Literary Perspectives* edited by M. Yamani. New York: New York University Press.

Aghajanian, A. 1986. "Some Notes on Divorce in Iran." *Journal of Marriage and the Family*. 48(November): 749-755.

Anderson, N. 1976. *Law Reform in the Muslim World*. London: University of London/ The Athlone Press.

Al-Azmeh, A. 1993. *Islams and Modernities*. London: Verso.

Al-Fanar Report. 1995. "Developments in the Struggle against the Murder of Women against the Background of So-called Family Honour." *Women Against Fundamentalism* 6: 37-41.

Alloula, M. 1986. *The Colonial Harem*. Minneapolis: University of Minnesota Press.

Amin, S. and S. Hossain. 1995. "Women's Reproductive Rights and the Politics of Fundamentalism: A View from Bangladesh." *Women Against Fundamentalism* 7: 8-12.

Arberry, J. 1964. *The Quran Interpreted*. London: Oxford University Press.

Atkinson, K. 1992. "Tales of Desire." *New Observations* 88(March/April): 5-9.

Badran. M. 1995. *Feminists, Islam and Nation*. Princeton, NJ: Princeton University Press.

————. 1993. "Independent Women: More than a Century of Feminism in Egypt." Pp. 129-148 in *Arab Women: Old Boundaries New Frontiers*, edited by J. Tucker. Bloomington and Indianapolis: Indiana University Press.

Bagley, F.R.C. 1971. "The Iranian Family Protection Law of 1967. A Milestone in the Advance of Women's Rights." Pp. 47-64 in *Islam and Iran*, edited by C.E. Bosworth. Edinburgh: Edinburgh University Press.

Bamdad, B. 1979. *From Darkness into Light: Women's Emancipation in Iran*. Hicksville, NY: Exposition Press.

Bouhdiba, A. 1985. *Sexuality in Islam*. London: Kegan Paul.

Bullough, V.L. 1976. *Sexual Variance in Society and History*. Chicago: The University of Chicago Press.

Chaudhry, M.S. 1991. *Women's Rights in Islam*. Delhi: Adam Publishers.

Chaumeil, M. 1995. Algeria: Defend Intellectuals! *Women Against Fundamentalism* 6: 42-43.

Cheriet, B. 1992. "The Resilience of Algerian Populism." *Middle East Report* (January-February): 13-14.

Colla, E. 1993. "Silencing is at the Heart of my Case." *Middle East Report* (November-December): 27-29.

Coulson, N.J. 1969. *Conflict and Tensions in Islamic Jurisprudence*. Chicago & London: The University of Chicago Press.

Cummins, A. 1995. "Taslima Nasreen and the Fight Against Fundamentalism." *Women Against Fundamentalism* 6: 53-56.

Doi, A.R.I. 1989. *Woman in Shariah*. London: Ta-Ha Publishers Ltd.

Dragadze, T. 1994. "Islam in Azerbaijan: The Position of Women." Pp. 152-163 in *Muslim Women's Choices: Religious Belief and Social Reality*, edited by C. Fawzi El-Solh and J. Mabro. Oxford: Berg Publishers.

Graham-Brown, 1988. *Images of Women: The Portrayal of Women in Photography of the Middle East 1860-1950*. London: Quartet Books.

Gruenbaum, E. 1992. "The Islamist State and Sudanese Women." *Middle East Report* (November-December): 29-32.

Guardian Weekly. 1996. August 18.

Haeri, S. 1989. *Law of Desire: Temporary Marriage in Shi'i Iran*. Syracuse, NY: Syracuse University Press.

Hale, S. 1996. *Gender Politics in Sudan: Islamism, Socialism, and the State*. Boulder, CO: Westview Press.

Haq, F. 1996. "Women, Islam and The State in Pakistan." *The Muslim World* LXXXVI(2, April): 158-175.

Hatem, M.F. 1994. "Privatization and the Demise of 'State Feminism' in Egypt." Pp. 40-46 in *Mortgaging Women's Lives: Feminist Critiques of Structural Adjustment*, edited by P. Sparr. London: Zed Press.

Hoodfar, H. 1993. "The Veil in Their Minds and On Our Heads: The Persistence of Colonial Images of Muslim Women." *Resources for Feminist Research* 22(3 & 4): 5-18.

Jorgensen, C. 1994. "Women, Revolution, and Israel." Pp. 272-296 in *Women and Revolution in Africa, Asia and the New World*, edited by M.A. Tetreault.

Landau, D. 1993. *Piety and Power: The World of Jewish Fundamentalism*. London: Secker and Warburg.

Lazreg, M. 1994. *Eloquence of Silence: Algerian Women in Question*. New York & London: Routledge.

Lowe, L. 1991. *Critical Terrains: French and British Orientalisms*. Ithaca, NY: Cornell University Press.

Khomeini, Ayatollah R.1980. *Clarification of Questions*. Qum: Sadra Publisher.

Malti-Douglas, F. 1991. *Women's Body, Women's World: Gender and Discourse in Arabo-Islamic Writing*. Princeton, NJ: Princeton University Press.

Mabro, J. 1996. *Veiled Half-Truths: Western Travellers' Perceptions of Middle Eastern Women*. London & New York: I.B. Tauris & Co. Ltd.

Mayer, A.E. 1995. *Islam and Human Rights*. Boulder, CO: Westview Press.

Mernissi, F. 1991. *The Veil and the Male Elite: A Feminist Interpretation of Women's Rights in Islam*. Reading, MA: Addison-Wesley.

———. 1985. *Beyond the Veil: Male-Female Dynamics in Modern Muslim Society*. London: Al Saqi Books.

Minai, N. 1981. *Women in Islam: Tradition and Transition in the Middle East.* New York: Seaview Books.

Moghissi, H. 1995. "Public Life and Women's Resistance." Pp. 251-267 in *Iran After The Revolution: Crisis of An Islamic State*, edited by S. Rahnema and S. Behdad. London & New York: I.B. Tauris.

________. 1994. *Populism and Feminism in Iran: Women's Struggle in a Male-Defined Revolutionary Movement.* London: Macmillan Press.

Momeni, D.A. 1972. "The Difficulties of Changing the Age of Marriage in Iran." *Journal of Marriage and the Family* 34(3): 545-551.

Motahhari, M. 1979. *The System of Women's Rights in Islam* (*Nezm-e Huquq-e Zan Dar Islam*). Qum: Sadra Publishers.

Mumtaz, K. and F. Shaheed. 1987. *Women of Pakistan: Two Steps Forward, One Step Back?* London: Zed Books.

Sabbah, F. 1984. *Women in the Muslim Unconscious.* Elmsford, NY: Pergamon Press.

Sahgal, G. and N. Yuval-Davis, eds. 1992. *Refusing Holy Orders: Women and Fundamentalism in Britain.* London: Virago Press.

Sanasarian, E. 1983. *The Women's Rights Movement in Iran: Mutiny, Appeasement, and Prepression from 1900 to Khomeini.* New York: Praeger Publishers.

Tannahill, R. 1989. *Sex in History.* London: Sphere Books Ltd.

Tessler, M. and J. Jesse. 1996. "Gender and Support for Islamist Movements: Evidence from Egypt, Kuwait and Palestine." *The Muslim World* LXXXVI(2): 200-217.

Zan-e Rooz, Tehran, July 3, 1993.

MUSLIM FAMILY LAW:
ARTICULATING GENDER, CLASS AND THE STATE

Nahla Abdo

INTRODUCTION

Under the influence of cultural imperialism and Orientalism, Islam, in much of Western literature, has been viewed as a homogeneous and immutable entity resistant to change. Islam is often essentialized and perceived as a monolithic force residing outside of history. On the other hand, some writings by feminists, both Western and Eastern, have tended to conflate Islam and patriarchy and present the two as an inseparable phenomenon. Whereas the first approach presents Islam in a fundamentally ahistorical manner, the second approach fails to historicize and contextualize gender relations.

This paper will argue that Islam as an ideology, Muslim laws or the Shari'a, and social customs or practices conducted in the name of Islam are not one and the same, nor are they experienced similarly in all Muslim societies. This paper builds on the historical fact that Islam emerged and developed in already existing civilizations; that it spread into other societies already having their own social structures and ideological and cultural norms. While Islam may have influenced these societies, they, in turn, have also left their imprint on

International Review of Comparative Public Policy, Volume 9, pages 169-193.

ISBN: 0-7623-0268-2

Islam. Islam, in other words, will be viewed as the changer and the changed. It will be dealt with as a force operating within history and not despite or outside of history.

The present paper will apply this historical approach to the analysis of the conditions, status, and role of women in Muslim societies and particularly within Muslim Law(s). The paper contends that gender relations, in general, and the role, status, and position of women, in particular, are not determined by Islam alone. Without denying the role of Islam, particularly in its recent politicized form(s), this paper maintains that women's productive and reproductive lives are largely molded, shaped, and reshaped within very real material forces of a social, economic, political, and cultural nature, and further that these forces are constantly in flux.

It is ironic that most discussions on women's issues in the Middle East and in other Muslim countries are basically confined to the contexts of the family and religion, hence, the primacy of family laws, perceived of as the only sphere for women's growth or stuntation (Hijab 1988). Muslim women's liberation, as a result, becomes predicated on their liberation from their own religion.

In contrast to the above position, this paper will examine the heterogeneity and in fact flexibility of Islam through a comparative analytical approach of the family laws in various Middle Eastern and Muslim states. It will demonstrate that a proper understanding of women in Muslim countries necessitates going beyond the sphere of ideology. One of the means of accomplishing this will entail conducting an enquiry into feminist debates within the Muslim world.

FEMINIST LEGALISTIC OPPORTUNISM: A CRITIQUE

Discussions on women in Muslim societies, not unlike general discussions and concerns regarding Islam, have been influenced by specific historical developments at both the national and international levels. For the past two decades, the Middle East has been undergoing a major identity crisis, as the region tries to come to terms with the fast changing pace of globalization and the internationalization of capital. The economic crises and political turmoil that have been gripping the region for quite some time have had major impacts on the Arab and other Muslim states in general, and on gender relations in particular. At both the national and local level, Muslim societies have also had—and continue to cope with—internal conflicts based on gender, ethnicity, religion, and nationalism. It is not surprising, therefore, that studies and debates on Muslim women have mushroomed in recent decades, nor is it unexpected to find that most discourses and debates are reactions to a perceived threat from Western culture.

It is understandable that under specific conditions, priorities, urgencies, and preferences lead to shifting emphases between and even within these contexts.

This is particularly true with regard to women's participation in national liberation movements, where their primary allegiance is often focused on the wider cause. Thus, for example, during the Palestinian Intifada, women's massive political participation, especially during the first three years, was mainly geared toward resisting Israeli colonialism, therefore prioritizing national liberation over gender equality and women's own rights. On the other hand, there have also been shifts emphasizing the "internal," namely social-gender forms of oppression, and a privileging of these over "external" forms of oppression. Women of the Intifada demonstrated this shift in the last year or so of their uprising, and Algerian women's ongoing struggle is another indication of this change.

Whether the emphasis is placed on one context or the other, what needs to be asserted is that all debates on women in Islam are themselves the product of political-economic factors. Therefore, the debates on Muslim women, as well as any critical reevaluation of these debates, must be located within these broader settings. Such a contextualization (i.e., one that accounts for the particular social, economic, political and cultural circumstances) helps to avoid an entrapment in a culturalist and ethnocentric approach, or conversely, from falling into the monolithic and homogenizing methodology or paradigm of the Orientalists. The imperative of adopting a systemic, structural, and contextualized approach to gender relations in Islam is more urgent today than ever before—partly due to the proliferation of literature on gender and Islam, and partly because of the hotly debated and quite controversial positions taken by various Muslim scholars and cultural advocates.

The following discussion will shed some light on the main conceptualizations of women living under Muslim laws, highlighting the variety of opinions and positionalities. To capture such diversity, one needs to go beyond scholarly published works into the realm of popular culture, as the most controversial positions are often expressed in artifacts of popular culture. A review of two videos dealing with these issues will illuminate this point.

In a video titled *Women in Islam*, produced by the Islamic Media Services in 1991, women interviewed presented a position defending Islam and all of its assumed laws and regulations, including veiling and early marriage. An almost unanimous voice of a basically subservient position of women in most matters, including polygamy, was strongly articulated by the heavily veiled women of this production. Most of the interviewees appeared to be of middle- and upper-class backgrounds, some were professionals while others were students. All shared similar opinions on Islamic symbols or aspects of the Family Laws. Thus, for example, the veil was viewed as a symbol of purity, shielding women from objectification, while polygamy was appreciated as a form of security to women.

The discourse in this video was primarily constructed in opposition to Western values, and particularly to the notions of democracy, individual freedom, and feminism. Feminism was essentialized and homogenized to denote anything and everything immoral in women's behavior in the West. This was quite clear in the approach taken by most of the women interviewed to the issue of polygamy. Polygamy in the West, the women argued, happens in very "immoral" ways, "under the table." Islam, in contrast, has legalized and legitimized polygamy, protecting women from becoming mistresses.

The Islamic discourse presented above is quite popular among Muslim fundamentalists or revivalists and is found in most of their popular literature (Afshar 1994). Such approaches are based on the gender ideology that assumes women's minds and bodies are weakened by natural processes, and that women must accept positions in society within a very specific sphere of economic and social activity involving submission to their fathers and husbands (Maududi 1974, pp. 114-137).

In contrast to the almost unanimous discourse mentioned above, a video titled *Veiled Revolution*, produced by Elizabeth Fernea in 1992, provides a diversity of opinions on the most controversial symbol of Islam, namely, the veil. Here, most "newly" veiled Egyptian women tend to see the veil as a means of protecting their place in the public sphere. Their argument suggests that the veil forces men to take women as serious partners rather than as sex symbols and enables women to enter into the masculine world without disturbing society's order. The hijab (veil), in this discourse, takes on both a political and a practical meaning rather than an ideological one. Similar arguments, as shall be further discussed, are prevalent among an increasingly popular type of scholarship.

In the discourse of *The Veiled Revolution*, the veil is not unanimously embraced as a religious axiom. To the contrary, it acquires a controversial character as non-veiled Muslim women resent the interpretation of the "new Islamists." In fact, some argue that the "reinvention" of the veil is a backward phenomenon that threatens the historic achievements of Egyptian feminism.

Within Muslim feminist scholarship, writings on women and Islam fall largely into two categories: those who adopt a secular(ist) approach to women by contextualizing their lives, conditions and struggles within the wider framework of the state, class, social, gender and cultural forces (Hammami and Rieker 1988; Sa'adawi 1983, 1988; Ghossoub 1987; Moghadam 1994; Hijab 1988); and those who locate their analysis within a cultural Islamic framework, suggesting that it is the misuse or misinterpretation of Islam that must be blamed for women's status (Ahmed 1992; Mernissi 1988, 1992a, 1993, 1994; Hoodfar 1993, 1996; Dirie 1996; Hessini 1994). While scholars in the first category are identified—by self and other—as secularists, scholars embracing the religious Islamic approach are not necessarily religious or Islamist themselves. Ironically, the most eloquently articulated studies on Muslim women's resistance within the veil or within Islam are those espoused by scholars who identify themselves as secular, modernist, or non-religious.

The following discussion will focus on the second (i.e., religious cultural) approach. This approach is of particular concern because of the popular appeal it claims to command, and also because of what I suggest is the problematic nature of its current argumentation. Scholars in this category may adopt different social science approaches and may focus on different points in explaining issues of women in Islam. For example, Leila Ahmed (1992) takes an interesting historical approach in order to revisit the role of Muslim women since the inception of Islam. Hoodfar (1991, 1993, 1996), on the other hand, takes an anthropological ethnographic approach, focusing on daily life experiences and forms of resistance espoused by Muslim women. Whether the case study is Iran or Egypt, Hoodfar's conclusions appear to describe a particular trend: veiled women, even if veiling was imposed, do not see themselves as victims. To the contrary, Hoodfar identifies "Islamic feminists" as women who are active, conscious, and capable of resistance and self-empowerment.

Authors in this category deal with various forms of Islamic laws, such as polygamy and divorce, nonetheless, veiling remains top on their agenda. Feminists subscribing to this approach unfalteringly consider the veil a tool of patriarchy. However, this tool is presented as a double-edge sword; while imposed as a patriarchal and oppressive tool, the veil, these authors insist, can also be used by women to "subvert patriarchy." Leila Hessini's description of how Moroccan Muhajjabat (i.e., veiled women) utilize the veil speaks well to this issue. According to Hessini, "by wearing the Hijab women can enter into traditional spaces without disturbing society's equilibrium; this also demonstrates a woman's choice to belong to a group which professes ethical standards in tune with Muslim values" (Hessini 1994, p. 51).

"Subverting patriarchy" or changing women's position is, in this literature, believed to be possible either through female interpretation of the Muslim family laws (e.g., Mernissi's position), or through incremental gains achieved by women who challenge patriarchy through their own daily life experiences (e.g., Hoodfar's analysis). The latter position does not offer any definition to these gains, which can refer to anything from using the veil as a coping mechanism to a much more subtle and passive mode of resistance in the street or in the bus (Hoodfar 1993, 1996).

Common to scholars adopting this position is what I refer to as *Feminist legalistic opportunism.* Feminist legalistic opportunism is the claim that women's problems, oppression, and victimization, as well as their solutions and emancipation, can primarily be attained within the confines of the Muslim laws and the boundaries of the legal system. In other words, this approach entrusts all discussions of women's problems and prospects in the legal system, trusting the latter to be the ultimate savior or champion of women's emancipation. Additionally, this approach confines women's rights to religious jurisprudence, albeit with a call to reinterpret the Shari'a.

Fixation on the legal sphere is quite characteristic of the work of the prolific scholar Fatima Mernissi, particularly in her book, *The Veil and the Male Elite: A Feminist Interpretation of Women's Rights in Islam* (1992a). While Mernissi's approach is more of a social-psychological one that, among other things, is heavily concerned with men's or the masculine perception of women's bodies, her solution to women's oppression within Muslim laws is similar to that adopted by other scholars in this approach. For Mernissi, the main problem for Muslim women is that the Quran, the Shari'a, and the Hadith have been interpreted by men in order to serve their own interests. To change their conditions, Muslim women must engage in the work of reinterpreting the Family Laws, diverting it from a male-centred into female-centred interpretation.

The ethnographic or micro-level-analyses characteristic of this literature can be interesting and illuminating, particularly as this approach tries to challenge Orientalist perspectives by restoring a human face onto veiled, "faceless" women. Equally illuminating and quite important is the work of Mernissi when she challenges Islamic laws as immutable. For Mernissi, the power of these laws comes from the male and patriarchal character of their inventors or interpreters, and not from a divine unchallenged source. Nevertheless, I believe that an approach of feminist legalistic opportunism remains highly problematic and, sometimes even dangerous.

To begin with, this approach gains its strength and perhaps even popularity mainly through the strong claim made by its propagators who identify themselves as secular and present a religious (Islamic) thesis based on their claim that "you need to speak the language of the masses in order to reach them." This position presents an ethical problem, namely, that of accepting secular life for oneself while advocating a religious one for the other. A more important issue here is that the scholarly or scientific community has yet to demonstrate that the masses of Muslim women do indeed prefer an Islamic solution over other forms of solutions. Moreover, from a tactical perspective, it is difficult to maintain that an Islamic solution will lead to women's liberation (and more generally to human liberation). In the absence of any research on the true needs or wishes of the masses, the basis of such popularity becomes difficult to sustain. In reference to the Pakistani case, for instance, Fauzia Gardezi (1990, p. 21) warned against applying ideal factors to explain women's oppression, while ignoring, in the process, the real material life conditions and experiences of women.

Before leaving this argument, I would like to point out that privately, among political grass roots activists, another discourse is taking shape. Here, Islam, reinterpreted and employed by women to achieve certain (particularly political) gains, becomes a tactical device. It is, as one Sudanese activist confided: "fighting men in power with the very same weapon they use against us." In this discourse the Islamic shield is used as a temporary working device to politicize the "poor, the illiterate and the disempowered." However, the ultimate goal or long-term strategy is articulated in terms of the separation between religion and state. Intellectualizing for public (and mainly Western) consumption is not a concern for these activists. Their concern is with effecting political changes in their daily lives. As much as one can appreciate such an experiential and activist position, the privacy surrounding what is tactical and what is strategic makes it difficult to engage in any discourse or debate.

Confining one's analysis to the legal sphere leaves room for other lacunae. Various women's organizations throughout Muslim countries, including "The Centre for Women's Counselling and Legal Aid" in the West Bank and "Women Living Under Muslim Law" with its two branches in Lahore and Paris, have been working on reforming the family laws for several years. These organized women see their work as empowering to women living within Muslim laws, as it raises their consciousness about their rights and challenges some misconceptions around the laws. However, these organizations are increasingly coming to the realization that there is a limit to reform within the status quo. Short of rejecting the divinity of the Quran, how can, for example, any feminist analysis feminize the following verses—let alone convert them into emancipatory principles:

> If you fear that you cannot treat orphans with fairness, then you may marry other women who seem good to you: two, three or four of them. But if you fear that you cannot maintain equality among them, marry one only, or any slave girl you may own. This will make it easier for you to avoid injustice (IV:2).

Or, consider the following verses from the same *Sura*:

> [M]en have superiority over women because Allah has made the one superior to the other, and because they spend their wealth in maintaining them (IV:34).

Or,

> Good women are obedient. They guard their unseen parts because Allah has guarded them. As for those from whom you fear disobedience, admonish them and send them to beds apart and beat them. Then if they obey you, take no further action against them (IV:34).

Feminist legalistic opportunism, as this discussion reveals, faces problems at various levels. Take for example the Muslim inheritance law that affords women the right to inheritance—albeit a lesser portion than that of the male. In practice, women have not benefited from this law. Historically, most women have been deprived of their right to property. Even if activists or feminists strive for and succeed in reforming this law to give women, in theory, an equal share to that of the male, there is no guarantee that women will, in practice, inherit property. Customarily, inheritance has been tied to traditional cultural norms of landed and other forms of property, around which the whole unit of the family, clan or tribe have been maintained and reproduced. Challenging the inheritance law would necessitate challenging the traditional cultural norms and not only the legalistic aspects of these laws. Having said this, however, this argument should not be construed as an anti-reform position. Quite to the contrary, legal reforms and changes are necessary, particularly in the face of the fast changing material life that all Muslim societies have been undergoing. The point here is that confining women's emancipation primarily to the legalistic sphere is no guarantee for true change.

Clearly, a one dimensional, single sphere (legalistic) analysis is too narrow an approach. Among other things, it fails to account for the role of culture and tradition. This is evident in the way scholars tend to conflate Islam with patriarchy. Islam is often essentialized and discussed as a monolith rather than as an ideological force that is often sustained and reproduced under particular social, economic, and political conditions. Islam is frequently used interchangeably with culture, thus masking the differences between, for instance, Arab, Iranian, and Indian or Pakistani cultures. All of these cultures become, miraculously, one and the same.

Furthermore, there is a potential danger in adopting such approaches. Such peril lies in the failure of these approaches to demonstrate the connection between women's conditions and the wider context within which these conditions arise, develop and are reproduced, namely, the nature and character of the state and the class structure. Few and far between are the studies that attempt to perceive social-gender problems as institutionalized systemic problems that are the product of the specific historical junctures of patriarchy, the religious elite, and the state. In fact, any form of state criticism in almost all of this literature is surprisingly and shockingly absent, as if the state does not matter. Yet, as will be shown shortly, the relevance of the state is central to any analysis or understanding of women's conditions. It is not incidental, one may note, that throughout the Middle Eastern region only Tunisian and Turkish women currently enjoy the abolition of polygamy. Such legal transformation could not have been possible had it not been for the direct intervention of the state. Short of abolishing such a law, it is difficult to imagine what type of gender equality any reform or feminist interpretation could bring about.

Finally, considering the multiplicity of male hegemonies reinforced by the recent tide of Islamization, such scholarship—appearing even implicitly to condone acts restricting women's movement—might feed into the already existing patriarchal systems of the religio-political parties.

ISLAM, PATRIARCHY AND THE STATE: AN ALTERNATIVE APPROACH

The above critique is intended to provide a basis for an alternative approach to the study of women and Islam. An alternative approach to studying women and Islam, or women and family law, must be contextualized within the specific history of the particular region or geographical locality under investigation. The merit of such an approach is that it avoids a monolithic essentialist approach to Islam. Moreover, by contextualizing Islam within the history of a particular culture, such an approach can show the relationship between culture, patriarchy and Islam as an ideology and practice. Such studies cannot be generalized: they are historically specific.

Whether one employs the concept of "classical patriarchy" (Kandiyoti 1991) or "neo-patriarchy" (Sharabi 1988) to describe the male nature of power and control that characterizes most states and institutions throughout the Middle East, it remains pertinent to realize that this maleness or patriarchy is not a purely abstract ideological construct. Rather, it is the product of very specific historical junctures of socioeconomic and political configurations, with deep roots in real historical material culture. It is within this material base of the modes or forms of production of class structures and states that women's conditions must be understood.

When first introduced in the Hijaz, Islam encountered a highly patriarchal structure on which it built its main tenets. In pre-Islamic Arab (tribal) society, a marriage contract strongly resembled a bill of sale according to which the female became the husband's possession. Esposito describes some characteristics of pre-Islamic law. First, the female's tribe received the dower (*mahr*) from the husband for the bride; in return, the tribe relinquished all claims to the woman and to her future children. Esposito notes that the female's worth at that time was deemed to be her dower and that in her life, chastity and fidelity were encouraged and emphasized. Further, women were denied any rights to inheritance because this would mean the transfer of property from one tribe to another, which was inconceivable. Women had no say whatsoever in the contracting of the marriage, nor in its dissolution; upon marriage, the female became subject to her husband's control. It was also a common, legitimate practice for female children to be buried at birth because they were devalued to such a large degree (Esposito 1982).

It is not surprising therefore that Islam retained some aspects of the already existing gender practices, while abolishing some, such as the "*wa'd*" (burying of female children at birth) and altering others, such as including women in inheritance. More research is needed to illustrate how Islam was received or changed by cultures that practiced matrilineal, rather than patrilineal or even patriarchal forms of relations. There is some indication suggesting that the introduction and imposition of Islam on, for example, parts of Kerala in India and among the Qabyle Berbers in Algeria, where matrilineal practices predominated, took different forms than in the overwhelmingly patriarchal societies of the Muslim world. In some parts of Muslim India, such as Kerala, it was not until the state (both the British colonial and the Indian) enforced Islamic laws that a patriarchal system began to dominate.

Control over private property and inheritance are basic mechanisms of power used by all states. Such control, as Souad Joseph has correctly observed, is a fundamental element in the reproduction of gendered citizenship throughout the Middle East (Joseph 1996). Yet, in order to understand this aspect of Muslim family laws, one cannot resort to ideological interpretations alone. It is historically plausible and logically conceivable that inheritance acquires an important status in the transition to private property. In order, for example, to comprehend why some Muslim women did historically inherit property (i.e., among upper-class women in late nineteenth-century Egypt [Tucker 1985]), and why others did not, one must look at, among other things, pre-capitalist class structures and forms of marriages. One may note here that inheritance or lack of it varies not only between countries but within the same country between different regions. In tribal, often exogamous marriages, the value placed on property is different from when landed property is at stake. Peasant societies place a special value on landed property, since land is their primary means of survival, as well as their major source of wealth and social prestige. It is not incidental, for example, that in pre-capitalist Palestine endogamous marriages, particularly among first cousins, were encouraged as such marriages were expected to keep property within the family, *Hamula* or clan (Abdo 1987).

An important question to be addressed here is why, despite the destruction and disappearance of most pre-capitalist forms of production and social life, ideological forms of social-gender organizations continue to survive? What contributes to the longevity of traditional forms of marriage and inheritance despite the destruction of the material basis that gave rise to them in the first place? To simply state that an ideology expressing the articulation of traditional customs and religious beliefs dies hard is not sufficient. Other factors are equally, if not more, important. These include political and economic forces, such as the level of economic development or the state of economic dependency under which most previously colonized countries found themselves after independence. Another important factor lies in the role of the state in maintaining and reproducing various forms of gender inequality. As demonstrated elsewhere (Abdo 1987), both the British and the Israeli colonial states have played a crucial role in fostering and further reproducing gender inequality among the Palestinians. "The different political projects of modern nation-states," as Kandiyoti observes, "...account for deep and significant variations in policies and legislation affecting women. These variations find concrete expression in the degree of access that women have to education, paid employment, social benefits, and political participation" (Kandiyoti 1991, p. 38).

The analysis in this paper focuses primarily on the level of the Canon versus reality/experience. There is, however, another important level of analysis, the inclusion of which improves our understanding of Islam. That is religion versus politics, or the distinction between Islam the religion (i.e., the relationship between the individual and God) and Islamization as a political movement. The latter, whether referred to as fundamentalism or revivalism, represents a political movement that bases itself on existing or re-invented religious codes. Political Islam, expressed in the various Islamicist movements and operating in almost all Muslim countries, seeks state power with a somewhat common agenda. This is a highly conservative, patriarchal agenda which, in terms of its social gender implications, is also reactionary.

It is in this context, namely, when Islam acquires a basically political face, that one begins to comprehend how the state in the Middle East walks a tight rope between secular nationalism and Islamism. This is particularly so in the context of international/global capitalism.

Globalization often pushes the state toward further Westernization and modernization. In this regard, the state can be viewed—particularly by its female citizens—as the site for their protection and, perhaps, even emancipation. However, when faced with internal resistance by Islamists (political Islam) and in a bid to control and maintain the status quo, the state does not hesitate to abandon its "secular" outlook adopting more rigid and often oppressive measures, particularly in the domestic, private sphere. The role of the Algerian state throughout the past decade and a half provides an example: on the one hand, the state has tried to pacify the FIS (*Front Islamique De Salut*), while on the other, under economic pressures and a strong women's resistance, has also attempted to not alienate the secular "modern" forces. The state's ambivalent role was most obvious during the introduction of the controversial Family Code of 1984, which instead of addressing the injustices against women in Algeria, reinforced the existing patriarchal grip over them.

During nation-state building, particularly in periods of economic and political crises, the state often seeks alliances with different ideological-political movements. Such alliances, particularly with political Islam, often work to the detriment of women's equality and against improving their status. In a study of the Indian state, Amrita Chhachhi cites examples of how the state, at different points during its nation-building, aligned itself both officially and unofficially with Hinduism, thus alienating Muslims. She also explains how in the 1950s, the government sought to convince Muslims within India that it was committed to creating a nation-state that would grant all religions the same rights, by allowing the continuation of Muslim personal law rather than pushing for equality of all women before the state law. Muslim women in India, as Chhachhi concludes, became twice oppressed by virtue of their religion and their sex (Chhachhi 1991).

A similar situation was observed in Malaysia where Malay women and men continued for centuries to follow the Law of *Adat* (customary laws). Both sexes enjoyed "equal rights to property (including land) and divorce, and were subject to the same codes of sexual behaviour, promiscuity being condemned in both sexes" (Agarwal 1988, p. 17). Yet, as in the Indian and other experiences, the intervention of the state, influenced by increasing Islamization, has begun to threaten the customary non-Islamic cultural traditions, particularly as they relate to gender-sex relations.

The state's dualism or vacillation between the secular and the religious is becoming increasingly characteristic of Middle Eastern states such as Egypt, Algeria, and the Sudan (prior to the take over by the Inqath Salvation government). The ambivalent role of the Egyptian government is a case in point. On the one hand, fearing a tide of Muslim Brotherhood (*al-Ikhwan*), a recent law was issued banning school girls from wearing the Hijab. Yet a contradictory position is demonstrated through the widely publicized case of the two Cairo University scholars who were forced to divorce each other because of the Azhar's verdict of alleged blasphemy committed by the husband. In fact, even the most secular state in the region, Turkey, could not escape such dualism. Since the military intervention in 1980, the Turkish state has increasingly been struggling to maintain a balance between its nationalist secular legacy and the pressure of Islamization that threatens the country's political stability from within.

In its constant effort to balance the major contending tendencies, the state often places women in an equally ambivalent position with a penchant for viewing the state as both the protector and the oppressor, depending on the particular configuration of alliances struck during any particular period. However, what women often tend to forget is that secularism, particularly in the national(ist) context of the Middle East, is not necessarily emancipatory or libratory, especially when issues of gender equality are at stake. The ideology of nationalism is a patriarchal, gendered ideology based on the assumption that women are (or should be) primarily mothers and housewives and secondarily workers. As history has shown, women's mass public participation and political activism during the upsurge of national movements have yet to be translated into change in the movements' gender ideology. Addressing this dilemma within the Iranian context, Afsaneh Najmabadi notes that: "with the consolidation of the Pahlavi state, women became objects of contestation of loyalty, as they became crucial to the state's project of social transformation, while remaining central to the Islamic claims of moral community" (Najmabadi 1991, p. 70).

The complex, multifarious and often pernicious role played by the State, to use Agarwal's words, is not confined to the Middle East. It is also evident throughout the countries of Asia where political Islam is gaining increasing momentum. In these countries the role of the state, similar to that in the Middle East, has been extended into the spheres of people's reproductive choices, religious beliefs, and even interpersonal relationships, dress and behavior patterns (Agarwal 1988, p. 1).

WOMEN AND FAMILY LAWS IN ISLAM

The articulation of traditional culture, the role of the state (whether colonial or nation-state), class structure, and women's organized resistance are crucial to understanding the gendered Muslim family laws. Three areas in the Personal Status Law, also known as the Family Code, will be traced in the following section. These are: marriage, polygamy and divorce.

At the outset of this discussion on Family Laws it is important to point out that there are different schools in Islam that provide varying interpretations to the laws. While the scope of this paper inhibits any detailed account on these schools, changes espoused by the four schools of Sunni jurisprudence, as al Hibri affirms (1992), are minor as the differences between the schools are more cosmetic than fundamental.

Marriage

Traditionally, marriage is considered one of the most important, if not the most important, event in the life of the community in Muslim Middle East. Children, mainly girls, are often prepared at a young age for such an event. Because of the transitional form it takes, namely, between the extended family and the smaller, more nuclear living arrangement, marriage is assumed to be the road toward asserting individuality and even securing a level of freedom. This individuality, however, does not translate itself into individual freedom at the legal level of state and citizenship. Throughout the Muslim countries of the Middle East citizenship is the prerogative of the man/father/husband and not the woman/mother/wife. Children are typically included in the passport of the father, consequently depriving the mother of the right to travel with her children if she chooses to do so. Women cannot give citizenship to their children, only men, even if married to non-Muslims, can confer citizenship to their children. The alliance constructed here between state and religion enforces the exclusion of women from true citizenship in the nation-state. In the Middle East, the female, viewed as an appendage or dependent, derives her citizenship through her relation to the man, be he father or husband (Joseph 1996).

The institution of marriage in most Muslim countries in the Middle East has undoubtedly undergone major changes due to increased urbanization, proletarianization, and the nuclearization-individualization of the traditional extended family. Exogamous marriages, particularly in urban areas, are increasingly replacing endogamous marriages. Also changing in urban areas, particularly among middle-class people, is the concept and practice of arranged marriage which is being gradually replaced with love-marriage or marriage of the choice of the couple. Despite these changes, traditional forms of marriages remain the norm.

One area in which traditional Muslim marriage is slow to change is in mixed or inter-faith marriages. This phenomenon to which I refer as "female religious endogamy versus male exogamous rights," remains a point of contestation for scholarship on women and Islam. Al-Hibri (1992) takes up this point by showing how the Hanafi school, considered by her as the more lenient school, deviates on the issue of mixed marriage from the *Sunna*. The Sunna, according to al-Hibri, indicates that marriage should be based on piety as the preferable reference point. However, the Hanafi school of jurisprudence has established that sex, rather than piety, be the basis of its jurisprudence on mixed or intra-religious marriages, therefore making sameness in religion the basic condition for Muslim women's marriages. This restriction is imposed on women only; Muslim men in all schools of jurisprudence are allowed to marry outside of Islam (al-Hibri 1992, p. 3). Nevertheless, the assertion made by al-Hibri by no means suggests that restrictions on mixed marriages are limited to the Hanafi school. In fact, all Muslim schools of jurisprudence share these constraints. Even within Islam such restrictions make it very difficult for a Sunni Muslim to marry a Shi'i Muslim.

Mixed marriage is an issue in the various urban centers of the Middle East, yet very little attention has been paid to it. Scholars who write about countries having a predominant Muslim presence, such as Pakistan, the Gulf states and Morocco, tend to treat this as a non issue. In other parts of the Arab Middle East, such as Tunisia, Palestine, Lebanon, and even Jordan, women appear to be more interested in discussing and debating this form of inequality, partly because of the principle of injustice involved in this "law" and partly because of the prevalence of such practices. In most of these countries, mixed marriages have been enacted not only among lay women and men but also among the political-national elite. While in most of the elite practices marriages have been conducted between Muslim men and non-Muslim (mainly Christian) women, mixed marriages involving Muslim women and non-Muslim men are not unheard of.

Although Islam prohibits Muslim female mixed marriages, it is interesting to note that class can mediate that prohibition. Muslim women who can afford to have the right to choose their partners often have a particular class backing that allows them to engage in such relations. The presence of such a class background is crucial for providing some social, moral, and economic security for the woman who finds herself isolated and, in many cases, shunned by the community. The class backing in such relations is all the more important when one considers the role of the state or lack thereof. No religious state in the Middle East (including Israel) has any mechanism to support such choices by its citizens. In fact, mixed marriages, like other gender issues vaguely and arbitrarily assumed to be in the realm of "family honor," are areas where the state refuses to interfere. Secular Muslims who choose such marriages can place their lives at risk while the state turns deaf ears and blind eyes. There is no legal means to formalize or legitimize inter-faith marriages. Except for Turkey, where family laws have been secularized, civil marriages are virtually non-existent throughout the Middle East. Even Israel, a country that prides itself for having a strong civil society, confines all family matters within the boundaries of religious courts and rules (Jewish, Muslim, and Christian jurisprudence). Among Palestinians, for example, a couple wishing to engage in a mixed marriage would have to leave Palestine/Israel, obtain civil marriage outside and return home. Civil marriages, while not performed in Israel, are still accepted as legal marriages.

Marriage restrictions imposed on Muslim women have far-reaching gender implications, particularly in the area of citizenship. Muslim countries will have to eventually deal with the thousands of women who, while living in their home countries, cannot give their nationality to their children because they are either married to foreigners, divorced, abandoned, or widowed. In many cases, as Nahid Toubia observes, the children become stateless persons with no educational, health, or employment rights. This situation occurs even if the father is from another Muslim country—an increasingly common occurrence with the expansion in labor migration to the Gulf countries (Toubia 1994, p. 56). On the other hand, the phenomenon of immigration and labor migration that characterizes many Muslim countries does not seem to be abating. Quite to the contrary, further globalization and integration within international capitalism will most likely reinforce and strengthen Muslim migration to the West. Consequently, it is expected that mixed-marriages may become more of a reality in the new living conditions for both sexes in Muslim communities. Such a reality, for example, was crucial in the Tunisian government's decision to abolish restrictions on inter-faith marriages. It was also a major consideration in the Lebanese government's decision in 1994 to pass a nationality law that would give a woman the right to grant her nationality to her children as well as to her foreign-born husband.

A second related issue is the age of marriage. The age of marriage in Islam is in fact a clear testimony to the diversity of rules and practices among different Muslim communities. It is generally held that according to the Shari'a, the age of marriage for the female is when she reaches puberty or once menstruation begins. This can be as low as 11 years of age. In Iran, for example, prior to the 1979 Revolution, the age of marriage was set at 13 for females. The Islamic Republic changed the rules and decreed the legal age of marriage to be 9 for girls. After some dissent and debate, the Revolutionary Council brought back the old rule and accepted marriage for girls at the age of 13. The declaration of the new age of marriage, however, did not cancel the existing rule (Hoodfar 1996). In practice, parents can still marry off their daughters at the age of 9 without facing any legal repercussions.

In most Arab countries a minimum age of marriage has been set at 16 for girls and 18 for boys. Yet, in many rural and poor urban communities girls are still married off under the legal age (Toubia 1994, p. 24). Some countries, such as the Sudan, Yemen, and Bahrain have no standard age of marriage. Reaching puberty or menstruation is considered legitimate. Except for Egypt, which considers marriage under the age of 16 as illegal, most Arab countries sanctify marriages under the minimum age as long as a court order is obtained. Obtaining a court order, as studies have revealed, is not a difficult task: all that is needed is a physician's certificate asserting the girl has reached puberty.

Marriage age is heavily influenced by economic conditions, levels of literacy, and cultural traditions or customary laws. For example, in rich Arab countries, such as the United Arab Emirates, the rate of adolescent or minor marriages is estimated to account for 55 percent of all marriages. In this case tribal cultural norms, rather than economic well-being, appear to predominate. High rates of minor or adolescent marriages have also been reported in some of the very poor countries; Sudan has a rate of 41 percent of minor marriages. In this case it is more likely that economic conditions along with a high rate of illiteracy are the major contributors to the phenomenon of early marriages (Toubia 1994, p. 25). A combination of poverty with illiteracy or constrained access to education, together with social and political repression can have regressive effects on the age of marriage. In the Gaza Strip, for example, although the Egyptian law has been followed—where marriages are set at the age of 16 for the female and underage marriages are declared illegal—during the Intifada (which produced harsh economic and political realities, and a consequent high dropout rate from school), girls as young as 13 were reportedly pulled out of schools and forced into arranged marriages (Hammami 1992).

Another contested issue in Muslim marriages is that of the *wali* (the male guardian), where consent is a precondition for a marriage contract. The idea of having a wali is based on the gender ideology that assumes the inferior status of women, depicting them as weak, emotional, and incapable of making rational decisions on their own (Mernissi 1992b; al-Hibri 1992; Agarwal 1988). Despite slight differences between the schools of Islamic jurisprudence on the role of the wali, most Muslim countries follow similar policies of requiring the wali.

The only exception is, once again, Tunisia, which abandoned *wilaya* (male guardianship) as a condition for marriage and gave mature women the right to choose whether they want to contract their marriages by themselves or relegate it to a wali (Perkins 1986; al-Hibri 1992). In various countries, such as Syria, Egypt and Algeria, reforms to the role of the wali have been attempted at various times. However, as al-Hibri points out, such changes were of hardly any significance. The wali continues to have the right to prevent marriage if he finds it "not in the interest" of the female who is often his daughter or blood relative. For example, in Egypt, where women can marry without the consent of the guardian, the father/guardian can request the annulment through court on the basis of incompatibility (Toubia 1994, p. 57). The negligible change in the Algerian Code on this matter was from its "not profitable" form as it existed in the 1981 Code, to "not in the interest of the daughter" as expressed in the 1984 Family Code (al-Hibri 1992, p. 235).

The role of the guardian, however, goes beyond issues of marriage. Whether a father, a husband, or a brother, the guardian often controls women's movement. Sudan, for example, commands the most restrictive measures on women's movement through the wali. Based on an administrative regulation, the Numeiri regime and all governments that followed, it enforced the law that requires a woman, regardless of her age or marital status, to have the consent of her male guardian if she wishes to obtain a passport. A guardian can be a teenage boy who is given authority over mature older sisters. In both Egypt and Morocco, married women need the consent of their husbands to obtain a passport. In many Muslim countries women also need the consent of the husband to take a job. As Toubia indicates, Syria, Tunisia, and Bahrain have the least restrictive laws, but they still discriminate according to marital status: "Unmarried women must have the consent of their fathers if they want a passport, but a wife who applies for a passport does not need her husband's consent, and he cannot stop her from travelling unless he secures a court order" (Toubia 1994, p. 50).

Polygamy

Polygamy, or the right to marry more than one person, is the prerogative of men throughout the Muslim Middle East, with the exception of Turkey and Tunisia where polygamy was abolished in the 1956 Personal Status Code. Muslim jurists have different opinions on this issue. Some jurists, particularly among the new Islamists, defend polygamy as an indisputable Muslim rule that, among other things, is seen as necessary for the propagation of the Muslim *Umma* (nation). Other jurists differ slightly, arguing that polygamy does not have to be an absolute right; the potential wife has the right to place a condition in the marriage contract to prohibit her husband from taking another wife. In cases where some reform has occurred, such as in Libya, Syria, and Morocco, polygamy, while still sanctioned by the law, is permitted only in court. In these countries the husband must inform or obtain the approval of the existing wife before marrying another (Toubia 1994, p. 60). In fact, as al-Hibri (1992) has observed, many Muslim countries, particularly Arab countries, have recognized women's right to intervene in the marriage contract around the issue of polygamy. Such intervention (real or potential), however, must be understood within the context of power relations between husbands and wives.

The poignant reality for most Muslim women is the fact that for traditional and customary reasons, women are often a marginalized party in the marriage contract, leading, in most cases, to women forfeiting their rights of negotiation.

Numerous reasons may account for this, ranging from the relatively high rate of illiteracy among women, particularly rural, poor and young women who are forced into arranged marriages, or by women unwilling to take the risk of being rendered socially outcast, or to others who realize the futility of challenging male hegemony in the overwhelmingly male-dominated structures and institutions.

Women's education and their consciousness about their right to interfere in, and perhaps shape, their marriage contract, while necessary, is not sufficient. Raising women's consciousness about their legal rights in marriage has become an integral part of women's activities in many Muslim countries. For instance, the Women's Legal and Counselling Center in Jordan, the Lahore based Women Living Under Muslim Law, and the Center for Legal Aid and Counselling for Women in the West Bank have all been involved in education and consciousness raising. Yet, for consciousness-raising efforts to work, there needs to be not only mass-scale grass-roots organized work, but also legal mechanisms and enforcement policies. In the absence of enforcement laws, of legal and institutional recourse and state commitment to enforce stipulations in the contract, this exercise and the efforts of many women's organizations remain futile.

The reluctance of the state to intervene in the issue of polygamy, or for that matter, in other aspects of Family Laws, leaves the Islamic/Shari'a courts in a position of full control over the lives of women.

Polygamy, like other forms of religiously based legal measures used to control women's bodies, has more symbolic meaning at the national-political level than it does in terms of its practical relevance. As noted by Perkins, the abolition of polygamy in Tunisia had no major implications for the conditions of women, since less than 3 percent of the Tunisian population was involved in multiple marriages at the time (Perkins 1986, p. 124). In Iraq, on the other hand, after the Iraq-Iran war, Saddam Hussein encouraged polygamy and prohibited the use of contraceptives in order to meet the regime's insatiable need for soldiers (Helow 1996, p. 18).

Encouraging polygamy as a means of controlling women's reproductive activities by confining them to the domestic sphere was the basis for the uproar in the Algerian political quarters in 1984. Despite the fact that multiple marriages as a practice was far from being widespread among the increasingly individualized and urbanized Algerian family, the state's attempt at restricting polygamy was met with fierce resistance from the religious elite. Thus, according to Article 8 of the Family Code of 1984:

> It is permitted to contract marriage with more than one wife within the limits of Shari'a if the motive is justified, the conditions and intentions of equality provided, and after the consultation of the preceding and future wife. Either wife may take judicial action against the husband or request divorce should he ignore her refusal to consent (Cheriet 1996, p. 24).

However, as Cheriet observed, "conservative delegates swiftly and violently denounced the Article as 'heretic'...Out of 60 interventions reported...45 opposed placing conditions on polygamy" (1996, p. 25). While reasons for such opposition varied, the most startling, according to the author, is the one which held that:

> Polygamy is not to be disputed, whatever the case, for a Muslim state is one based on *Jihad*, and this calls for the involvement of men alone. To whom will women be left in the case of Jihad, and how will society be protected from subsequent depravity, if widows cannot find parties to marry? Polygamy is therefore a must (Cheriet 1996, p. 25).

The logic of the Algerian Islamicists appears to differ little from that of Saddam Hussein's encouragement of a controlled version of Islamicization after the Gulf war: leaders in both countries use women's bodies to promote their political-military agendas.

Divorce

Not unlike other aspects of the Shari'a that pertain to women's status, the issue of divorce is contested among various schools of jurisprudence. All Muslim jurisprudence, while not questioning the right of the man to divorce his wife or wives, accept the view that the woman may have the right to divorce her husband if, and only if, this right is clearly established as a condition in the marriage contract. However, as al-Hibri points out, this condition can be violated and even nullified through other legal and extra-legal recourse reserved for men. According to the Hanafi school, a limited monetary payment to the wife can revoke the significance of such a condition, while in the Maliki jurisprudence, if a judge learns about the existence of a stipulation granting women the right to divorce, he can annul the whole marriage (al-Hibri 1992, p. 235). In other words, the liberal appearance of the Hanafi school over that of the Maliki is exactly that: in form or in theory alone, with little practical or real meaning.

There appears to be a uniformity in rule over the issue of marriage by all schools. This uniformity stands against the intent of the *Sunna*. "The Prophet," al-Hibri asserts, "regarded conditions in the marriage contract as most worthy of satisfaction" (al-Hibri 1922, p. 235). In fact in many Arab countries, the husband still has repudiation rights over his wife. The extent to which the Shari'a rules, despite attempts at reform by some of the more liberal countries, is evident. In Egypt, Lebanon (the Sunni population), and Syria, countries that accept the right of the wife to initiate divorce—albeit under strict conditions—nevertheless continue to uphold the husband's right to repudiation (Toubia 1994).

In a situation where the wife is able to initiate her own divorce, provided this condition is included in the marriage contract, divorce for her is neither necessarily a gain, nor does it indicate an improved status. Quite to the contrary, she frequently remains the main victim. A divorced woman in a traditional patriarchal society is, to say the least, an unwelcome phenomenon. In a society where the discourse on women is framed either in the context of "Woman as sexual being" or in the context of "Woman as male/family honor," virginity remains the most sacred quality imposed on women; a non-virgin is less than desirable.

Although variations in divorce experiences exist, in many Muslim societies a divorced woman has little, if any choices to make. In some Arab countries, for example, if she remarries she knows she is entering into a new contract with inferior status. If she remains single, which is often the case as society constructs her fate, her new status will be that of an unfavorable member of the community. No matter how mistreated, abused, or victimized she was during marriage, the woman is often blamed for the breakdown of marriage. It is not incidental that popular culture, at least in several Arab countries, has constructed fixed images of the divorced woman: a woman who breaks the first marriage is capable of breaking the second; or, had she been fitting for marriage she would not have been divorced in the first place!

Breaking up marriages in cultural contexts where dowry or bride price (*mahr*) is a significant part of the contract, renders women not only socially undesirable, but perhaps in danger of being abused or even, in some cases, killed. Pakistan provides a case in point. Various reports indicate that a woman suffering sexual abuse by her husband or a member of her marital family rarely finds support from her own family. As one report states:

> Marrying a woman off involves a significant financial commitment which makes her relatives committed to making sure that she remains with her husband. Furthermore, a woman's leaving her husband's house or exposing what is seen as a private matter between husband and wife is seen as a source of shame to her family...A common social response to a woman who leaves her husband's home is that she is a bad, incapable or dangerous woman and that other women related to her must have the same qualities (Special Bulletin 1995, pp. 47-48).

Moreover, divorce becomes particularly poignant in Muslim countries because of the role, or rather the absence of the role of the state in this arena. The inability of the state to act in favor of divorced women—by devising certain mechanisms, such as shelters, protection rights and means of survival for women on their own or when with children, may be interpreted as a result of the power of the religious institutions. However, there is evidence that supports the claim that the state can, under specific conditions, curb the power of the religious institutions. The Egyptian crackdown on al-Azhar is just one example. The state's reluctance to take responsibility for family or domestic issues that have implications on gender relations can be better explained in the patriarchal or neo-patriarchal nature of most of these states. Lack of interference, which in this case means keeping a tight lid on a potentially revolutionary force—the women—helps maintain the status quo. It is not incidental that so far no Muslim country has had the courage to officially and legally deal with issues of domestic sexual abuse, incest, or rape. To date, while some Muslim states, such as Pakistan, sanction "killings for honor," known otherwise as *karo kari* crimes, many Muslim states are complacent when it comes to "honor killings".

Changes in family codes or personal status laws and the consequent changes in women's status in the family and in society, must simultaneously occur in both the structural and cultural realms. While structural changes can lead to legal changes, ideological and legal changes can also influence social relations. Changes to the Divorce Act in India provide a case in point. Pressured by women's demands and the concern that Muslim women were converting to other religions in order to attain the divorces denied them under the existing Muslim personal laws, the *ulema* in India pushed for the passing of the Dissolution of Marriages Act (Chhachhi 1991, p. 160). Similarly, the firm stand on family or personal status laws adopted by the consecutive Tunisian governments has no doubt had an empowering effect on the improved status of women in this country.

CONCLUSION

The above analysis on family laws in Muslim countries has demonstrated that a diversity of Islams and a multiplicity of laws, regulations, and rules prevail throughout the region. The paper has argued that despite attempts at homogenizing Islam by both the outside forces of the Orientalists and the inside forces of the fundamentalists, the former remains a diverse and complex ideology.

Specifically, the paper has undertaken a critique of *feminist legalistic opportunism* as a means to analyze the complexity of family laws in the Middle East. The paper has attempted to expose the potential dangers of employing this approach, arguing that while its emphasis on legalities provides an important and useful tool for analysis, it risks an overemphasis on the legal aspect while neglecting other important issues. Rather, this paper has contended that state initiated laws, as well as Islamic or Shari'a laws, are heavily influenced by social, economic, political, and cultural factors. The state can play a crucial role in maintaining, reproducing or even abolishing repressive laws against women. The reluctance of most Muslim states to do so must be located in the ability or inability of the state to represent its citizens as a *sovereign state*. The background of patriarchy together with the economic and political dependency of most of these states on the West, further combined with their ineffective ways of dealing with internal class contradictions, the urban-rural divide, ethnic and religious problems, as well as the increasing danger of political destabilization they face from the Islamist movements, render most Middle Eastern states incapable of protecting even the simplest human rights of their female citizens.

Under relatively stable economic and political situations—very rare in today's Muslim countries—a commitment on the part of the state to liberalize or Westernize the economy can simultaneously lead to openness around social gender relations, through legal changes. Turkey and Tunisia have been clear examples. Yet, when economic crisis and political instability are the lot of a country, neither liberalization, nor attempts at progressive changes can survive; as witnessed, for example, in the unsuccessful attempts at liberalization of economy and laws in the late 1970s by Egypt's Sadat.

It is not insignificant that in the cases of both Turkey's secular laws and Tunisia's most progressive laws affecting women, the personal intervention and commitment of the head of state was involved. Ataturk and Bourguiba were personally involved in encouraging women's rights. While the issue of personality, particularly at the level of leadership, might play a role, the true measure for the success of social-gender changes must be located elsewhere, namely, in the realms of structural and cultural changes. For, despite the success of both Turkey and Tunisia in promoting progressive legal structures, one cannot but observe the vulnerability of both systems in today's international context. Such vulnerability is not the sole work of "outside" forces such as the emergence of strong Islamicist movements. It is also the result of the failure of both governments to accompany the legal changes with concrete structural changes to sustain their economies: rural populations in both countries remained largely removed from the process of change.

Finally, this paper has asserted that Muslim women must not be seen as passive recipients of their victimization. Through personal agency and organized forms of resistance, women throughout the Muslim Middle East have and continue to resist their subjugation. Women's organizations and feminist activism in the form of centers for women and feminist studies, legal-aid to women, counselling on women's legal rights, and a multiplicity of other activities, have mushroomed in the past decade and a half. In addition to local and national organizations, a number of international Muslim women's solidarity organizations and networks have also emerged in the past decade. The Lahore and Paris based network, Women Living Under Muslim Law (WLUML) is one such organization. Among its aims, WLUML seeks to provide information for women and women's groups from Muslim countries and communities; support women's struggles from within the Muslim countries and communities and make them known outside; and provide channels of communication among women from Muslim countries and communities (Helie-Lucas 1990, p. 73).

It is nevertheless important to remember that women are not of one class, nor is the concept of gender of a unitary or homogeneous character. While most women may agree on the fact of their lower status in the family and society, they disagree on the means to achieve progress or gender equality. As part of the real world in which they live and work, Muslim women's political activism is influenced by the class and political ideological movements to which they belong. Moreover, family, clan and tribal allegiances can also influence women's activism or lack thereof. Finally, not unlike the heterogeneous movement of feminism in the West, feminism in Muslim countries is also heterogeneous. The two most influential groups are those who struggle for the secularization of family laws, the separation of state and religion, and the eventual democratization of the state, on the one hand, and those who choose the Canon (Quran and Sunna) as their point of reference and orient their struggles and energies toward a reinterpretation of the text, on the other. In either case women's resistance efforts must be accompanied with changes at both the structural-institutional as well as the cultural levels.

REFERENCES

Abadan-Unat, N. 1991. "The Impact of Legal and Educational Reforms on Turkish Women." In *Women in Middle Eastern History: Shifting Boundaries in Sex and Gender*, edited by N. Keddie and B. Baron. New Haven, CT: Yale University Press.

Abdo, N. 1991. "Women of the Intifada: Gender, Class and National Liberation." *Race and Class* 32(4): 19-35.

________. 1993. "Race, Gender and Politics: The Struggle of Arab Women in Canada." Pp. 71-98 in *And Still We Rise*, edited by L. Carty. Toronto: Women's Press.

________. 1987. *Family, Women and Social Change in the Middle East: The Palestinian Case*. Toronto: Canadian Scholars' Press.

Afshar, H. 1994. *Why Fundamentalism? Iranian Women and Their Support For Islam*. University of York, Department of Politics, Working Paper No. 2.

Agarwal, B. 1988. *Structures of Patriarchy: The State, the Community and the Household*. London: Zed Books.

Ahmed L. 1992. *Women and Gender in Islam: Historical Roots of a Modern Debate*. New Haven, CT: Yale University Press.

Al-Hibri, A. 1992. "Marriage Laws in Muslim Countries: A Comparative Study of certain Egyptian, Syrian, Moroccan, and Tunisian Marriage Laws." Pp. 227-244 in *International Review of Comparative Public Policy*, Vol. 4, edited by N. Mercuro. Greenwich, CT: JAI Press.

Chhachhi, A. 1991. "Forced Identities: The State, Communalism, Fundamentalism and Women in India." Pp. 144-175 in *Women, Islam and the State*, edited by D. Kandiyoti. London: Macmillan.

Cheriet, B. 1996. "Gender, Civil Society and Citizenship in Algeria." *Middle East Report*, No. 198 (January-March): 22-26.

Dirie, F. 1996. "Female Circumcision and Islam." Seminar presented at the Canadian Arab Women's conference, Toronto, March 1st.

Esposito, J. 1982. *Women in Muslim Family Law*. Syracuse, NY: Syracuse University Press.

Gardezi, F. 1990. "Women's Movement in Pakistan." *South Asian Bulletin* 10(2): 18-24.

Ghossoub, M. 1987. "Feminism—Or the Eternal Masculine—in the Arab World." *New Left Review* 161: 3-18.

Hammami, R. 1992. "Women in Palestinian Society." Pp. 283-311 in *Palestinian Society: A Survey of Living Conditions*, edited by M. Heiberg and G. Ovensen. (A FAFO Report)

Hammami, R. and M. Rieker. 1988. "Feminist Orientalism and Orientalist Marxism." *New Left Review* 120: 93-106.

Helie-Lucas, M.-A. 1994. "Report: Women Living Under Musllim Law." *South Asia Bulletin* 10(January): 73.

Helow, J. 1996. "Women's Court in Beirut." *Middle East Report*, No. 198 (January-March): 18.

Hessini, L. 1994. "Wearing the Hijab in Contemporary Morocco: Choice and Identity." In *Reconstructing Gender in the Middle East: Tradition, Identity and Power*, edited by F.M. Goecek and S. Balaghi. New York: Columbia University Press.

Hijab, N. 1988. *Womanpower: The Arab Debate on Women at Work*. London: Cambridge University Press.

Hoodfar, H. 1991. "Return to the Veil: Personal Strategy and Public Participation in Egypt." In *Working Women: International Perspectives on Labour and Gender Ideology*, edited by N. Redclift and M.T. Sinclair. London and New York: Routledge.

————. 1993. "The Veil in Their Minds and On Our Heads: The Persistence of Colonial Images of Muslim Women." *Resources for Feminist Research* 22(Fall/Winter): 5-18.

————. 1996. "Reflections on Gender and Religion in the Middle East: Can Muslim Women Be Feminists?" Lecture, Carleton University, March 15.

Joseph, S. 1996. "Gender and Citizenship in the Middle East." *Middle East Report*, No. 198 (January-March): 4-10.

Kandiyoti, D. 1991. "Islam and Patriarchy: A Comparative Perspective." In *Women in Middle Eastern History: Shifting Boundaries in Sex and Gender*, edited by N. Keddie and B. Baron. New Haven, CT: Yale University Press.

Maududi, A.A. 1974. *Purdah and the Status of Women in Islam*. Delhi: Markazi Maktaba Islami.

Mernissi, F. 1986. *Beyond The Veil: Male-Female Dynamics in Modern Muslim Society*. London: al-Saqi.

————. 1988. "Muslim Women and Fundamentalism." *Middle East Report*, No. 153(July-August): 8-11

————. 1992a. *Islam and Democracy: Fear of the Western World*. Reading, MA: Addison-Wesley.

————. 1992b. *The Veil and the Male Elite: A Feminist Interpretation of Women's Rights in Islam*. Reading, MA: Addison-Wesley.

————. 1993. *The Forgotten Queens of Islam*. Minneapolis: University of Minnesota Press.

————. 1994. *Dreams of Trespass: Tales of a Harem Girlhood*. Reading, MA: Addison-Wesley.

Moghadam, V., ed. 1994. *Gender and National Identity: Women and Politics in Muslim Societies*. London: Zed Books.

Najmabadi, A. 1991. "Hazards of Modernity and Morality: Women, State and Ideology in Contemporary Iran." In *Women, Islam and the State*, edited by D. Kandiyoti. London: Macmillan.

________. 1984. *Marxism and the Muslim World.* New York: Monthly Review Press.

Perkins, K.J. 1986. *Tunisia: Crossroads of the Islamic and European Worlds.* Boulder, CO: Westview Press.

Sa'adawi, N. 1983. *Women at Point Zero.* London: Zed Press.

________. 1988. "The Political Challenges Facing Arab Women at the End of the 20th Century." Pp. 8-26 in *Women of the Arab World*, edited by N. Toubia. London: Zed Books.

Sharabi, H. 1988. *Neopatriarchy: A Theory of Distorted Change in Arab Society.* New York: Oxford.

Special Bulletin, 1995. *Women Living Under Muslim Law.* Lahore: Shirkat Ghah.

Toubia, N., ed. 1994. *Arab Women: A Profile of Diversity and Change.* New York: The Population Council.

Tucker, J. 1985. *Women in Nineteenth-Century Egypt.* Cambridge: Cambridge University Press.

Turner, B. 1994. *Orientalism, Postmodernism and Globalism.* London: Routledge and Kegan Paul.

Video I: Women in Islam. 1991. Directed by Ahmad Abdul-Rahman, produced by the Islamic Media Services in two parts.

Video II: The Veiled Revolution. 1993. Produced by Elizabeth Fernea.

ISLAM, ISLAMISTS, AND LABOR LAW

Marsha Pripstein Posusney and Karen Pfeifer

INTRODUCTION

Labor law is a window to some of the most important aspects of a country's political economy. It enables us to envision the nature of relationships between employers and employees and the formal constraints under which each group operates, such as how much leeway managers have in hiring and firing employees, and how much freedom workers have to organize collectively. These relations are the core of the productive processes that sustain society. Further, through the presence or absence of specific codes regulating the employment of women and/or children, labor legislation provides clues to the economic underpinnings of gender and family relations within a country. And with employees (as opposed to employers and the self-employed) making up a sizeable proportion of a country's work force, labor law also provides an indication of how a government views its responsibilities toward a large and economically vulnerable section of its adult citizenry.

To be sure, the picture one sees through the labor law window may distort reality. Various aspects of the codes may be unenforced, completely or selectively by sector or firm size; alternatively, certain "laws" may operate in practice, even though they are never codified. Still, there is a value to studying

International Review of Comparative Public Policy, Volume 9, pages 195-223.

ISBN: 0-7623-0268-2

the formal legislation as an indication of the image a government seeks to project to its own citizens, as well as to the international community.

This paper explores the degree to which Islam has influenced the prevailing labor legislation in predominantly Muslim countries. Toward that end, the following section examines what Islamic jurisprudence prescribes in the realm of labor relations. Next, we survey the evolving relationships between Islamic regimes and labor policy in three Middle Eastern countries. Then, we consider the likelihood that changes in labor law would result from Islamic groups coming to power in countries where secular governments now rule. Our case study concerns the position taken by certain Islamic forces toward the secular labor legislation now in effect, but under debate and contest, in Egypt.

While these issues are of positive interest, we approach them here with a normative concern. Though there are of course variations in degree, it is generally true throughout the Muslim world today that laborers suffer conditions of poverty and degradation. Thus the question that underlies this study is: Would implementation of Islamic labor laws improve the lives of wage workers?

ISLAMIC PRECEPTS AND LABOR RELATIONS

Most of the writing on Islamic doctrine as it pertains to labor relations has been generated by adherents of "Islamic economics," who believe that Islam constitutes a "third path" between capitalism and socialism (Pfeifer 1997; Behdad 1992a, p. 77). By organizing society according to Islamic precepts of fairness and justice, they argue, Islamic economics will avoid both the degradation and exploitation of workers, and the vast gaps in wealth, that are associated with capitalism, as well as the denial of individual talent and creativity ascribed to the communist countries.[1]

Beyond this general premise, there are considerable differences among "Islamic economists" in their policy prescriptions. In large part this stems from the fact that the texts used to derive their interpretations date from the seventh century, and thus cannot directly address the complex labor issues which arise from modern productive processes. Paying heed to these disagreements, we begin by discussing different conceptions of an Islamic view of work and how it confers rights, and imposes obligations, on both employers and employees. We then turn to the implications of this for the role of the state.

Islamic Views of Work

While it is common in the West to speak of "the Protestant work ethic," Muslims can rightly trace the association between labor and godliness to their own scriptures from several centuries earlier. The sanctity of labor in Islam derives from the belief that all of the earth is God's property, entrusted to mankind to be used for the good of human society. Human effort is required to extract the fruits of the earth, and is rewarded by their enjoyment. God, moreover, is understood to have endowed humanity with creativity to be used in this production.[2]

It follows that Islam encourages its adherents to work, so that the Muslim community as a whole may benefit from this labor. Sloth is frowned upon; although Muslims are enjoined to give charity to the poor, begging by those who are able to labor is discouraged (Cummings et al. 1980, pp. 42, 229; Said 1972, p. 64). Furthermore, those capable of working are instructed to seek employment at a distance if none is available locally (Said 1972, p. 101).

Some interpret these injunctions to mean that labor, although it is not one of the five pillars of the religion, is a duty of all able-bodied citizens, and an additional form of worship (Abd al-Salam 1993, pp. 25-26; Al-Faruqi 1980, pp. 79-80). However, such an interpretation may be based on a misreading of the basic Islamic texts. Noting the multiple interpretations of the Arabic word for labor (*'amal*), Bayat argues convincingly that it is 'amal as good deeds, not 'amal as work, to which Islam ascribes holiness. Comparing Islam to Protestantism, he maintains that Protestantism regards work as a religious duty and promotes laborers who "work with the interest of over-satisfying their daily needs for moral, material, and social reasons." Islam, in contrast, encourages moderation in working, which it sees as a mundane concern rather than a spiritual enterprise.[3]

But even if not a pillar of worship, manual work in Islam is held in high esteem. Typically, in Western capitalist discourse, workers are denigrated as having less intellect and/or initiative than entrepreneurs. In contrast, Islam considers the worker to be "beloved of God," and entitled to the same status as the employer. Employers, therefore, are expected to treat their employees with dignity. Islam holds that mistreatment of employees will be judged harshly by God, as reflected in the *hadith*, "The owners who mete out evil treatment toward their servants shall find the gateway to Paradise shut in their faces" (Moore and Delener 1986, p. 75; Shabon 1981, p. 274).

By the same token, employers are not viewed in Islam as less productive than, or morally inferior to, workers. Whereas Marxism portrays employers as owners of capital who exploit labor while performing no productive work of their own, the Islamist thinkers studied here tended to paint employers as managers who contribute organizational and intellectual skills. While these skills should not be viewed as superior to those of workers, they must also not be discounted (Al-Faruqi 1980, pp. 79-81; Yusuf 1988, pp. 74-75; Moore and Delener 1986, p. 75).[4]

Employers' and Employees' Rights and Responsibilities

Once hired, workers are expected to perform their jobs faithfully. There are numerous *ahadith* (pl.) which point to this; according to Said, contemporary *fiqh* scholars have determined that employees who fail to work to the best of their ability are guilty of *tatfif*, or shirking, a moral transgression. This injunction stems not only from workers' obligations to their employers, but also because their production is seen to benefit society as a whole (Said 1972, pp. 58-61). Similarly, workers must not steal from employers or damage their property, and must be honest in all other aspects of the employer/employee relationship.[5]

In exchange for their diligent labor, Islam bestows various rights on workers, which are a mirror image of the obligations of their employers. The injunction against mistreatment of workers would prohibit forcing them to undertake labor they do not wish to do, or punishing them for equipment damages that are not their fault. In addition, employers may not overtax their employees. The working day should be limited, and the labor itself should not exhaust the worker. Workers are entitled to periods of rest during the day as well as vacation, in addition to the weekly Sabbath. Working conditions should be safe from accidents and should not damage the employee's health (Abd al-Salam 1993, p. 32; Mannan 1986, p. 89; Chapra 1979, p. 17; Moore and Delener 1986, pp. 75-76; Rahman 1990, p. 151; Aziz 1992, p. 54; Said 1972, pp. 100-102).

Some scholars go further, finding in the Shari'a an obligation for employers to provide their workers with housing, medical facilities, transportation, education, and meals (Said 1972, p. 85), or to sell them the products of their labor at subsidized prices (M.A. Khan 1983, p. 108). Vocational education is also considered an employer's obligation by some, both because it contributes to the development of the individual's productivity, thereby enhancing the overall society's productive capacity, and because it connotes a form of respect from the employer for the laborer (Chapra 1992, p. 253; Aziz 1992, pp. 48-49).

The Islamic texts are widely understood to prohibit the economic exploitation of workers. However, most Islamic economists reject, implicitly or explicitly, the Marxist notion that exploitation is inherent in capitalist labor relations.[6] Rather, exploitation is to be prevented by requiring employers to pay workers a "just" or "fair" wage. This should be set by mutual agreement—a written or oral contract—before the work is undertaken. Again, Muhammad is quoted as having said that the employer who fails to pay workers their due shall meet the displeasure of God (Chapra 1979, p. 17; Abd al-Salam 1993, p. 33).

Exactly what constitutes a fair wage, however, remains a source of ambiguity and dispute. Choudhury holds that it is equivalent to the average product of labor (Choudhury 1989, p. 8). However, for Mannan,[7] Aziz (1992, pp. 51-52) and Abd Al-Salam (1993, p. 33), it is the equilibrium wage that results when demand equals supply—which should equal the marginal, rather than average, product of labor—that constitutes fairness.

A unique distributional argument is offered by Yusuf, based on the Islamic belief that all property ultimately belongs to God. If all agricultural land, pasture, standing timber, and mineral deposits remain public property, he maintains, then all individuals will be able to live off of their own labor. They would agree to work for another only if the wage is greater than what their own labor on free land would produce, and this would be a just wage. Thus, it is private ownership of natural resources by some that compels others to undersell their labor, and there is no need for the state to impose any minimum wage so long as it ensures equal access to these resources. By the same token, employers are not obliged to worry about the subsistence or comfort of their employees; wages should be determined strictly by the productivity of the worker and the utility of his services (Yusuf 1988, pp. 74-79).

There is a clear basis in Islam for the promotion of profit sharing schemes. Its roots lie in the ancient system of *mudaraba*, in which one individual supplied the financing for a venture, while the other performed the work, and the proceeds were divided between them according to prior agreement. Losses are likewise borne by both parties. Muhammad himself is said to have participated in such arrangements. Proponents of Islamic economics concur that the modern day equivalent of *mudaraba* would be some type of cash distribution of profits or employee stock ownership plan (ESOP), perhaps combined with worker representation in management (WRM) decision making (Choudhury 1989, pp. 321-324; Aziz 1992, pp. 52-53).

Islamic economists disagree, however, on how widespread these profit sharing systems should be and on the method of dividing profits. The strongest advocate is Chapra, who maintains that every firm in an Islamic economic system should be required to do profit sharing. Some of the proceeds would be distributed to workers as a cash supplement to wages, and the remainder would be used to finance workers' necessities such as housing, daycare, and medical expenses (Chapra 1992, pp. 254-255). Said holds that mudaraba is "the best way of doing business according to Islam," but does not suggest that it be mandatory. In his interpretation, profits or losses would be divided equally between owners and employees (Said 1972, p. 68). Afzal-ur-Rahman praises profit-sharing but does not claim that Islam prefers such schemes, and in his view any division of proceeds between owners and workers, even if unequal, would qualify as mudaraba (Afzal-ur-Rahman 1990, pp. 230-233).

No prohibition on the employment of females was suggested in the works studied here. On the contrary, Aziz argues that the Quran specifically enjoins employers to hire the person best suited for the job, eschewing discrimination based on gender, race, religion, or national origin. The only restriction Islam imposes on female employment, he claims, is that women must dress modestly and men must lower their gaze when approaching women (Aziz 1992, pp. 50-51). However, a woman needs permission from her husband or male guardian (father, brother) to work. Abd al-Salam maintains that, in addition, Islam proscribes the intermingling of male and female employees, and prohibits women from working at night (Abd Al-Salam 1993, p. 41). Iran recently passed legislation entitling women to wages for labor in the home; these must be paid by her husband prior to a divorce. Though derived from the Islamic texts and consistent with the Islamic Republic's civil and constitutional codes, the law is controversial within the religious community there.[8]

The Role of the State

We shall argue below that there is a kind of corporatist conception of the role of the state that comes into play in those Middle Eastern societies with established Islamic regimes. In Islamic economic theory, the government is charged at minimum with enforcing the rights and responsibilities of employees and employers, however defined. There is debate, though, around whether the state has additional obligations to workers. Controversy also surrounds the possibility of government regulation of collective action by employees or employers.

Numerous scholars suggest that workers may be unable on their own to negotiate a "fair" wage. Acknowledging that labor supply generally exceeds demand, depressing wages, both Mannan and Abd al-Salam would support government intervention to elevate them; Chapra and Said as well endorse minimum wage laws (Abd al-Salam 1993, pp. 31, 42-43; Chapra 1979, p. 17; Said 1972, pp. 98, 109). Abd al-Salam also calls for employers to raise wages in accordance with workers' experience, the cost of living, and the company's profits, but implies that this can be achieved through contractual negotiations if both parties act according to religious precepts (Abd Al-Salam 1993, p. 35).

Behdad argues that there is no basis in the Shari'a for these minimum-wage policies, nor for any government intervention so long as the contractual agreement between employer and employee is fulfilled (Behdad 1992a, pp. 89-90). Nevertheless, the proponents of minimum wage laws generally call for additional forms of government intervention as well. If it is an individual's duty to be productive, they maintain, then it is up to the state to find employment for those unable to do so on their own, or provide them with unemployment benefits (Abd Al-Salam 1993, pp. 23-27; Chapra 1979, pp. 1, 12; Said 1972, p. 75; Aziz 1992, p. 52). Vocational education is also seen as an obligation of the government, either in place of, or in addition to, employers (Said 1972, p. 103; Chapra 1979, p. 18; Aziz 1992, p. 49).

According to Khan, the first four caliphs believed it to be incumbent upon the state to provide a minimum standard of living, including food, clothing, and shelter, for all citizens, To be true to their teachings, he argues, a contemporary government must provide medical care and education as well. Thus he posits, in essence, the modern welfare state (M.A. Khan 1983, passim). Likewise, Abd Al-Salam holds that "the Islamic society is responsible for filling the gaps between workers' wages and the [costs of the] necessities of life" (Abd Al-Salam 1993, p. 35) and this was also the position of the Muslim Brotherhood in pre-1952 Egypt (Beinin and Lockman 1987, p. 377). For Chapra and Shabon, however, the state's obligations extend only to the elderly and infirm, while Said would also include widows and orphans (Chapra 1992, pp. 251-252; Shabon 1981, p. 276; Said 1972, pp. 104-107). Wages should be sufficient to enable the able-bodied and employed to provide life's necessities for themselves.

Uniquely, al-Faruqi holds that Islam's emphasis on production also obligates the state to ensure that all necessary work be done. This entails planning, including assigning people in their adolescence to different career tracks, in order to ensure that they are properly trained for the functions they will ultimately fulfill. Recognizing that all individuals have different talents, he argues, an Islamic society should seek to channel individuals into the careers for which they are best suited. And while ideally the faithful would do this willingly, "the state is empowered to coerce its citizens to undertake the jobs which are necessary for its welfare, just as it is entitled to conscript them into the military for its defense" (Al-Faruqi 1980, p. 83).

As noted above, the Shar'ia, as interpreted by Islamic economists, clearly specifies that employment must be undertaken by contract, with a wage agreed upon in advance. But the relevant sources assume a situation where an individual worker negotiates with a lone employer, typical of the master/servant relations of ancient times. How can these translate into the modern day, where hundreds if not thousands of workers can perform similar jobs in the same establishment? In particular, does Islam have implications for collective bargaining?

Although some countries claiming to apply the Shari'a, such as Saudi Arabia, have outlawed trade unions, the works studied here find no textual basis for this prohibition. Muhammad Akram Khan implicitly assumes unions' existence, and both Mannan and Abd al-Salam specify that there is no Islamic injunction against them. Joined as well by Said, the latter two writers further concur that Islam does not explicitly prohibit either strikes or lockouts, though they argue that its precepts imply objection to both. Whether initiated by employers or workers, work stoppages are thought to impede production and thereby harm society. More generally, clashes between workers and employers are viewed as injurious to the fabric of society, so both parties are expected to strive for amicable relationships and, in upholding Islamic principles, to minimize industrial conflict; from this perspective the Muslim Brotherhood in pre-1952 Egypt actively discouraged labor militancy (M.A. Khan 1983, passim; Mannan 1986, pp. 92-94; Abd al-Salam 1993, p. 32; Said 1972, pp. 106-109; Beinin and Lockman 1987, pp. 383-384). Where disputes do arise, adherents of this position expect the state to resolve them fairly (Shabon 1981, pp. 275-276). Thus, Mannan maintains that in an Islamic world workers would have no need and feel no desire for unions.[9]

Thus, there is no uniquely "Islamic" system of labor/state and labor/business relations. It is evident that Islam promotes the idea that workers be treated "fairly," but because the concepts of fairness and justice are so vague, and the modern work environment so different from what was prevalent in Muhammad's time, a wide range of different interpretations as to how the basic texts of Islamic jurisprudence apply to contemporary society is possible. In particular, both a minimalist state and one that intervenes extensively in labor markets can be construed as Islamic.

ISLAMIC REGIMES AND LABOR LAW

The three societies discussed in this section, Saudi Arabia, Pakistan, and Iran, are the best-established of the formally-declared Islamic states in the Middle East. They tend to be corporatist in ideological structure. As we use it here, "corporatism" refers to a worldview in which society is conceived as one giant enterprise in which three groups of actors cooperate to produce the national product, to reproduce the citizenry and to contribute to growth and "development." The three partners are business (owners of capital, enterprise and land), the state (the political elite and bureaucracy), and the mass of the citizenry. As used here, the term "corporatism" does not refer to specific political structures linking the state to particular classes, but rather to a philosophy that is intended to bind the society together around the common purpose of development.[10]

Labor is usually a junior partner in the corporatist conception, for example being underrepresented in positions of power where important decisions are made. But the arrangement often includes programs that benefit labor and other parts of the citizenry, public education for example, which comprise a kind of social contract. The state provides these functions because its role is not simply to serve capital in an instrumental way (capital using the state to promote its own narrow interests), but rather to promote and protect the system as a whole, integrated structure. In a developing capitalist economy, where it may be difficult for capital to control labor just through production technologies and labor market mechanisms, the state may take it upon itself to perform two seemingly contradictory tasks. On the one hand, the state presumes to regulate labor on capital's behalf through a labor code it can enforce through its judicial and policing functions. On the other hand, the state claims to provide protections for workers from the worst vagaries of the labor market, for example via the institutions of unemployment compensation and social security insurance. The state takes on its latter role both to promote social and political quiescence in the short run and to ensure the reproduction of the workforce, and thereby the economy, over the long run.

There is no close correlation between Islamism as a political movement and any specific agenda for labor policy. All three cases below, Saudi Arabia, Pakistan, and Iran, illustrate that when Islamic regimes promote a corporatist philosophy, labor law tends to be used to tightly restrict the rights of workers to organize their own unions and to exercise forms of protest like the strike. On the one hand, workers cannot expect Islamism to unequivocally deliver them from their prior oppressors. On the other hand, Islamism will not necessarily become a greater oppressor either. Two of the three cases illustrate that Islamism is not synonymous with either etatist economic policies or authoritarian controls on societal groups. There have been periods when Islamic regimes have been associated with more democratic policies, namely Pakistan in the years immediately following independence and the first few years under Zulfiqar Ali Bhutto, and Iran immediately after the revolution. These periods provided broader scope for workers to openly and successfully exercise their organizing and bargaining rights.

The Case of Saudia Arabia

In Saudi Arabia, the law of the land is the Shari'a, as interpreted by the Hanbali school. It has been supplemented in the twentieth century by a growing body of royal decrees, which have the force of law, to handle those areas of modern life not covered by the Shari'a. The Hanbali school of Islamic thought is close to laissez-faire in economic policy, allowing wide latitude to private business. However, in Saudi Arabia it is tempered with a paternalistic expectation regarding the government's responsibility toward its citizens' welfare, *maslaha* (Nomani and Rahnema 1994, p. 150). This corporatist framework posits that an Islamic economic system is "based on a one-class society rejecting any kind of class struggle or class distinction," where the government is charged with responsibility for warding off class conflict (Shabon 1981, pp. 273-276; Moore and Delener 1986, p. 76).

Nevertheless, class conflict surfaced in Saudi Arabia soon after capitalist elements, in the form of a consortium of international firms extracting and exporting oil, penetrated and transformed what had been a pastoral, agrarian and trading economy. As modern capitalist relations were introduced into Saudi society, the government elaborated mechanisms for managing industrial and class conflicts. The first oil-worker strikes took place in 1935 and the first formal law governing compensation of workers in the technical (oil-related) sector was decreed in 1937. Many more strikes occurred in the subsequent decades, and the government responded by outlawing strikes and unions in 1956.

A full-fledged labor code dealing with contemporary capital/labor issues was decreed in 1969. The code provided for such modern features as formal labor contracts, paid vacations, regulation of working hours and overtime, safety standards, and medical treatment on the job, but it also imposed yet stronger penalties for union and strike activity (Shabon 1981, p. 68; Nomani and Rahnema 1994, p. 151). The monetary penalties imposed on employees for violations of the labor code ranged from 1000 to 10,000 rials (in the 1970s, when there were about 3.5 Saudi rials to the U.S. dollar) and possible jail terms ranged from one month to 6 years. In contrast, no threat of incarceration was defined for employers and their monetary fines ranged from 100 to 1000 rials for each violation or each worker (Council of Ministers 1970, pp. 39-40),[11] though code violations by employers were routine, especially for foreign workers.

About 60 percent of the labor force in Saudi Arabia was expatriate in 1990 (Abdel Jaber 1993, p. 149). Their situation is not highlighted in the labor code. The code does not specify that foreign workers be excluded from any of its provisions, except to define what permits and qualifications they need have and to stipulate preference for hiring Saudis whenever possible. The code does specify that employees of contractors working under concession (that is foreign-owned companies) be treated the same as employees of other firms (Council of Ministers 1970, pp. 12, 30).

From other evidence it appears that these provisions have not been used to protect expatriate workers. Some discrimination became tied directly to citizenship, such as foreign workers not being able to own real estate, corporate stock or other commercial property without a Saudi partner and not being able to bring their families to live with them in Saudi Arabia. Occupational segregation by ethnic group became common, with the non-Yemeni Arabs concentrated in the professions, the Yemenis in the trading and artisanal positions, and the Asians in construction, domestic service, health service work and other less desirable posts. Even controlling for occupation, however, there appeared a discriminatory sliding scale of wage payments based on national origin: in the late 1980s, a Saudi professional or technical worker was paid 1.4 times as much as a non-Saudi Arab, or 3.6 times as much as an Asian worker, for the same work. The ease with which Saudi Arabia expelled up to one million Yemenis and Jordanians during the 1990-91 Gulf crisis, often with great financial hardship to the deportees, bespeaks the insecurity experienced by even the more accepted of foreign workers (Abdel Jaber 1993, pp. 150-157; Stevenson 1993, p. 15).

Obtaining reforms of the limited sort embodied in the 1969 labor code entailed the dethroning and exile of King Saud and the installation of the "rational reformer," Faisal, as the new monarch in 1964.[12] With King Faisal's endorsement, a social insurance system was created at the same time as the labor code and was considered to be one of the most generous in the Arab World in its day. It was expected to play an important part in further averting labor conflict, providing an additional economic inducement for workers to rely on the government rather than on their own organizations for their welfare (Shabon 1981, p. 74). The high volume of oil revenues in the 1970s and early 1980s provided the largesse with which to fund these benefits.

Islamic corporatism, in the cultural as much as the religious sense, affects social relationships between employers and employees and also affects family life directly and indirectly.[13] Religious injunctions about work, the worker's relation to God, and the employer's obligations to the employee are frequently intoned in everyday worklife (Moore and Delener 1986, pp. 75-79). Shaw quotes the Saudi labor regulations as enjoining "the employer to 'treat his workmen with due respect and refrain from any work or deed that may affect their dignity or religion'" (Shaw 1986, p. 98). More concretely, in Saudi Arabia it is common for employers to provide non-mandatory benefits to employees, such as payment of bonuses just before the major Islamic holidays, cash loans to workers (usually for a wedding or other special event), and shortened workdays during the month of Ramadhan (Shabon 1981, p. 71).[14]

The labor code decreed in 1969 formalized some of these features as they apply to the employment of women. While the code makes it legal for women to work, it requires that they be physically separate from men on the job, which rules out some occupations entirely, and specifies stricter limits on hours worked. It also mandates accommodations for pregnant and breastfeeding women, including medical care, maternity leave, shortened hours of work, and time off from work to feed the child during the day (Council of Ministers 1970, pp. 35-37; Shabon 1981, p. 74).

Islamic corporatism plays a role in the institutions set up by the labor code of 1969 to adjudicate disputes between employers and employees. The code allows disputes to be settled either by arbitration or by a hearing panel. An arbitration procedure is conducted by a mediator chosen by the disputants and approved by the Ministry of Labor, and whose decision is binding and enforceable by the same ministry.

Alternatively, a hearing may be conducted by one of three lower ("preliminary") three-person panels appointed by the Ministry of Labor, in which two out of the three members must have a law degree in the Shari'a. After closed deliberations, the announced ruling may be appealed to a High Commission at the national level, made up of appointees of the ministries of labor, commerce and industry, and petroleum and mineral resources. The ruling at the lower level or, if appealed, at the higher level is binding and enforceable by the Ministry of Labor. While all these adjudicating bodies are enjoined to make their judgements in accordance with the Shari'a, nowhere in the process is there a role for advocates of labor (Lerrick and Mian 1987, pp. 316-319, 326, 328-329; Shabon 1981, pp. 78-80; Nomani and Rahnema 1994, p. 151).

A similar top-down decision-making process governs the determination of the wage structure. There is no advocate of labor, save a representative appointed by the ministry of labor, on the committee that recommends changes in the minimum wage to the Council of Ministers (Shabon 1981, p. 70). This may explain why the minimum wage is not systematically enforced and why the labor code does not apply to those sectors where it could make the biggest difference for low-wage, often casual, workers, namely in domestic service and agriculture (Shaw 1986, p. 101; Nomani and Rahnema 1994, pp. 150-151). It is noteworthy that these two sectors are among the most likely to have a high representation of women and foreign workers.

One of the most significant negative features of this framework from labor's point of view is that in most countries of the Peninsula, including Saudi Arabia, Bahrain, Oman, Qatar, and the United Arab Emirates, labor organizations and strikes are illegal. Even in those Peninsular countries in which unions are technically legal, namely Kuwait and Yemen (before and after unification), the unions are closely controlled, sometimes even suppressed, by the government and strike activity is tightly regulated or even prohibited (Shabon 1981). This has not prevented workers from organizing and even striking illegally, often over political as well as economic issues, but it does mean that the governmental response is usually harsh and repressive.

Saudi Arabia has a history of frequent labor activism, especially among oilfield workers, going back to 1935. The government's response has routinely been to fine or imprison Saudi workers, for as long as six years, and to deport foreign workers (Shabon 1981, pp. 74-77; Council of Ministers 1970, pp. 39-40). Workers' only recourse in cases of ill-treatment or discrimination, besides throwing themselves on the mercy of the Islamic moral sentiments of their employer, is to file a grievance with one of the commissions appointed by the Ministry of Labor to hear disputes. They have no autonomous voice in the corporatist scheme.

The Case of Pakistan

Pakistan is a country in which labor policy has clearly had an Islamic and paternalistic cast. However, despite intermittent claims otherwise, there has been no clear and consistent association of Islamic thinking with either a pro-labor program or a pro-business program. Indeed, in a corporatist fashion, Islamism has at times been associated with one or the other and sometimes with both simultaneously. For example, one Pakistani authority on labor law quotes approvingly a speech given in Geneva in June 1954 to the International Labor Organization Conference by the leader of the Pakistani delegation, who said that his country was trying to build a society "...where there is [the] widest possible distribution of property and income, where there is no hoarding or usury, and, above all, where duties and responsibilities are taken as a moral trust" (A.H. Khan 1994, p. 79).

The Pakistani industrial workforce is reasonably well organized even by Western standards. In 1990, about 2.5 million people, or 20 percent of the labor force, worked in industry (United Nations 1994, p. 162). Forty per cent of industrial workers belonged to unions. However, this potentially great strength is diluted by the complex structure of the union movement, by the large number of unions (8,300 in 1986), and by the practice of each union bargaining with its enterprise individually (*Far Eastern Economic Review* 1986, pp. 63-64).

The strongest unions are in the public sector, particularly in textiles. However, unions are organized by enterprise, not by industry, and there is often more than one union per enterprise. The individual unions are then grouped into a large number of federations, which are allied with political factions or parties. The eight largest federations have about two-thirds of total union membership, but they compete fiercely among themselves, being allied to rightwing or leftwing or Islamic (or some combination of the above) political formations (*Far Eastern Economic Review* 1986, p. 64).

These labor organizations played an important part in the independence movement and the issue of labor relations was addressed early on in the development of the Pakistani state. The first phase of labor policy development, "Islamic" in orientation, lasted from 1947 to 1958. In 1949, for example, the first Pakistani labor conference was opened by Prime Minister Liaquat Ali Khan with a statement proclaiming that the founding Islamic principles of Pakistan impelled the abolition of exploitation and the guarantee of justice to labor. Labor policy in this initial phase officially assured freedom of association for workers and for employers, freedom to "negotiate the terms and conditions of employment," and the right to strike and to lock out, respectively. It was also stressed that the main aim of economic policy should be to raise living standards and, that aim requiring economic growth, to promote industrial peace (A.H. Khan 1994, pp. 55-65, 74).

Under the regime of Ayub Khan and then Yahya Khan, from 1958 to 1971, labor policy took a more corporatist turn, shifting to stress economic growth and industrial peace more intensely, and downplaying the rights and needs of workers. Indeed labor policy was used to suppress "industrial conflict so much that in 1968 it erupted to shake the very foundation of the socio-economic order" (A.H. Khan 1994, p. 56). Militant unionism, strike activity, and political resistance to the military regime reached its peak in 1968. In response, the regime enunciated a "New Labor Policy" in 1969 that granted concessions to labor and promised fewer restrictions. These promises bore fruit under the next regime.

A new era, ostensibly more favorable to labor, was then ushered in with the end of martial law and the election of Zulfiqar Ali Bhutto and his Pakistani People's Party (PPP) to office. Like his predecessors, Bhutto laid claim to the Islamic tradition as the framework of his policies, even though the policies were considered "socialistic" (A.H. Khan 1994, p. 58). This new program was based on the idea that, although "industrial conflict is inherent in the factory [capitalist] way of production," the level of conflict can be lowered if the rights of workers to organize and bargain collectively are respected.

The emphasis in labor policy shifted away from industrial peace and back to "distributive justice," the latter both for its own sake and as a stimulus to productivity growth. It entailed the assertion of collective bargaining as both "a modus operandi of the industrial relations system" and "a problem-solving institution" for society (A.H. Khan 1994, pp. 56-58). In a series of pronouncements on a number of different occasions, the labor ministry made clear its intention to use labor policy to redress distributional inequities and exploitation. It was only in this period that labor policies expressly claimed the objectives of establishing a just social order and giving workers a role in participatory management.

Thus, workers' bargaining power, wages, benefits and job security all improved during Bhutto's watch, which also promoted profit-sharing and participation schemes. However, Bhutto's regime became more and more corporatist in practice. It maintained at best a patronizing, and at worst an authoritarian, relationship to labor. The Pakistan People's Party (PPP) forcibly took over the leadership of unions, while the state security service repressed those labor leaders who resisted, leaving a mixed legacy of both "Islamic" and "socialistic" labor policy (Nomani and Rahnema 1994, p. 122).

Claiming the mantle of Islam directly, the military regime of General Zia-ul-Haq, from 1977 to 1990, returned to overt corporatism. This regime imposed a long list of Islamic economic reforms on Pakistan, abolishing the payment of interest, requiring payment of zakat, imposing a land tax, and establishing profit-and-loss-sharing procedures in banking. Other Islamic reforms affected the educational system, enforced bans on drinking of alcohol and gambling, encouraged fasting during Ramadhan and observance of daily prayers, encouraged modest dress for women, and imposed the classical punishments for theft, robbery and adultery (Esposito 1980, pp. 152-155).[15]

None of the reforms touched directly on labor policy, leaving the official labor policy of the Bhutto era on the books (Kennedy 1990, p. 63). Nevertheless, this phase was the most self-consciously "Islamic" with regard to labor policy, mainly by exhorting workers to work harder. Not unlike the military regime of the 1960s, it emphasized the urgency of higher economic growth rates and reform of the economy. While taking care to restate the "Islamic principles" that all participants in an economic activity be treated justly and paid fairly, the policy stressed the "Islamic" obligation of all parties, employers and employees alike, to promote the objective of economic growth, based on "the right Islamic attitude to work and property" (A.H. Khan 1994, pp. 57-59).

General Zia's rule was based on martial law, which was used to circumvent "temporarily" various features of the labor law favorable to labor (from 1977 to December 1985). The ability of unions to bargain collectively, or even to be recognized, was abridged. Some employers successfully refused to recognize union agency. Strike activity was restrained or forbidden and in some industries unions were banned altogether, such as in the airlines and in television and radio broadcasting. There were even mass firings in public sector companies such as the PIA, Pakistan International Airlines (*Far Eastern Economic Review* 1986, pp. 63-64). In these years, Islamism was used as a rationale to curb the rights of labor.

After the lifting of martial law and the return to electoral politics in the late 1980s, observers expected a return to the more pro-labor practices of the earlier years (*Far Eastern Economic Review* 1986, p. 64). Benazir Bhutto's first electoral success promised an "Islamic welfare state," a promise repeated by her successor, the business leader Nawaz Sharif and then again by Benazir Bhutto. However, the phase begun in 1990 has essentially promoted private enterprise instead and has not produced a new labor policy or new welfare initiatives (A.H. Khan 1994, p. 59). Indeed, the beneficial aspects of the labor-policy legacy from earlier regimes came under attack from the world financial community via a stabilization and austerity reform program in the 1990s, similar to the experience of many other countries in Asia, Africa, Latin America and the Middle East (Nomani and Rahnema 1994, pp. 121-134). Measures such as privatization of public enterprises, for example, threaten to abrogate job security provisions in the Pakistani labor law by asserting the employer's right to fire workers. The similarity to the case of Egypt described below is striking.

The case of Pakistan illustrates the flexibility of Islam in serving as the framework for various, and sometimes contradictory, socioeconomic or labor policies. Whereas Zulfiqar Ali Bhutto was a modern secular social democratic politician until he came to power in 1971, after 1971 he was able to appeal to Islam and advocate the "Islamisation of the country" in order to promote his "socialistic" economic and political agenda. Islam was a handy umbrella to hold over otherwise conflicting classes in order to manage their conflict. However, Bhutto was criticized for using and abusing Islam to these ends (Esposito 1980, p. 160). So, too, was General Zia-ul-Haq criticized for using and abusing Islam to promote a pro-business social and economic agenda unfavorable to many other groups in Pakistan.

Pakistan has failed thus far to furnish an example of an Islamic system which consistently provides for "the abolition of exploitation and the guarantee of justice to labor" as promised at the first labor conference of independent Pakistan in 1949. It also struggles to provide a clear sense of what constitutes a contemporary "Islamic" state (A.H. Khan 1994, pp. 70-79). As Esposito summed it up for an earlier period, so it still is in the mid-1990s that "beneath Pakistan's Islamic and nationalist rhetoric and terminology is a vacuum, an unresolved identity crisis which promises to be a source of continued unrest" (Esposito 1980, p. 160).

The Case of Iran

Prior to 1979, Iran was a corporatist monarchy like Saudi Arabia, characterized by authoritarian governmental supervision of labor and industrial relations. However, the Pahlavi monarchs, first Reza Shah and then his son Mohammed Reza Shah, were more secular in orientation than the monarchs in Saudi Arabia and used the state significantly more to actively lead industrial development. After the restoration of the Shah to power in the 1953 coup d'etat, following a two-year period of nationalist rule, industrial development, with the participation of direct foreign investment, took off (Bayat 1987, pp. 22-24). While the majority of manufacturing firms remained very small, state-run enterprises were responsible for more than 70 percent of employment in large industrial firms (those having 10 or more workers) and 79 percent of their output in 1979 (Rahnema 1992, p. 74).

The working class, though fragmented by ethnic and other fissures, grew in tandem with industry, so that by 1978 more than half of the economically active population had become wage workers and one-third of the total labor force were working in industry (Bayat 1987, pp. 25-26; Rahnema 1992, pp. 75-81). Workers and their organizations seem to have played a major role in the political transformations of Iran in the twentieth century and were just as frequently curbed or outlawed. A trade union federation was organized as early as 1921, for example, only to be suppressed. A similar and larger federation existed in the 1940s, was suppressed, then resurged under the nationalist government only to be banned again when the Shah was reinstated. With the promulgation of the new labor code in 1959, unions were made legal again, but only insofar as they fit within the regime's warm and tight embrace (Rahnema 1992, p. 71).

The Shah's regime sought simultaneously to win workers over with a paternalistic provision of benefits and to intimidate workers who might try to slip out from under the state's aegis. In the 1960s and early 1970s, the Shah offered unfulfillable promises of much higher wages for everyone, universal employer-provided housing, profit-sharing and employee-stock-ownership schemes (Ladjevardi 1985, pp. 237-246). With a secular orientation, this approach carried a corporatist conception of the relationship between workers and employers reminiscent of Saudi Arabia's, "the idea of a community founded upon the cooperation of all its members (workers, bosses, state agents)," with the addition, when ideology did not suffice, of manipulation of union leadership and the threat of dismissal (or worse) to keep workers cooperating with management (Bayat 1987, p. 60). This labor policy was consistently similar to its Saudi counterpart in strictly prohibiting both political and strike activity (Bayat 1987, pp. 59-60).

Despite an oppressive legal apparatus, Iranian workers engaged in both political and strike activity against the Shah's regime. During the emergence of the revolutionary movement in the late 1970s, and especially in the fall of 1978, a new form of labor organization, the *showra* or council developed. It became an important vehicle at the production site for mounting economic resistance and fomenting the political struggle for the overthrow of the Shah (Bayat 1987, pp. 77-81; Rahnema 1992, pp. 71-72). In the period immediately following the revolution of February 1979, these bodies then restored production in those firms where the owners or managers had fled and struggled to press their demands for improved conditions and worker control in those firms where management remained (Bayat 1987, pp. 100-107).

However, as the Islamic regime consolidated its control of the new republican state, the secular and left-wing leaders of the councils were removed (at first by political means, later by more violent methods) and the councils became Islamicized (i.e., taken over by Islamic Republican Party [IRP] operatives). With the purging of the left-wing and liberal secular allies of the IRP from the national political structures in the summer of 1981, independent workers' organizations were crushed and the workers' councils were replaced by "Islamic Associations" composed of management, selected workers' representatives and appointees of the government (Ladjevardi 1985, pp. 246-254). This represented effectively an

"Islamic" reincarnation of the corporatist attitude toward labor prevalent under the erstwhile Shah's regime. Once again, labor policy was to be enforced by a combination of Islamic corporatist ideology, paternalistic promises of benefits, and suppression of dissent (Bayat 1987, pp. 183-190; 1991, pp. 72-88).

In the national political sphere, especially in the new parliament of the Islamic Republic, the first few years evidenced a tendency for the alleged "Third Way" of an Islamic economy to dissolve into the left-right debates common to virtually all modern societies.[16] The debate over a new labor law, first proposed in 1982, did not escape this implicit class conflict. The core argument focussed on whether Islam allows only a one-to-one contractual relationship between an employer and an employee, in which the former is simply renting something from the latter and in which the government should not intervene, or whether Islam enjoins the government to intervene in this relationship to set the minimum conditions for the contract and to protect the employee (Bayat 1987, pp. 185, 192). A parallel and larger debate between the supporters of laissez-faire economics and the promoters of a more interventionist economic policy raged at the same time, resulting in a paralysis that crippled the economy until after the war with Iraq ended. In the larger economy in the 1990s, the laissez-faire wing appears to be in the ascendance (Nomani and Rahnema 1994, pp. 169-172, 182-184).

The labor code finally enacted in 1987 included the standard interventionist clauses on working age, working hours, social insurance, and paid leave. Save for a few improvements, like a limited unemployment compensation provision and a modest social security system, and for new restraints on female employees, this code seems similar to the one prevailing under the Shah, including the feature that prohibits labor's right to form independent unions or to strike. As noted by Nomani and Rahnema: all non-Islamic labor organizations from the pre- and post-revolutionary era have been dissolved and "state-run Islamic labour councils and Islamic societies [associations] constitute the only legal labor organizations" (Nomani and Rahnema 1994, pp. 174-175).

Indeed some students of contemporary Iran argue not only that the "Third Way" of the Islamic economy has disintegrated into the same old two-way, capital/labor class struggle, but also that the struggle has been resolved for the foreseeable future in favor of capital, that in this case Islamism does not end class exploitation but legitimizes it (Moore 1992; Parvin 1993; Vakili-Zad 1992).

Thus do the countries with the most self-conscious history of formal Islamic rule, Saudi Arabia, Pakistan, and Iran, illustrate the pitfalls of Islamic corporatism for labor law and labor's well-being.

ISLAMIC MOVEMENTS AND LABOR LAW CHANGES

It follows from the above that what might be labelled as Islamic labor law will vary from country to country, subject to the particular constellation of Islamic and other forces in, or contesting, power. Furthermore, many predominantly Muslim countries have secular systems of labor law. Thus the changes that would occur when an Islamic regime replaces a secular one cannot necessarily be foretold accurately.

However, at least some Islamic forces have been quite specific about the type of labor law they envision. In Egypt, which has a secular labor law, proposed changes to the labor laws have engendered sustained public criticism of the Mubarak regime since 1991. This marks the first time since the 1940s that Islamic forces have played a significant and public role in the dialogue about labor law.[17] Some Muslim Brothers were involved in the labor controversies that developed during the initial year of Free Officers rule in 1952/3, but these took place in closed-door committees removed from any public spotlight; shortly thereafter, Islamic forces were removed from politics by widespread arrest and imprisonment. Labor law was more sharply and publicly contested during the mid-1970s, but although Sadat had released the Muslim Brothers from prison by this time, they did not engage in these debates.[18]

Until 1987, the Islamic forces in Egypt appeared to have ignored workers as an organizational category. That year, however, the Muslim Brotherhood (*al-Ikhwan al-Muslimin*) entered into an alliance with the Socialist Labor Party (*Hizb al-'Amal al-Ishtiraki*, hereafter SLP), and contested both parliamentary and trade union elections. Since then the party's newspaper (*al-Sha'b*) and its other propaganda have been understood to reflect the views of these "mainstream" Islamic forces, who until recently were tacitly tolerated by the regime. However, the Alliance's perspective on labor, as reported below, may in fact reflect only the views of a somewhat left-leaning segment of the Brotherhood.[19]

Egypt's Proposed New Labor Law

The current controversy stems from the regime's desire to amend the existing labor code, law 137 of 1981, and other laws specific to public sector workers, in order to facilitate Egypt's economic reform program. While Egypt's current and proposed labor laws are both secular, they differ in that the existing law and its predecessors were shaped by Egypt's earlier flirtation with socialist ideas and thus predicated on an interventionist and provisionary state. Law 137 is supplemented by special codes pertaining only to, and bestowing privileges upon, public sector workers. The new law, which seeks to move the country toward a more liberal economy, would override these codes.

Taken together, the existing laws make it very cumbersome for employers to fire full-time, permanent workers individually, and prohibit mass workforce reductions without the express permission of the government. Many of Egypt's private sector employers have historically evaded these provisions by, for example, falsely classifying their workers as part-time or temporary. In the public sector, however, the laws have for the most part held sway. Public sector workers enjoy other privileges as well, most notably access to public housing in the large industrial centers. Mainly because of these disparities, proposals to dismantle the country's public sector have been controversial, particularly among organized labor, which successfully combatted many of them in the 1970s and 1980s. Egypt's trade unions did agree to support the privatization legislation designed by the regime in 1991, but only when it was modified to specify that all firms which were sold under the auspices of this law continue to abide by the existing labor legislation. The sales agreements that have been concluded since then have contained clauses guaranteeing that the firms' workforces would not be reduced (Posusney 1992, 1995a).

After the privatization law was passed (law 203 of 1991), the multilateral lending agencies supporting Egypt's reform program, mainly the International Monetary Fund and the World Bank, complained that these restrictions inhibited the success of the privatization effort by making the parastatals less desirable for purchase.[20] Western economists frequently argued as well that the existing labor laws were an impediment to private sector development in general. Therefore, within a few months, the government convened a committee to develop new labor legislation. It consisted of representatives from: the Egyptian Trade Union Federation (ETUF), Egypt's umbrella labor confederation; businessmen's organizations; the Ministry of Labor; the local legal community; and the International Labour Organization (ILO), which provided funding for the endeavor.[21]

The proposed law gives both private employers and public sector managers far greater leeway in hiring and firing.[22] Although all firms must still obtain government approval for mass layoffs, the new law makes such permission more likely by stating explicitly that it is an employer's right to adjust the workforce according to economic conditions. Individual dismissals are facilitated in numerous ways. All workers must still be employed under written contract, which can be of temporary or permanent ("indefinite") duration, but the new law widens the range of serious worker infractions for which employers are entitled to break a contract.

The 1981 legislation requires that employers consult a tripartite committee representing management, unions, and the Ministry of Labor before firing a worker.[23] Although the decisions of this committee were only advisory, dismissed employees could appeal its rulings before turning to the courts, which in turn could require an employer to continue paying the worker until the case was settled.[24] The new legislation—at the insistence of union leaders—makes committee decisions to reinstate workers binding. However, it expands the committee to include two judges, which will probably make rulings in favor of workers less frequent.[25] While a worker can still appeal a negative committee decision to the courts, there are no longer specific provisions for ongoing salary payments, thus making it unlikely that workers will be financially able to appeal dismissals.

While law 137 specifies that a temporary contract, if renewed, acquires indefinite status, the new law permits multiple renewals. This makes it unlikely that any temporary worker will ever achieve the security of permanent status, or that any new worker will be hired indefinitely.[26] The law also enables employers to effectively force workers to resign by empowering them, "for economic reasons," to lower the contractual wage and/or require employees to perform different jobs. This article, included at the insistence of the business organizations which participated in the committee, nullifies the very essence of the work contract concept. In addition, the proposed new law gives public sector managers far greater leeway in setting wages and benefits for future employees, by nullifying the uniform wage scales previously obtained. Minimum wage laws will remain in effect, with a national committee meeting every three years to consider cost of living adjustments to it.[27]

Finally, in an explicit quid pro quo for the "right to fire," the draft law legalizes strikes, which have been banned since 1952. However, the new provisions legalizing strikes require the approval of two-thirds of an industrial *federation's* (as opposed to the individual local's) leadership for walkouts at individual plants. Furthermore, strikes are still prohibited while contracts are in effect, during mediation and arbitration, and in vital services. Moreover, union leaders are required to give employers and the government 15 days notice, and a justification, before striking.

The Islamists' Critique of the Law

Although the labor law committee's work was intended to be secret, the opposition parties soon learned of the effort, and later were able to publish details of its negotiations. By the middle of 1994, the thrust of the proposed new legislation, although it was not yet finalized, was well known, and both secular and Islamic opposition forces were actively challenging it. The SLP began forming "committees for the defense of workers' rights" in industrial areas, and convened a labor conference in April to denounce the law. Shortly thereafter, the party published a detailed critique of the draft law, entitled *La li-Qanun al-'Amal al-Muwahhid* ("No to the Unified Labor Law," Al-Sakhawa 1994, pp. 5-6; *al-Sha'b*, April 19, 1994).

The SLP's perspective puts them solidly in the camp of Islamists favoring an interventionist state. In fact, beyond opposing the neoliberal revisions to the existing laws, they would in some ways extend the protections for workers inherent in them. Their critique argues that the state has primary responsibility for preserving social peace, and is obligated to provide insurance against unemployment, illness, and industrial injuries (Al-Sakhawa 1994, pp. 17-21).

Islam, according to the SLP, sees work not only as a duty, but also as a *right* of all individuals. Islam therefore obliges those who have capital to use it for the benefit of society by providing employment opportunities. It follows that the party is highly critical of those provisions of the draft law that would cause job loss. The SLP also maintains that the law is biased toward employers, thus violating the Islamic doctrine of fairness, and treats workers as a commodity rather than as members of humanity who are "beloved of God." Decrying the draft as "presented by the IMF in the name of the government and the Egyptian Trade Union Federation," the SLP charges that it would give a "free hand to exploitative capitalists" (Al-Sakhawa 1994, pp. 5-6, 14-21, 32-33; *al-Sha'b*, April 19, 1994).

The party objects to the draft law's recognition of mass layoffs and temporary plant closings as a response to market conditions. Instead, the SLP proposes to maintain Article 9 of Law 137, which specifies that the sale or lease of a firm obligates the new employer to the existing workforce, and to strengthen this clause by stipulating that neither bankruptcy nor permanent closure of a plant be permitted to diminish its workers' rights in any way. Thus, the government is called upon to provide unemployment insurance and free worker retraining. The absence of such protections and benefits, the party argues, would harm society by causing unemployment and/or suffering to the worker and his/her dependents (Al-Sakhawa 1994, pp. 32-35, 50-51).

To protect workers from arbitrary firing, the SLP would restore the provisions of Law 137 that obligate employers to continue paying workers during the appeals procedure, and which exempt workers from paying the costs of taking firing cases to court. Likewise, the party rejects those provisions that make it easier for an employer to terminate or revise an individual work contract. The SLP would maintain those articles of the current law which specify that a temporary contract becomes permanent upon renewal, and would make the maximum period for a temporary work contract one year (Al-Sakhawa 1994, pp. 32-35, 50-51).

The SLP's position on wage-related issues reflects the perspective that an Islamic society must ensure that a worker's take home pay be sufficient to support his/her family. This translates into a political critique of those aspects of the proposed law which threaten to shrink paychecks in the short term, or will gradually reflect reduced purchasing power. Thus, the party objects to the clause that would allow employers to lower workers' wages because of "economic circumstances." Furthermore, while they do not condemn the removal of the uniform pay scales in the public sector, the Islamists believe that the three-year interval for meetings of the minimum wage committee is too long, and will cause real wages to erode due to inflation; they argue instead for annual meetings of the committee (Al-Sakhawa 1994, p. 34).

Additionally, the party objects to provisions of the new law that would allow up to 50 percent of an employee's pay to be deducted for repayment of debts.

Fearing that this would cause suffering to the worker's family, they call for a ceiling of 25 percent. The Islamists also object that the draft law's provisions for overtime pay are inadequate and could lead to undercompensation of workers. They call for the application of international standards for overtime pay, specifying time-and-a-half for day-shift workers, and double-time for back-shift workers. Double-time pay is also sought for work on the weekly day off or on holidays (Al-Sakhawa 1994, pp. 37-39).

Finally, the SLP's critique of the law reflects their view that it is incumbent upon the state to supervise the functioning of private capitalists, to ensure that they are in fact behaving in the best interests of society. Toward this end, the party calls for the new law to require employers to submit regular performance reports to the government. This stipulation, which has been incumbent only on public sector firms under the auspices of law 48 of 1978, is not contained in the new draft. The Islamists would also charge the government with supervising the use of machinery, to ensure that private entrepreneurs do not allow their capital to deteriorate prematurely through overuse and that dilapidated equipment is replaced (Al-Sakhawa 1994, pp. 47-48).

The SLP does not support the interpretation of some Islamic scholars and governments, such as Saudi Arabia's, that trade union activity is un-Islamic. On the contrary, the party faults the existing unions for not being aggressive enough. During their April conference on the law, party leaders denounced the current president of the ETUF for his "failure to defend workers," and called for his removal. The party advocates reform of the trade union movement, which they claim is plagued by corruption (Al-Sakhawa 1994, 21, 48-49; *al-Sha'b*, April 5 and 19, 1994).

The same SLP documents object vociferously to those provisions of the draft law that restrict the right to strike. However, leftists point out that the practice of Islamists elected to union office—most of whom are associated with the Alliance—is different from their written word. They claim that the Islamists neither initiate mass struggle around labor issues among workers, nor support labor protest that is generated by others, and cite this as one reason why numerous Islamists who captured union office in 1987 lost their seats in the 1991 elections (Al-Shafi'i 1994, pp. 203, 206-209). Moreover, even the aggressive language of the SLP's labor coverage and electoral campaigns may reflect differences within the Islamist camp. A 1989 factory occupation at the Helwan Iron and Steel plant, which was brutally repressed by the Egyptian government, was condemned by the Muslim Brotherhood. Their magazine, *The Banner of Islam* (*Luwa' al-Islam*) warned of the social dangers of labor militancy and praised the role of the police in suppressing the protest.[28]

These differences notwithstanding, the published perspective of the SLP Islamists closely resembles that of Egypt's secular left. It should be clear, however, that while the two movements share the belief in an interventionist, provisionary state, the Islamists do not join in the Marxists' fundamental hostility to capitalism. Thus, whereas the left has historically been outright opposed to the privatization of the parastatals, the Islamists have argued only for certain conditions to be imposed on the sales. These conditions are that the buyers be Egyptian or Arab, that the sale price reflect the actual value of the company, and that no diminution of workers' rights occur as a result of ownership transfer. The SLP has also argued for workers to be able to purchase a larger share of the stock in privatized parastatals than the 10-20 percent currently being proposed by the government (*Al-Sha'b*, April 19, 1994).[29]

The proximity between the positions of the SLP-associated Islamists and the secular left is not surprising when one considers that both groups are largely defending the status quo ante, which took shape during the Nasir years. Nasir molded the expectations of workers, especially those in the public sector, into a moral economy.[30] Arguably his emphasis on reciprocal rights and responsibilities between workers and the state had underpinnings in the Islamic notions of fairness and justice, and was able to gain widespread acceptance for this reason. In any case, it would be difficult for the Islamists today to gain credibility among workers by challenging this established understanding. Political expediency dictates instead that the Islamists mold their interpretation of Islamic precepts about labor relations to conform to it.

CONCLUSION

The answer to the question, "Would the application of Islamic labor law improve the lives of wage workers?" can only be a resounding, "It depends." An Islamic framework is not a clearly defined pattern on which to construct a labor law or build a labor movement. Labor law must address the actual conditions of labor in a particular society at a given time in its development.

As we have seen here, Islamic precepts regarding labor are open to competing interpretations. Some would offer workers little or no protection from the vagaries of the capitalist labor market. Others support a high degree of government regulation of business that would guarantee to workers employment with at least subsistence wages, protection from overwork and/or arbitrary dismissals, and the provision of basic social services. Yet others lead to pro-labor laws on the books that are honored more in the breach than in the observance.

The only way to assure that labor law effectively promotes the welfare of workers is for the workers to be organized in unions with a vision of progressive social change and a political strategy for promoting that vision. Labor cannot rely on paternalistic politicians, no matter how well-intentioned or how imbued with moral or religious precepts, to protect their interests for them. They must be independently organized and mobilized to protect and promote their own interests.

What changes to secular labor law would occur when Islamic forces come to power in a country, then, will depend on the initial perspective of these forces themselves, and the evolution of debate among them. The shaping of this perspective does not occur in a political or historical vacuum. The prior and ongoing struggles of workers against employers and/or the state help to shape the overall political/ideological environment in which Islamic forces operate. A labor movement can effectively ally itself with the Islamists only as long as the alliance represents a two-way commitment. An Islamic movement that has not made such a commitment cannot be relied upon to serve or protect labor's interests in any future government in which it plays a part. But, like any other political force, an Islamic movement that has a base in the working class, and a compatible doctrinal basis, will have a good and rational reason to promote legislation favorable to labor if and when it comes to power.

ACKNOWLEDGMENTS

The authors would like to acknowledge the support of their respective institutions and, in particular, the provision by Smith College of the assistance of Rebecca Hanson '97. Part of Pfeifer's research was assisted by a grant from the Joint Committee on the Near and Middle East of the Social Science Research Council and the American Council of Learned Societies with funds provided by the National Endowment for the Humanities and the Ford Foundation.

NOTES

1. See especially Mannan (1986, pp. 80-88).
2. See especially Said (1972, p. 61).
3. This contrast between Protestantism and Islam was first proposed by Max Weber, as an explanation of why capitalism developed first and most successfully in the West. Challenging Weber, Maxime Rodinson (1981) held that Islam is perfectly compatible with capitalism. Bayat's argument supports Weber's views but only, he is careful to clarify, if it can be shown that today's Islamic societies adhere to the correct interpretation of the original texts (Bayat 1992, esp. pp. 6, 18-22).
4. See Behdad (1994, p. 798) on some differences among Islamist thinkers regarding the exploitation of wage labor and the relationship between labor and capital. The more radical, such as Abu Suleiyman, Peyman and Shariati appear to have significant overlap with Marxist thinkers, in contrast to the Islamic thinkers cited here.
5. See also Mannan (1986, pp. 88-91) and Afzal-ur-Rahman (1990, pp. 150-151, 168).
6. For a counterexample, see Haque (1992, p. 122).
7. Based on the critique of his book by Behdad (1992b, pp. 228-229).
8. Mehranguiz Kar (1996, p. 37).
9. Private communication from Sohrab Behdad (July 11, 1996).
10. The concept of corporatism here is somewhat distinct from the usual usage of the term insofar as it refers to a vision of society as a sort of grand patriarchal family and because it tends to be associated with non-democratic governments. The more

common usage is found in a broad and rich body of literature in which corporatism is a type of political arrangement where the class-defined parts of an economy, in particular capital and labor, bargain with each other in a centralized format, and where the government participates in negotiations between these groups as part of social and economic planning and guidance. Thus conceived, corporatism is compatible with either democratic or authoritarian governments. For further discussion, see Posusney (1997).

11. Council of Ministers (1970) contains the complete text of the labor code, in English translation. Discussion of the various provisions of the code can be found in Shabon (1981) and in Lerrick and Mian (1987).

12. Nomani and Rahnema (1994, pp. 142-143), review the whole set of political and economic reforms discussed at the time, not all of which were adopted.

13. See especially Moore and Delener (1986), for a discussion of how cultural factors, including Islam, affect work practices and relationships.

14. One of the authors, Pfeifer, found this to be true of both self-defined Islamic and non-Islamic business firms in Egypt in 1994 as well.

15. The punishments are flogging, amputation and stoning. Esposito (1980) argues forcefully that these reforms were either insignificant or, in the end, unenforceable, and therefore made little difference to the socioeconomic trajectory of Pakistan.

16. Analyses of how and why this occurs in Islamically defined societies, as in others, are offered by Behdad (1994), Valibeigi (1993), and Zubaida (1989).

17. For more on the role of Islamic forces in the labor movement during the pre-coup period, see Beinin and Lockman (1987, pp. 363-394) and Goldberg (1988, passim).

18. For more on these earlier debates, see Posusney (1997, chaps. 1 and 2).

19. Comments by Ahmed Abdalla, seminar on The Moral Economy of Islam, University of California at Berkeley, April 26, 1996.

20. The information in this and the succeeding three paragraphs is drawn from Posusney (1995b).

21. See *al-Ahali* (December 2, 1992), *al-Ahram* (March 5, 1993), *al-Gumhuriyya* (October 21, 1993), and *al-Tadammun* [a newsletter of opposition unionists] (No. 1, December 1993, pp. 14-16). The ILO insisted on the inclusion of the right to strike, making its role problematic for the government. However, although the organization has historically served as a liberal counterweight to the orthodox multilaterals, some Egyptian leftists cite evidence that it has now allied with them in promoting structural adjustment. See Freeman (1992) and al-Hilali (1994, pp. 10-11).

22. This discussion is based on the draft published in *Al-Ahali* (November 30, 1994).

23. Public sector workers enjoyed the protection of these tripartite committees since 1962.

24. The courts were empowered to order reinstatement of the worker only if it was established that the dismissal was not punishment for trade union organizing; otherwise, they could mandate only financial compensation. However, the burden of proof that the firing was not union-related fell on the employer. The new law maintains this provision.

25. Employers complained that the tripartite committees were an obstacle to firing because both government and union representatives usually sided with the worker. In contrast, the courts have upheld dismissals in most cases. See Al-Muhami (1994, p. 20).

26. Notably, this controversy coincided with the government's effort to liberalize economic relations in the countryside. As part of the latter, the regime had the Shaykh al-Azhar issue a fatwa that permanent land tenancy contracts are un-Islamic (Hinnebusch n.d., p. 6). We are grateful to Ragui Assaad for calling this to our attention. During discussions in the panel on "Labor-Business and Labor-State Relations in the Contemporary Middle East," at the Middle East Studies Association annual conference (Washington, DC, December 1995), Assaad suggested that this fatwa might also be used by the regime to prohibit permanent work contracts.

27. However, Article 2 of what was reported at the end of 1994 to be the final version of the law, which was expected to passed by parliament in late 1995 or early 1996, maintains Law 137's stipulation that all workers receive a minimum 7 percent annual raise, and actually removes a previous cap on the pound value of this increase. In addition, Article 4 specifies that workers' current wages and benefits must not be reduced by the law's application. Thus only new public sector hires should be affected by the removal of the uniform pay scales, though the language of Article 4 is ambiguous enough that this interpretation could be contested in court.

28. Sana Al-Masri, *Al-Ikhwan al-Muslimun w'al-Tabiqa al-'Amila al-Masriyya* [The Muslim Brotherhood and the Egyptian Working Class] (Cairo, n.p. 1992, p. 96, cited in Al-Shafi'i 1994, p. 208).

29. For more on the leftists' position on the labor law and privatization, see Posusney (1995a, 1995b).

30. For an expansion of the moral economy argument, see Posusney (1993, 1997).

REFERENCES

Abd al-Salam, J.H. 1993. "Haqq al-'Amal fil-Islam." *Majallat al-Mu'amalat al-Islamiyya* 1: 13-45.

Abdel Jaber, T. 1993. "Inter-Arab labor Movements: Problems and Prospects." Pp. 145-162 in *Economic Development of the Arab Countries*, edited by S. El-Naggar. Washington, DC: International Monetary Fund.

Afzal-ur-Rahman, M.A. 1990. *Economic Doctrines of Islam*, Vol. 1, 3rd ed. Lahore: Islamic Publications Ltd.

Al-Faruqi, I.R. 1980. "Islam and Labour." Pp. 79-101 in *Islam and a New International Economic Order: The Social Dimension*. Geneva: International Institute for Labour Studies.

Al-Hilali, A.N. 1994. *Hatha al-Mashru' Lan Yamirr*. Helwan: Dar al-Khidamat al-Niqabiyya Bihilwan.

Al-Muhami, S.S.A. 1994. "Al Harika al-'Ummaliyya wi al-Nidal al-Qanuni." *Al-Badil* (August): 20.

Al-Sakhawa, M. 1994. *La li-Qanun al-'Amal al-Muwahhid*. Cairo.

Al-Shafi'i, U. 1994. "Al-Quwa al-Siyasiyya w'al-Intikhabat al-Niqabiyya al-'Ummaliyya." Pp. 199-218 in *Al-Harika al-'Ummaliyya fi Ma'rakat al-Tahawwul*. Helwan: Trade Union Services Center.

Aziz, A. 1992. "Firm Level Decisions and Human Resource Development in an Islamic Economy." Pp. 45-59 in *Economic Growth and Human Development in an Islamic Perspective: Proceedings of the Fourth International Islamic Economics*

Seminar, Issues in Contemporary Islamic Thought, edited by E. Ahmed. Herndon, VA: The International Institute of Islamic Thought.

Bayat, A. 1992. The Work Ethic in Islam: A Comparison with Protestantism. *Islamic Quarterly* 36: 5-27.

________. 1991. *Work, Politics, and Power: An International Perspective on Workers' Control and Self-Management*. New York: Monthly Review Press.

________. 1987. *Workers and Revolution in Iran*. London: Zed Books.

Behdad, S. 1994. "A Disputed Utopia: Islamic Economics in Revolutionary Iran." *Comparative Studies in Society and History* 36(October): 775-813.

________. 1992a. "Property Rights and Islamic Economic Approaches." Pp. 77-104 in *Islamic Economic Alternatives*, edited by K.S. Jomo. London: Macmillan.

________. 1992b. "Islamic Economics: A Utopian-Scholastic-Neoclassical-Keynesian Synthesis! A Review Essay." Pp. 221-232 in *Research in the History of Economic Thought and Methodology*, Vol. 9, edited by W. Samuels. Greenwich, CT: JAI Press.

Beinin, J. and Z. Lockman. 1987. *Workers on the Nile: Nationalism, Communism, Islam, and the Egyptian Working Class, 1882-1954*. Princeton, NJ: Princeton University Press.

Chapra, M.U. 1992. "Invigorating the Human Factor" and "Reducing Concentration of Wealth." Pp. 251-261 and 262-281 in *Islam and the Economic Challenge*. Herndon, VA: The International Institute of Islamic Thought, Islamic Economics Series, 17.

________. 1979. *The Islamic Welfare State and its Role in the Economy*. Leicester: The Islamic Foundation.

Choudhury, M.A. 1989. "Manpower Exchange and Social Security System for Islamic Economic Cooperation." Pp. 314-351 in *Islamic Economic Co-operation*. New York: St. Martin's Press.

Council of Ministers, Kingdom of Saudi Arabia. 1970. *Labor and Workmen Regulations*. Dhahran: Arabian American Oil Company.

Cummings, J.T., H. Askari, and A. Mustafa. 1980. "Islam and Modern Economic Change." Pp. 25-47 in *Islam and Development: Religion and Socio-Political Change*, edited by J.L. Esposito. Syracuse, NY: Syracuse University Press.

Esposito, J. 1980. "Pakistan: Quest for Islamic Identity." Pp. 139-162 in *Islam and Development, Religion and Sociopolitical Change*, edited by J.L. Esposito. Syracuse, NY: Syracuse University Press.

Far Eastern Economic Review. 1986. April 3, pp. 63-64.

Freeman, R.B. 1992. "Labor Market Institutions and Policies: Help or Hindrance to Economic Development?" Pp. 117-144 in *Proceedings of the World Bank Annual Conference on Development Economics*. Washington, DC: World Bank.

Goldberg, E. 1988. "Muslim Union Politics in Egypt: Two Cases." Pp. 228-243 in *Islam, Politics, and Social Movements*, edited by E. Burke, III and I.M. Lapidus. Berkeley: University of California Press.

Haque, Z. 1992. "Islamic Perspectives and Class Interests." Pp. 105-116 in *Islamic Economic Alternatives*, edited by K.S. Jomo. London: Macmillan.

Hinnebusch, R.A. n.d. "Completing De-Nasserization: The Reversal of Egypt's Agrarian Relations Law." Unpublished manuscript.

Kar, M. 1996. "Interview." *Middle East Report*, No. 198 (January-March): 37.

Kennedy, C.H. 1990. "Islamization and Legal Reform in Pakistan, 1979-1989." *Pacific Affairs* (Spring): 62-77.

Khan, A.H. 1994. "Pakistan." *International Encyclopedia of Labor Law* Suppl. 159(June): 46-281.

Khan, M.A. 1983. *Issues in Islamic Economics*. Lahore: Islamic Publications Ltd.

Ladjerardi, H. 1985. *Labor Unions and Autocracy in Iran*. Syracuse, NY: Syracuse University Press.

Lerrick, A. and Q.J. Mian. 1987. *Saudi Business and Labor Law*, 2nd ed. London: Graham and Trotman.

Mannan, M.A. 1986. *Islamic Economics: Theory and Practice*. Cambridge: Hodden and Stoughton.

Moore, J.M. 1992. "The Iranian Revolution Revisited." Pp. 83-102 in *Modern Capitalism and Islamic Ideology in Iran*, edited by C. Bina and H. Zangeneh. New York: St. Martin's Press.

Moore, R.M. and N. Delener. 1986. "Islam and Work." Pp. 71-87 in *Workforce Management in the Arabian Peninsula: Forces Affecting Development*, edited by G.S. Roukis and P.J. Montana. Westport, CT: Greenwood Press.

Nomani, F. and A. Rahnema. 1994. *Islamic Economic Systems*. Atlantic Highlands, NJ: Zed Books.

Parvin, M. 1993. "On the Synergism of Gender and Class Exploitation: Theory and Practice Under Islamic Rule." *Review of Social Economy* LI (Summer): 201-216.

Pfeifer, K. 1997. "Is There an 'Islamic Economics'?" In *Political Islam*, edited by J. Stork and J. Beinin. Berkeley: University of California Press.

Posusney, M.P. 1997. *Labor and the State in Egypt: Workers, Unions, and Economic Restructuring*. New York: Columbia University Press.

________. 1995a. "The Political Environment of Economic Reform in Egypt: The Labor Movement vs. Privatization Revisited." *Amsterdam Middle East Papers* No. 2 (September).

________. 1995b. "Egypt's New Labor Law Removes Worker Provisions." *Middle East Report*, No. 194/5 (May-August): 52-53, 64.

________. 1993. "Irrational Workers: The Moral Economy of Labor Protest in Egypt." *World Politics* 46(October): 83-120.

________. 1992. "Labor as an Obstacle to Privatization: The Case of Egypt, 1974-87." Pp. 81-105 in *Privatization and Liberalization in the Middle East*, edited by I. Harik. Bloomington: University of Indiana Press.

Rahman, A. 1990. *Economic Doctrines of Islam*, Vol. 1, 3rd ed. Lahore: Islamic Publications Ltd.

Rahnema, S. 1992. "Work Councils in Iran: The Illusion of Worker Control." *Economic and Industrial Democracy* 13: 69-94.

Rodinson, M. 1981. *Islam and Capitalism*. Austin: University of Texas Press.

Said, H.M. 1972. *The Employer and the Employee: Islamic Concept*. Karachi: Dar Al-Fikr Al-Islami.

Shabon, A. 1981. *The Political, Economic and Labor Climate in the Countries of the Arabian Peninsula*. Philadelphia: Wharton School.

Shaw, P.F. 1986. "Saudi Arabian Manpower Requirements." Pp. 95-112 in *Workforce Management in the Arabian Peninsula*, edited by G.S. Roukis and P.J. Montana. Westport, CT: Greenwood Press.

Stevenson, T.B. 1993. "Yemeni Workers Come Home: Reabsorbing One Million Migrants." *Middle East Report*, No. 181 (March/April): 15-24.

United Nations Development Program. 1994. *Human Development Report 1994*.

Vakili-Zad, C. 1992. "Continuity and Change: the Structure of Power in Iran." Pp. 13-48 in *Modern Capitalism and Islamic Ideology in Iran*, edited by C. Bina and H. Zangeneh. New York: St. Martin's Press.

Valibeigi, M. 1993. "Islamic Economics and Economic Policy Formation in Post Revolutionary Iran: A Critique." *Journal of Economic Issues* XXVII(September): 793-812.

Yusuf, S.M. 1988. *Economic Justice in Islam*. New Dehli: Kitab Bhavan.

Zubaida, S. 1989. *Islam, the People and the State*. London: Routledge.

COMMERCIAL LAW:

THE CONFLICT IN SHARI'A AND SECULAR LAW

William Ballantyne

> *Commercial law is, strictly speaking, merely one aspect of the law of contract, but its volume and the importance of its subject-matter, have elevated it to the rank of an independent branch of the law in general.*
>
> —Gutteridge (1949)

> *A legal tenet is considered to go to "public policy" (Ordre public: An-nitham Al'Aam) if its purport is to establish a public benefit, be it political or social or economic, touching upon the highest order of society, thus taking precedence over individual benefit....*
>
> *There is no firm tenet absolutely defining "public policy" applicable to every age and place, because public policy is a relative matter, thus it is only possible to set a flexible yardstick, viz, "the public weal, going to the highest regulation of society. The application of such yardstick in one civilisation leads to results different from those which we may find in another civilisation."*
>
> —Al Sanhuri (1954, Vol III, p. 81)

In this particular context of his brilliant comparative jurisprudential study, Al Sanhuri was dealing with the subject under the general heading of "Validity

International Review of Comparative Public Policy, Volume 9, pages 225-248.

ISBN: 0-7623-0268-2

of the object (objet: Mahall) of the (commercial) transaction in western jurisprudence." He then went on to consider the position at the Shari'a. It is as well to appreciate at the outset of this paper a fundamental difference.

> Where, in most Western systems, the law may be refashioned as a result of political considerations, this can never, in theory at any rate, be the case in Islam. For here the law is the eternal blueprint to which the structure of state and political policy must, ideally, conform. As a matter of abstract principle, therefore, Islamic political theory represents the rule of law in its most absolute form. From the provisions of the rigid and immutable Shari'a no one, neither the ruler himself nor any official of state, is by privilege exempt (Anderson and Coulson 1958, p. 929).

Thus, in essence, while that essential flexibility to which Al Sanhuri refers may be said to exist in secular legal systems, the same cannot be said of the Shari'a itself where, if any such flexibility is to be found, it must, in view of the religious origin of the Shari'a, be strictly circumscribed; indeed, it may be said that, in regard to immutable basic tenets of the Shari'a which stem from the Quran and Sunna, it does not exist at all, other than by recourse to often contentious differentiations of definition. To anticipate what follows: for example, that riba is forbidden, is indeed a great sin, is unarguable, however, what riba *is*, has been debated by jurists at great length.

In dealing with public policy in Islam as affecting commercial law, there were, it seems, two possible approaches. The first, and perhaps most obvious, was to examine the matter on an academic or historical basis, by reference to the essential attitude of the ideal Islamic state, or Caliphate as going to public or state policy in the matter of commercial law. The second approach was to treat the subject in the context of the modern world, with examples from contemporary legal systems, their constitutions, and their codes. As essentially a practicing lawyer, I have without hesitation chosen the second option, although in effect there is little difference between the two, where consideration of the immutable tenets of the Shari'a is in issue. Public policy imposed upon a state, whether the ideal Islamic state as originally envisaged or a modern Islamic state, by the tenets of Islam, hence the Shari'a, necessarily extends to the legislation of such a state and to commercial contracts. Thus, under the Shari'a, we cannot differentiate between public policy, meaning state policy, and public policy or public order (ordre public: An-nitham Al'Aam) in commercial contract. The principles are the same.

I shall, in this paper, use the term "public policy" to denote those principles of law which go to "ordre public" or An-nitham Al'Aam and which, being mandatory at the Shari'a, must be applied by a judge applying the law of Islam and cannot be avoided by contract.

Modern Islamic states have had to deal with the very difficult situation of trying to reconcile their religious and, usually, constitutional obligations with the practices of modern commerce. It is here that conflict arises.

It would, of course, be impossible to cover the whole complex of Islamic states worldwide. I have thus taken my examples largely from the jurisprudence of the States of the Gulf Cooperation Council (Kuwait, Saudi Arabia, Bahrain, Qatar, the United Arab Emirates—UAE, and Oman) with, inevitably, references to Egypt from whose codes and doctrine much that is relevant in the Gulf systems derives. Those states provide an excellent study in microcosm and this is largely due to historical circumstance. All were precipitated with startling rapidity from a condition of virtual tranquillity into one where they have been obliged to adapt to and participate in the contemporary commercial and economic world, due, of course, to the sudden cataclysmic creation of wealth derived from oil. These states, and others with them, are all Islamic states, whose very constitutions provide for the Shari'a as the, or one of the, principal sources of legislation. At the Shari'a, many of the usual provisions in a modern commercial contract are anathema; at the Shari'a, they offend principles of "public policy" and, thus, would lead to nullity. Therein lies the conflict. Does an Islamic state, at one end of the scale, pass secular legislation in defiance, or defeasance of the Shari'a; or does it at the other end, continue to pay lip service to the Shari'a while in practice ignoring it; or does it adopt a grey middle position and endeavor to make its codes comply as far as possible with Shari'a principles, in a new *ijtihad*?

Therein lies a real and increasing crisis of conscience.

Dealing with this problem by reference to the Arab States:

> The problem is that the Arabs have, to a greater or less degree, in wishing to adopt the existing international world of commerce, come face to face with the classic situation: an irresistible force against an immovable object.
>
> As is not uncommon in these circumstances (not by any means only in the Arab world) the question has been begged on all sides. It will be, to say the least, interesting to see for how long and to what extent this apparent anomaly can continue (Ballantyne 1980).

The difficulty was propounded with his customary lucidity by Sir Norman Anderson:

> The Shari'a, or common law, of Al Islam...still, officially at least, reigns supreme in Saudi Arabia; that it constitutes only one among severally mutually conflicting sources from which the comprehensive and progressive Kuwait codes have been derived; that it still prevails throughout the whole area, largely unchanged, insofar as the "personal" or family law of Muslims is concerned; and that it is difficult to say whether Saudi Arabia, on the one hand, and Kuwait and Bahrain on the other, will continue to pursue their present line of policy. In all probability they will; but it is far more difficult to foresee whether the references to the Shari'a which are included in almost all the constitutional enactments in the other States in the peninsular will prove to have been pointers to the law which will actually be applied in matters which are still wholly uncovered by legislation—or, indeed

> to the way in which the courts will interpret, or fill the lacunae, in those enactments which have already been, or will soon be, promulgated—or whether, instead they will turn out to have been little more than pious acknowledgements of an undoubted religious allegiance (Anderson 1980, p. viii).

Al Sanhuri's general definition, above, of public policy retains validity, but there is a world of difference between what constitutes public policy at the Shari'a and at secular law. Modern commerce, in the secular non-Islamic sense, demands of its laws concepts of public policy different from those enjoined by the Shari'a. It is those differences that must be examined, together with the difficulties caused by them in theory and in practice and how modern legislators have sought to deal with the problem. Commercial law is predominantly a matter of contract (above). Thus, we examine at the outset the question of the sanctity of contract, the theory that "*pacta sunt servanda.*" Here the path is clear. At the Shari'a, as in secular systems, the binding force of contract is a basic precept. At the Shari'a, this emanates from the highest authority. Sura V of the Quran reveals: "Oh ye who believe, fulfil your contracts (bonds)."

Another revelation equally implores the Muslim to observe his obligations: "Nay, but (the chosen of Allah) is he who fulfils his covenant" (Sura III:76).

There are countless further injunctions as to the fulfillment of obligations, and the Prophet himself commented thereon: "Muslims are bound by their stipulations/agreements" (Schacht 1959, p. 139).

All authority emphasizes that:

> a party to an agreement cannot free himself from his obligations thereunder by his will alone, even if he had resolved such faculty for himself at the time the contract was concluded, unless such stipulation is specifically authorised by the Divine Legislator (Abu Zahra 1938, p. 218; Sanhuri 1968, p. 50; Mahmasani 1955, p. 79 et seq.; see also *Al-Majallat al-Ahkam ash-Shar'ia*—Definition, Articles 162-165).

Further, the great Hanbali jurist Ibn Taimiyya, in propounding the theory of the sanctity of contracts, opined that transactions not expressly forbidden at the Shari'a are permissible: "The following rule shall be obeyed: men shall be permitted to make all the transactions they need, unless those transactions are forbidden by the Book or by the Sunna" (LAOUST 1948, P. 167).

It may be fairly submitted that in modern circumstances, this is a view which would be generally accepted. However, even in this light, there are, as will be seen, prohibitions at the Shari'a which affect public policy and would cause severe embarrassment to, or indeed defeat, many modern commercial contracts.

At the Shari'a, contracts must be performed under Divine sanction; but of what may such contracts consist?

The problem reared its head, inevitably, early in the history of modern international commercial arbitration. In the *Aramco Arbitration* (*Arabian*

American Oil Company (*Aramco*) v *Kingdom of Saudi Arabia*, 1958: 27 ILR 117), the arbitrators had no difficulty in holding that the proper law of the concession agreement was the law of Saudi Arabia, which they found to be the Shari'a in the Hanbali school; they had, correspondingly, no difficulty in finding that the rule of pacta sunt servanda was strictly applied at the Shari'a. However, when they came to consider the substance of the particular concession contract, they ran into the brick wall of public policy at the Shari'a—had the contract to be interpreted strictly at the Shari'a, many of its provisions must be held to be void. They thus found it necessary to hold that the regime of oil concessions had remained embryonic in Muslim law, and to "fill the gaps in the law of Saudi Arabia" by resorting to "world-wide custom and practice in the oil business and industry, world case law and doctrine and pure jurisprudence." They also managed to hold that all the circumstances of the case demanded that the lex arbitri be public international law.

Similarly, in the earlier case of *Petroleum Development* (*Trucial Coast*) *Limited* v *the Shaikh of Abu Dhabi*, Lord Asquith rejected the Islamic law as applied in Abu Dhabi as not being competent to regulate a modern commercial instrument and, accordingly, applied "principles rooted in good sense and common practice of the generality of civilised nations—a sort of modern law of nature."[1]

Further, in *Ruler of Qatar* v *International Marine Oil Company Limited*,[2] the arbitrator, again, found the law of Qatar to be the proper law of the concession, and to be the Shari'a, Hanbali school, but also found that the latter was inappropriate to govern a modern oil concession. To quote from the Award:

> There are at least two weighty considerations against that view (i.e., the adoption of Islamic law). One is that, in my opinion, after hearing the evidence of the two experts in Islamic law, Professor Anderson and Professor Milliot, there is no settled body of legal principles in Qatar applicable to the construction of modern commercial instruments...I have no reason to suppose that Islamic law is not administered there (Qatar) strictly, but I am satisfied that the law does not contain any principles which would be sufficient to interpret this particular contract...Both experts agreed that certain parts of the contract, if Islamic law were applicable, would be open to the grave criticism of being invalid. According to Professor Milliot, the Principal Agreement was full of irregularities from end to end according to Islamic law applied in Qatar. This is a cogent reason for saying that such law does not contain a body of legal principles applicable to a modern commercial contract of this kind.

In effect, therefore, the Arbitrators in all three of these early arbitrations, faced with the situation that the proper law of the contract, the Shari'a, contained provisions according to which the contract was void—rendering themselves, presumably, ab initio ex officio (!)—chose not to apply the proper law. Instead, they invoked outside principles: where provisions of the proper law going to public policy are such as to avoid the contract, substitute another proper law, to avoid a situation of non licet. A strange solution, but perhaps inevitable as the only alternative to their packing their bags at the outset, with the dispute between the parties unresolved.

It should be said that were these cases to be adjudicated upon today, the Arbitrators would find in both Qatar and Abu Dhabi (now a constituent Emirate of the UAE) modern codes based upon the civil system. Saudi Arabia, however, on the other hand remains in much the same position as before.[3]

The problem inherent in the above examples was dealt with by Sir Norman Anderson and Professor Coulson in the following terms, referring to the Shari'a as:

> ...a system of law which attained its formal perfection centuries before the demands of modern technology, finance and marketing were dreamed of....But as we see it, it is preferable to recognise the typical economic development agreement (such as an oil concession) of today, with the detailed terms which the exigencies of the trade and the needs of the two contracting parties commonly dictate, as sui generis (Anderson and Coulson 1958).

It may be submitted that however much the inherent flexibility of an international arbitration may admit of such solutions, it would be very difficult, to say the least, for a judge in a forum whose conflict rules dictated the Shari'a as the proper law, to ignore its provisions in this way.

It is now necessary to examine, albeit briefly, those provisions of the Shari'a which may be said to be mandatory or going to public policy, after which some practical examples may be examined as to how some states have dealt with the undoubted difficulties.

Overall, there is an ostensible difficulty in that the classical fuqaha (jurists) did not endeavor to elicit in the fiqh any overall theory of contract; rather, they expounded on the basis of a series of nominate contracts representing those in most common usage. The first difficulty might be, therefore, whether the commercial lawyer is obliged to fit his transaction into one of these categories. We have referred to the opinion of Ibn Taimiyya in this respect. Al Sanhuri wrote:

> Thus, what the fuqaha (jurisprudents) maintained of nominate contracts, were only those contracts which formed the majority of transactions during their era, and if civilisation develops other contracts in accord with established principles of fiqh (jurisprudence), such contracts are lawful. The new Iraqi Civil Code is based upon this policy of the Shari'a. Article 74 thereof enumerates the types of contracts in accordance with their objects as set out in Murshid Al Hairan. That is then followed by a general provision, Article 75, as follows:
>
> > "The contract may deal with any other thing wherein the obligation is not forbidden by law or contrary to public policy or morals."

This principle is generally adopted by other Civil Codes of the Arab countries, and it may be confidently stated that it would now be unlikely that an Arab civil or commercial forum, called upon to apply the Shari'a, would take the restricted view that only if a transaction fell within one of the classes of contract dealt with by the classical jurists, would it be entertainable.[4] Rather, an attempt would be made to extract a basic theory of contractual obligations and apply it to the contract in question. Herein, however, as we have seen, lies the difficulty.

Sanhuri continues the above passage:

> However, the scope of public policy in the Islamic jurisprudence is wider than in western jurisprudence. The prohibition of contracts involving riba, and contracts of qarar greatly broaden such scope in the Islamic jurisprudence and similarly the theory of the associated condition in contract increases the widening of such scope to a great extent.
>
> Thus the basic principle in the Islamic jurisprudence is *freedom of contract within the limits of public policy, but the multiplicity of the principles which are considered as going to public policy narrow such freedom* (Sanhuri 1953, Vol. 1, p. 81; emphasis added).

Al Sanhuri thus narrowed the main difficulty to the three aspects: riba, gharar, and associated conditions.

These are difficult to define succinctly in English but may be broadly described as any unjustified accretion and upsetting of the equivalence of the contract—the obvious and most contentious example being the ban on interest on monies lent (riba); the element of risk or uncertainty (gharar); and, where you have extraneous conditions, a multifarity of transactions within one contract.

> You will observe that each of these three matters had a profound effect on detailed provisions to the extent that, in many such, the Islamic jurisprudence distanced itself from the requirements of the transaction. If it were possible to clarify our understanding of these three matters in accordance with the development of civilisation, then much of the reasoning which prevents the Islamic jurisprudence from going along with the requirements of such development would disappear (Al Sanhuri 1953, Vol. III, p. 14).

While it cannot be said that reference to these three matters alone suffices to include all the difficulties inherent in the application of the Shari'a in commercial law, Sanhuri's condensation continues to have considerable relevance.[5]

The concept of riba alone, embracing as it does in any acceptable definition an absolute embargo on interest on monies lent, or owing, is enough to constitute an insuperable obstacle to modern commercial transactions.

It is not possible in this paper to analyze in detail the extent of the difficulties imposed even by the three main obstacles mentioned by Al Sanhuri. To highlight their main purport suffices to demonstrate the difficulties, going in effect to invalidity of the object in a commercial contract.

It may be submitted that the main element is that of uncertainty. A judge called upon to apply the Shari'a, would be obliged to listen to considerable argument as to the actual meaning of gharar, riba, and extraneous conditions. For the purposes of this paper, we shall accept that all of these considerations contain severe obstacles to commercial contract. Riba is the one most commonly encountered, having been identified loosely with the prohibition on interest (although it goes much further than that) and, indeed, being the principle which, bracketed with gharar, has led to the creation of the whole structure of Islamic banking. As the early cases quoted above show, it is virtually impossible to envisage a modern commercial agreement which would not contain serious illegalities if construed in the light of Shari'a tenets which are mandatory, under Divine sanction leading to inevitable temporal observance. Can one envisage, for example, any loan agreement which did not contain provisions for interest, or a complicated contract for the supply of weapons, linked to the most complex and diverse financing provisions and with several volumes of conditions, which was limited to one transaction, containing equivalence of prestations? Such contracts make the early oil concessions look simple indeed.

It must be accepted that the uncertainties, difficulties, and illegalities exist. Thus, modern commerce and modern economics are irreconcilable with many mandatory provisions of the Shari'a, much of such policy being of Divine origin. This has inevitably led to a crisis which is not only juridical, but for Muslims, one of conscience. Al Sanhuri's avowed object of drafting the 1948 Civil Code of Egypt with due regard to the requirements of the Shari'a, in the hope that gradually the principles of the Shari'a would predominate, has been submerged in the sea of conventional economics.

It is now necessary to consider how in practice some of the Arab states have sought to deal with the situation of the "irresistible force against an irremovable object." The example of riba will be taken as that most commonly occurring and as constituting, in the context of conventional financing, the most obvious obstacle. Such a study, however superficial, reveals, in the undoubted reassertion of Islamic principles now taking place, the inevitable crisis of conscience.

We first consider the relevance of the Shari'a in the modern commercial laws of the Arab states, which consist, from Morocco in the West to Oman in the East, of 18 jurisdictions. All of them, with the strict exception of Lebanon, are Islamic states, whose constitutions provide that Islam is the religion of the state, with the Shari'a, in varying degree, as a source of legislation.

Due largely initially to the genius of Al Sanhuri, who drafted not only the Civil Code of Egypt of 1948 but also subsequently the Codes of Syria, Libya, Iraq, and Kuwait,[6] and also provided the monumental civil commentary "Al Wasit," Egypt has provided the source of codes and doctrines for many of the Arab jurisdictions, and continues to do so. In 1980, the constitution of Egypt, which until then had provided for the Shari'a as *a* principal source of legislation, was amended to make it *the* principal source of legislation.[7] This change led to a furor among the judiciary, puzzled as to whether its effects must be to oblige them to immediately apply the strict provisions of the Shari'a, in defiance of the modern commercial legislation and practices prevalent in Egypt. The Egyptian Court of Cassation took the unusual step of circularizing the courts to the effect that no change should be given effect pending clarification (Ballantyne 1986, Appendix XV, p. 50). Clarification finally came in the guise of the decision of the Constitutional Court discussed below.

A similar reference to the Shari'a as *the* principal source of legislation is to be found in the Constitution of Qatar. In Bahrain, Kuwait, and the UAE it is *a* principal source.[8]

In 1992, Saudi Arabia promulgated its new Constitution, providing in Article I:

> The Kingdom of Saudi Arabia is an Arab Islamic state, having full sovereignty: its religion is Al Islam and its constitution is the Book of God Almighty and the Sunna of his Apostle (God's prayers and peace be upon him); its language is Arabic and its capital Riyadh (*Umm al Qura*, Official Gazette, No. 3397, 6 March 1992).

The sole source of law continues to be Shari'a. Article 45 provides:

> The source from which a fatwa is to be derived in the Kingdom of Saudi Arabia is the Book of God Almighty and the Sunna of his Apostle (God's prayers and peace be upon him): the law shall provide for the set up and powers of the Body of Senior Ulema (jurists) and the Directorate of Scientific Research and Ifta.

and Article 46:

> The judiciary is an independent authority; there is no authority to impose on them, when adjudicating, other than that of the Islamic Shari'a.

and Article 48:

> The Courts shall apply, to matters brought before them, the provisions of the Islamic Shari'a in accordance with the precepts of the Book of God and the Sunna and the laws that are enacted by the King which are not inconsistent with the Book and/or the Sunna (Al Mohaimeed 1993, p. 30).

Thus, in Saudi Arabia the law paramount remains the Shari'a. "Regulations" (not "laws" which are of Divine origin) must conform to the Shari'a to avoid nullity.

The constitution of the Yemen Republic provides:

> *Article 2*: Islam is the religion of the state and Arabic is its official language.
> *Article 3*: The Islamic Shari'a is the principle source of legislation.

We thus find a broad spread of constitutional provisions, ranging from where, as in Saudi Arabia, the Shari'a remains the sole source of legislation and Regulations are valid only insofar as they do not infringe the Shari'a; to where the prescription is of the Shari'a as *the* principal source of legislation (Egypt, Qatar, Yemen); to where it is *a* principal source of legislation (Bahrain, Kuwait, UAE); to where the constitution is silent (Jordan): to where there is no written constitution (Oman).

The basic question for determination is to what extent can a state, within the ambit of its constitution as the paramount law, legislate in defeasance or defiance of Shari'a principles? In particular, to what extent may a given state legislate to replace those aspects of the Shari'a which are mandatory or go to public policy but which are found to be incompatible, nay irreconcilable, with the commercial and economic interests of the state?

At one end of the scale, in Saudi Arabia, where the Shari'a reigns supreme, the answer is simple. In many cases, notably where riba is involved, commercial necessity turns a blind eye to the difficulty, but this gives no satisfaction to the commercial lawyer required, for example, to give a clearing Opinion in a syndicated loan Agreement. Where such an Agreement was expressed to be subject to Saudi law, it would be "as full of irregularities from end to end" as were the old Concession Agreements (see above)—the solution being, of course, to provide, where possible, for another proper law for the contract, with jurisdiction vested in the courts of the lender and, in addition, assets outside the Kingdom against which any judgment may be enforced (Mallat 1988; Ballantyne 1995).

In other Islamic jurisdictions, the matter must be individually considered in each case by reference to the Constitution in the first place, then by reference to legislation promulgated downstream from the Constitution. To the Western lawyer—in particular, the Common lawyer—the necessity to go behind the laws "on the statute book" to look at the constitution appears strange. It is not, however, so strange in jurisdictions where there are Constitutional Courts and where the actions of the executive and even the constitutionality of legislation can be, and are, challenged.[9] Two examples suffice to illustrate this.

We have referred to the furor caused by the change in the Constitution of Egypt in 1980. The question that immediately arose was whether such a deliberate change in the prescription of the Shari'a from "*a*" to "*the*" principal source of legislation, immediately obliged the judiciary to have regard to the Shari'a, in defeasance of existing provisions in the laws which were contrary to the Shari'a. Here, our prime example of ribawi transactions provides a convenient yardstick.

The matter came to a head in 1985 in *Rector of Al Azhar University* v *The President of the Republic, The President of the Council of Ministers, The President of the Legislative Committee of the People's Assembly, and Atef Fouad Goudah*, before the Supreme Constitutional Court of Egypt.[10]

The essence of the dispute went to the constitutionality of Article 226 of Al Sanhuri's Civil Code of 1948. That Article provides:

> If the object of the obligation is a sum of money of a known quantity, and if the debtor is late in repaying it, then the debtor is obligated to pay the creditor interest at the rate of 4 percent in civil matters and 5 percent in commercial matters by way of compensation for the delay.

The Plaintiff pleaded that this text was contrary to the absolute prohibition of interest in the Shari'a, which prohibition must be immediately applied having regard to the change in the Constitution.

The lower court of Egypt, in application of that Article, had ordered Al Azhar to pay an amount of 592 Egyptian pounds *plus interest*, at the rate of 4 percent until date of payment, pursuant to Article 226. The Plaintiff argued that although the jurists differed in some details, they were practically unanimous in condemning interest as riba.[11] Thus, in the light of the change in the Constitution, Article 226, which offended against the Shari'a in this respect, was now unconstitutional.

It will be appreciated that the significance of this case went far beyond its immediate substance. Had the Court agreed with the basic proposition of the Plaintiff, this would not only have rendered Article 226 unconstitutional but would also have had the same effect upon many further provisions of the Civil Code, the Commercial Code, and generally the secular commercial laws (in particular, the banking laws) of Egypt. It would, in effect, have decreed the paramountcy of the Shari'a and the illegality of interest, with corresponding repercussions throughout the Islamic world. There can be little doubt that the Constitutional Court adverted to such weighty considerations in formulating its judgment.

In any event, the Court found no need to contribute to the great debate on riba. The Court based its judgment upon the principle of non-retroactivity, deciding that the amendment to the Constitution should not be given retroactive effect.

In its judgment, the Court referred to the 1979 Report of the Committee on the amendment to the Constitution:

> It enjoins on the Legislature to have recourse to the rules of the Shari'a in its quest for the rule of law which it is seeking, to the exclusion of any other system of law and in case it does not find in the Shari'a a clear ruling, then it should apply the approved methods of deducing legal rules from the authorised sources of endeavour in Islamic jurisprudence. These may help the Legislator in making a ruling which does not contradict the principles and general framework of the Shari'a.

The Court found that this principle was to be followed in future legislation but that the change in the constitution did not cover former legislative enactments. The Court quoted with approval further extracts from the Report, which are worth quoting here in full:

> The Constitution of 1971 was the first constitution in our modern history which explicitly provided that the Shari'a is *a* principal source of law. Then later on, the Constitution was amended in 1980 so that the Shari'a became *the* principle source of legislation. This amendment means that *it is no longer possible in the future to enact any legislation which contradicts the rulings of Islamic law. It also means that it is imperative to review the laws which were in effect before the application of the Constitution of 1971 and to amend these laws, in such a manner as to make them conform to the principles of Islamic law* (Emphasis added, see note 10).

The Report of the General Committee goes on to say:

> The departure from the present legal institutions of Egypt, which go back more than one hundred years, and their replacement in their entirety by Islamic law require patient efforts and careful practical consideration. Hence, legislation for changing economic and social conditions which were not familiar and were not even known before, together with the innovations in our contemporary world and the requirements of our membership in the international community, as well as the evolution of our relationships and dealings with other nations—all these call for careful consideration and deserve special endeavours. Consequently, the change of the whole legal organisation should not be contemplated without giving the law-makers a chance and a reasonable period of time to collect all legal materials and amalgamate them into a complete system within the framework of the Quran and the Sunna, and the opinions of learned Muslim jurists and Imams...

Thus, the Court decided that:

1. The change to the constitution did not affect existing legislation; however
2. All legislation enacted thereafter must conform to Shari'a principles, otherwise it would infringe on the constitution; and furthermore
3. Such finding did not free the Legislature from responsibility with regard to existing enactments, but imposes on it the duty of purifying them of principles which infringe the Shari'a.

If anything, therefore, this important judgment highlighted the dilemma of the "irresistible force against the irremoveable object," without suggesting any compromise solution: the existing social and a fortiori commercial and economic structure in Egypt requires, ratione materiae, secular laws which offend against the Shari'a; public policy at the Shari'a, reflected a fortiori since 1980 in the Constitution, requires the abrogation of such laws.

The conflict continues.

The Constitutional Court of Kuwait, at its session on 28th November 1992, issued an important judgment touching upon Article II of the Kuwaiti

Constitution of 1962, which provides that the Shari'a is *a* major source of legislation.

The point at issue was the legality under the Constitution of Articles 110 and 113 of the Kuwaiti Commercial Code which provide for the charging of interest (see below). The Court, after reciting the paramount nature of the Constitution and the necessity for ensuring always that the legislation had complied therewith, said:

> The second Article of the constitution provides: The religion of the State is al Islam and the Islamic Shari'a is a principal source of legislation and the passage in the Explanatory Memorandum to such provision reads: This Article does not limit itself in its limitation that the religion of the State is al Islam, but also that the Islamic Shari'a—meaning the Islamic fiqh—is a principle source of legislation, and in thus providing and directing the legislator to a basic Islamic approach without precluding him from giving effect to principles and other sources in matters in which the Islamic fiqh does not lay down any provisions, or where it is approved to develop such provisions with regard to giving effect to the necessities of natural development with the passage of time: in the provision, for example, which permits the adoption of modern penal laws together with the hadd punishments of the Islamic Shari'a, all of which would not be acceptable if it were said "and the Islamic Shari'a is *the* principle source of legislation," in that *the requirement of such a provision would also make it illegal to take from any other source on any matter dealt in by the Shari'a*, which subjects the legislator to an extreme dilemma where practical necessities move him to pause before accepting as obligatory the view of the Shari'a fiqh in some matters, in particular e.g. regulations going to companies, insurance and banks and loans and Islamic punishments etc.... It should also be noted in this connection that provisions set out in the Constitution, that "The Islamic Shari'a is *a* principle source of legislation" *only imposes upon the legislator a trust to adopt the Islamic Shari'a to the extent possible* and summons him expressly and clearly to such a path, thus the said provision does not preclude the adoption, sooner or later, of Shari'a provisions in toto in all matters if the legislator is of such opinion: this indicates that the constitutional directive to the legislator in the second Article aforesaid, in the light of its constitutional interpretation, is a political directive to the legislator to adopt the provisions of the Shari'a as far as possible for him to do so, considering them as a substantive source from which to deduce the legal principles to be applied by him in matters put to him and further, that they are not an official source for the law—otherwise there would have been an express provision to that effect. Furthermore, *the provision of the aforesaid Article in the form which provides "a principal source" without providing that it be "the principal source" has the effect of making the Islamic Shari'a a material source from among other sources of legislation without being the sole source*: thus, this effect of the constitutional provision is not to oblige the legislator to be limited in his choice of principles of legislation to the Islamic Shari'a alone, so that there is nothing which prevents him from drawing such principles from sources other than the Islamic Shari'a which he deems suitable for the requirements of the

moment without thereby falling into the position of contravening the Constitution, whereby the judge thus becomes obliged to apply it.

The Court accordingly gave judgment:

> As a result of the foregoing, whereas the foundation of this appeal is contravention by Articles 110 and 113 of the Commercial Code of Article 2 of the Constitution, based on the contention that the provisions of these two Articles contravene the Islamic Shari'a, which is created by Article II of the Constitution as *a* principal source of legislation—in that the said Article, as explained above, does not preclude the adoption of principles of legislation from sources other than the Islamic Shari'a, so as to nullify them because their provisions are in conflict with the Islamic fiqh—*whatever may be the opinion of the Islamic fiqh*, it is not acceptable. Thus judgment must issue not to accept the constitutional argument put forward.[12]

Here the Court, in a very tautologous judgment, virtually adopted the official commentary to the Constitution: the Shari'a was merely prescribed as *a* principle source; thus, the legislator has considerable freedom. If, however, the provisions were for the Shari'a as *the* principle source, the result would be otherwise. This reasoning would seem to accord with that of the Supreme Constitutional Court of Egypt (above). Although the reasons for not upsetting existing legislation were very different in both cases, the effect of the more restrictive wording of the Constitution was given virtually the same construction. (See, for example, the position in Saudi Arabia and the constitutions of Qatar and Yemen.)

It is interesting to compare here that part of the Commentary to the Civil Code which deals with Custom, as another prescribed source of law in Kuwait:

> Customs (*aadat*) which contradict the social, political, economic or moral principles upon which the community rests, can never be elevated to Custom (*urf*) even if of long duration: *in a country such as Kuwait, whose religion is by the Constitution Al Islam it is not possible that anything which contravenes one of the bases or established precepts of Islam should constitute Custom* (Ballantyne 1986, p. 52 et seq.; emphasis added).

Thus, semble, express legislation may contravene the Shari'a; Custom may not. What, then, of Custom which becomes codified in legislation?

Against the foregoing constitutional backdrop, it remains to consider some of the formulae adopted by States in prescribing their sources of law, with examples of how these mandatory provisions of the Shari'a have been dealt with in legislation. In so doing, we shall continue to concentrate upon riba as constituting perhaps the most intractable of the Shari'a principles in this respect, while again recognizing that even riba, wideranging as it is in its effect, does not have a monopoly of difficulty in this context.

Downstream from the Constitutions, we find prescriptions in the various codes of the sources of law upon which the judge is to rely in formulating his judgment.

In Egypt, the Civil Code provides:

> Provisions of laws govern all matters to which these provisions apply in letter or in spirit.
>
> In the absence of a provision of law which is applicable, the judge will decide according to custom and in the absence of custom in accordance with the principles of Moslem law. In the absence of such principles, the judge will apply the principles of natural justice and the rules of equity.

In Kuwait, the Civil Code gives preference to express legislation, failing which

> ...the judge gives judgment in accordance with the dictates of the Islamic Jurisprudence (fiqh) most in accord with the morality and interests of the country in the absence of which according to custom.[13]

In Oman, there is no written Constitution and no Civil Code; however, the Authority for the Settlement of Commercial Disputes is referred to:

> ...clauses and laws in force in the Sultanate: contracts between the litigants provided that such do not conflict with the laws, public order or morals; established and observed customs in the field of commercial activity; and that which will achieve justice between the adversaries and lead to stability in commercial transactions.

Further, the provisions of the Omani Banking Law appear to confer upon contracting parties the right to choose both jurisdiction and law (Omani Banking Law 1974—Ballantyne 1986, pp. 50, 163).

In Saudi Arabia, it is not surprising to find no Regulations which derogate from the Shari'a as the law paramount.

In Bahrain:

> In the event that the judge finds no provision of law capable of application, he shall deduce the basis of his judgment from the principles of the Shari'a and the provisions thereof, and in the absence of any such provision, Custom shall be applied. A particular custom shall be preferred to a general Custom, and in the absence of Custom the tenets of natural law or the principles of equity and good conscience shall he applied.

This formula of itself appears to confer a certain flexibility. The matter is, however, taken further by the Commercial Code (Law 7/1987), which not only further defines the sources of Bahraini law in Commercial matters but also, as we shall see, effectively grasps the nettle of riba. Article 2 of the Code provides:

Article 2

1. There shall apply to commercial matters that upon which the two contracting parties have agreed provided that such agreement does not conflict with mandatory legislative provisions.
2. If there is no special agreement the principles of commercial custom shall apply insofar as there is no relevant provision in this or any other law relating to commercial matters. Special Custom shall be preferred over general Custom.
3. If there is no commercial Custom the laws relating to civil matters shall be applied, in the absence of which the judge shall deduce the bases of his judgment from the principles of the Islamic Shari'a then the principles of natural law and justice.

It may be doubted whether these provisions have done much to clarify the jurisprudential position in Bahrain.[14] In view, however, of subsequent provisions in the Commercial Code referred to below, it may be submitted with some confidence that the Shari'a should not now be regarded as having an overriding mandatory effect in a commercial matter.

In the United Arab Emirates, against the constitutional backdrop of the Shari'a as a principal source of legislation, we find that the law setting up the Union Supreme Court provides:

> The Supreme Court shall apply the provisions of the Islamic Shari'a, Union laws and other laws in force in the member Emirates of the Union *conforming to the Islamic Shari'a.* Likewise it shall apply those rules of custom and those principles of natural and comparative law *which do not conflict with the principles of the Shari'a* (Law 10/1973, Article 75; emphasis added).

Further, the Union Law which set up the Union Courts in the first instance provided:

> The Union Courts shall apply the provisions of the Islamic Shari'a, Union laws, and other laws in force, just as they shall apply those rules of custom and general legal principles which do not conflict with the provisions of the Shari'a (Union Law 6/1978, Article 7; see also Law 17/1978 and Ballantyne 1986, p. 57 et seq.).

Further, the Code of Civil Transactions (Civil Code) of the UAE contains in Articles 10 et seq. basic principles of private international law (conflict) (Ballantyne 1986, p. 245 et seq.; Union Law 5/1985). In contract, the principle of "autonomie de la volonte," the right of the parties to choose the law which is to govern a commercial contract, is enacted (Law 5/1985, Article 19) However, a sweeping up provision at the end of this conflict section of the Code provides:

> It shall not be lawful to apply principles of law designated by the foregoing provisions if such principles are contrary to the Islamic Shari'a or public policy or morals in the UAE (Law 5/1985, Article 27).

Thus again the predominance of the Shari'a appears to be preserved in Union Law. By reference to Article 27, those mandatory provisions of the Shari'a which form an essential embarrassment to a commercial contract cannot, in the Union forum, be avoided by the choice of a foreign law to govern the contract. This raises further the extent to which a non-Union forum would apply the chosen law, in a contract of which the lex loci solutionis were entirely Union law, in defeasance of public policy provisions of Union law.

A final example may be taken from another Gulf State, Qatar, where the Civil and Commercial Code (Law 16/1971) provides:

> *Article 2*: In prescribing principles which apply to merchants and commercial acts regard shall be had to contracts recognised by law, which constitute the law of the contracting parties.
> *Article 3*: If there is no contract or if there is but it is silent as to (a particular) provision or the provision set out in the contract is void, the legislative provisions set out in this law shall apply.
> *Article 4*: In the absence of a legislative provision which can be applied, the judge shall adjudicate in accordance with custom, special or local custom taking precedence over general custom, and in the absence of custom the principles of the Islamic Shari'a shall be applied.

The foregoing provisions must of course be viewed against a constitution which provides for the Shari'a as *the* principle source of legislation. As to whether the Shari'a must in such circumstances be regarded as paramount, I have expressed the view that as the Shari'a is not expressed by the Constitution of Qatar to be the *sole* source of legislation, it might not be regarded as paramount, in defeasance of the other specified sources of law. However, a contrary view has been expressed in two constitutional court cases of Egypt and Kuwait (above) (Ballantyne 1986, p. 56).

With the foregoing examples in mind, we turn to consider how these Shari'a difficulties going to public policy, best typified in although not monopolized by, riba, have been dealt with in the contemporary legislation.

In the Kuwait Civil Code, loans are dealt with in Articles 543 to 551 inclusive. Article 547 provides:

1. Loans shall be without interest. Any condition to the contrary is void, without prejudice to the loan agreement itself.
2. Any benefit stipulated by the lender shall be considered interest.

The Explanatory Memorandum to the Code declares:

> Islam allows loans on the basis of human cooperation and brotherliness and the relieving of worries and the reverses of fate. It vehemently denounced those who practise usury, and threatened them with the severest kinds of threats and menaces. It was accordingly forbidden in Islamic jurisprudence that loans should

include a provision granting a benefit to the lender. Otherwise, the contract is void as to such provision (Kuwaiti Law 67/1980).

Such elevated sentiments were not, however, adverted to in framing the Kuwaiti Commercial Code, passed contemporaneously with the Civil Code (Law 68/80; see also Ballantyne 1986, p. 128 et seq.):

Article 102(1) The creditor has the right to interest in a commercial loan unless the contrary is agreed. If this rate of interest is not specified in the contract, the interest due shall be the legal interest of 7 per cent.

(*2*) If the contract contains agreement on the rate of interest and the debtor delays in payment then interest for delay shall be calculated on the basis of the agreed rate.

Article 104 of the Code provides that there can be no prepayment of a loan unless by express agreement, unless the interest for the whole period is paid to the creditor. Article 110 provides that where there is delay in payment of a commercial loan in a liquidated amount, interest shall be paid at the legal rate. Article 111 limits the rate of interest to that set from time to time by the Central Bank.

Despite some attempts at justification in the commentary to the Commercial Code, it cannot be contended that these provisions of the Code do not offend against the Shari'a. Thus, the strategy of paying service to the Shari'a in the Civil Code while legislating to the contrary in the Commercial Code, is in contravention to the Shari'a—hence, the application to the Kuwaiti Constitutional Court referred to above (Ballantyne 1986, p. 103).

This strategy has, faute de mieux, been adopted by other Islamic jurisdictions under consideration, of which the following can serve as examples[15].

Bahrain has as yet no Civil Code, the law of Contracts and Civil Wrongs (torts) being still enshrined in early codes inherited from the British extra-territorial jurisdiction, and thus based upon the common law[16]. As would be expected, these make no mention of Shari'a principles. However, the Commercial Code of Bahrain[17] in addition to codifying many of the incidents of modern commerce, has put the matter of interest in a commercial loan beyond doubt.

Article 76

1. There shall be considered as a commercial loan every loan concluded by a merchant in matters relating to his commercial activities.
2. It shall be presumed that the contracting parties have stipulated for interest on such loan unless it be proved that they intended otherwise.
3. If the activity of the merchant required a granting of loans to his fellow merchants or otherwise or the payment of amounts or expenses for their account then he may require from them interest thereon from the day of payment unless it be agreed otherwise.

> 4. Interest shall be calculated at the legal rate unless another rate be agreed upon provided that the same shall not exceed the limit provided for by law.
> 5. The Bahrain Monetary Agency shall from time to time prescribe by notice the legal rate and the maximum limit for interest as provided in the preceding sub-article.
> 6. Interest shall be payable at the end of each year where the loan is deferred for more than a year, and on due date where the period is one year or less, unless it be agreed or custom dictates otherwise.
>
> *Article 275*
> 1. A money deposit is a contract which grants to the bank possession of the monies deposited and the right to deal therein consistently with its professional practice with the obligation to return the equivalent to the depositor such return to be in the same currency as the deposit. Interest may be stipulated in a contract for the deposit of money and in such case the provisions of sub-paragraphs 4, 5 and 6 of Article 76 of this Law shall apply.

Again, the Civil Code of the UAE (Law 5/1985) pays lip service to the Shari'a in many of its Articles, following in many respects the Civil Code of Jordan. Article 714 stipulates that:

> If there be stipulated in the loan Agreement any benefit additional to the requirement of the contract other than security to the lender, the condition is void and the contract valid (CP Jordanian Code 640).

This at least makes clear that the principle of severance applies. However, the UAE code of Commercial Transactions (Union Law 18/1993) provides for bank loans in Articles 409 et seq.

> *Article 409(3)* The borrower is obligated to discharge the loan with interest to the bank at the times and on the conditions agreed upon.

Article 441 deals with the calculation of interest in cases of discounting; Articles 450 et seq. with loans secured by commercial paper; Article 457 with interest charged; and Article 475 et seq. cover negotiable instruments. Interest may be stipulated in bills payable at sight or a fixed period after sight; promissory notes are similarly treated (Article 594).

We may therefore highlight, within the constraints of this article, the essential difficulties which would inevitably arise were the Shari'a to be applied per se to a modern commercial transaction. The position in that respect remains basically as it was at the time of the early arbitrations referred to above.

The veto on riba alone, viewed even in its narrow sense as interest on money, ostensibly constitutes a direct opposition to the economic system of the capitalist world. Our world today is ruled by economics and finance. It is virtually impossible to contemplate a contemporary commercial contract in the international sector, which does not contain complex financing arrangements involving not only interest, but also broad elements involving gharar, and multifarious and complex conditions. The basic clash of ideologies inherent in this situation points up the dangers, along with the need to consider those ideological problems which are quickly becoming identified with the political aspects of confrontation.

The rationale of the Islamic position on interest was well summed up by the great Hanbali jurist, Ibn al Qayyim, who, following Shafi'i's doctrine, wrote:

> Dirhams and dinars are the units of valuation of articles sold, and constitute the standard by which the evaluation of property is recognized. It must therefore be fixed and regulated so that it does not go up or down, since were the price (of currency) to go up or down like commodities we would not have a standard with which to value the articles sold. Indeed, everything is commodity, and the people's need for a price by which to value articles sold is a general and compelling one. Such valuing is not possible, save on the basis of a rate by which to know value. This requires a standard on the basis of which things are assessed, which continues upon one state of affairs, and which is not itself assessed by reference to anything else. If it becomes a commodity which goes up and down, then the transactions of people will be impaired (Ballantyne 1988b, p. 1).

It is basically for these reasons, and because of the uncertainty (gharar) inherent in the conventional system, that "in Islam the relationship between work and capital has been based on a just partnership of the two factors in the profit and loss accrued, and the payment of a pre-determined profit has been forbidden" (Shirazi 1990, p. 10).

From these principles stems of course the whole still emerging system of Islamic Banking,[18] a system based essentially upon such partnership, to the exclusion of riba. It cannot be denied that it is in the banking and financial sector that the conflict between conventional Western methods and the public policy provisions of the Shari'a are most glaringly apparent. Further, in looking at the contemporary macroeconomic scene today, with its conventional banking system, it seems difficult to dismiss the Islamic precepts, which involve in essence equity rather than lending at interest, as inappropriate.

> Despite this (i.e., the disadvantages) some people in the West have begun to find the idea attractive. It gives the provider of money a strong incentive to be sure he is doing something sensible with it. What a pity the West's banks did not have that incentive in so many of their lending decisions in the 1970's and 1980's. It also emphasises the sharing of responsibility by all users of money. That helps to make the free market system more open: you might say more democratic (Ballantyne 1988a, 1988b; Shirazi 1990, p. 10; Ray 1995; Mirakhor et al. 1988).

In conclusion, the following propositions may be offered by way of summary.

1. Overriding considerations of public policy in al Islam as propounded in the Shari'a are incompatible with what man has made of international commerce.
2. Thus, contemporary commercial laws contain much that would be illegal if adjudged at the Shari'a.
3. Islamic states have dealt with this dilemma in a variety of ways but only effectively in legislation in defiance or defeasance of the Shari'a. With the reassertion of the Shari'a, the latter solution may not be tenable for much longer.
4. It is incumbent upon the Western lawyer, advising his client with regard to an Islamic jurisdiction, to consider from the beginning whether the Shari'a does or could apply, and if so, of the difficulties inherent in such an application and how, with reference to the particular jurisdiction concerned, it may be dealt with.

As to any overall solution, it may be restated that: What, if anything, can be done by the Western lawyer, other than to sit back and wait for an inevitable crash application of the Shari'a? First of all, realize the existence of the problem, and problem it is! Second, Western lawyers must show a great sympathy and understanding for this powerful religious and cultural background, and we have to devise where necessary (not as yet in all jurisdictions) our bridges—if indeed these can be built. There are many ifs. The Shari'a is a sacred subject and, apart from understanding on the Western side, any dialogue requires the cooperation of the Muslim side in what, in many quarters, is regarded as a forbidden subject. Certainly, let us never underestimate the importance of this subject at this time—a classical and rigid application of the Shari'a to modern economic agreements could be catastrophic, and it is sobering to reflect that this could occur in most of the Arab jurisdictions with no need for prior legislation. Let it not be thought either that such application is relevant only within a Muslim jurisdiction. What of the situation in a Western court directed by its conflict rules to apply to a contract the Shari'a law, proved by expert evidence to be the law applicable as the proper law of contract? Also, what of the enforcement of foreign judgments in the Muslim jurisdictions? These are grave matters. Perhaps what is required in today's climate is not just a restructuring of the Shari'a to fit Western economic concepts, but some restructuring of those concepts in order to meet the Shari'a—a study of just how far Western and Shari'a concepts are incompatible in the light of a new ijtihad. Insofar as dialogue is possible, let us pursue it. We must take the opportunity for discussion while wise and moderate men are available to participate in it. Really, this can only be done by lawyers and, in the final analysis Muslim lawyers at that—such is the difficulty where the law is identified with religion. The political implications are, of course, obvious. But for the issue to become an international political issue, or to be overtly treated as such, is, I hope, unthinkable—to say that doing so would be counterproductive is to understate the obvious. We are too apt to write learned memoranda based upon the material pros and cons of a situation. Despite the obvious difficulties, I am optimistic enough to believe that something may be achieved (Ballantyne 1988a, p. 317 et seq.).

NOTES

1. *International Law Reports* (ILR) 1951: Case No. 37; *International Common Law Quarterly* (ICLQ) 1951. It may be noted that in so deciding, Lord Asquith entirely overlooked, or ignored, the fact that at the time the British Crown exercised extraterritorial jurisdiction in Abu Dhabi, with laws prescribed by Order in Council based upon the Anglo Indian law.

2. ILR (20), 1953: 534: Once again the arbitrator (Lord Radcliffe) failed to advert to the extraterritorial position in Qatar.

3. As to Qatar, Law No. 16/1971: Civil & Commercial Code. As to the UAE, see the Code of Civil Transactions, Law 5/1985, and Ballantyne (1985a, p. 245 et seq.). See also the UAE Commercial Code, Law 18/1993. As to Saudi Arabia, see Al Muhaimeed (1993, p. 30).

4. Refer to the Codes of, for example, Syria, Jordan, Kuwait, Qatar, and the UAE; compare Bahrain Contracts Law.

5. A broad definition of the incidents of the Shari'a going to public policy and morals may be found in the distinction between Haqq Allah (the law of God) and haqqal'abd (the law of the slave—man). Precepts of the former are immutable, mandatory, and unavoidable by contract (Al Sanhuri 1953, Vol. III, p. 101 et seq.).

6. The 1961 Commercial Code of Kuwait, done by Al Sanhoury, has now been replaced by the Civil & Commercial Codes of 1981, Laws 67/1980 and 68/1980 respectively. For the index to both Codes, see Ballantyne (1986, Appendix XV).

7. Constitutional amendment dated May 1980. As to the predominance of the constitution as the law paramount, see Badawi (1971) and Barthelemy and Duez (1933, p. 183).

8. Ballantyne (1986, Appendix XI). As to the constitutional position in the Gulf States, see Ballantyne (1986, Appendix XI, p. 27 et seq.).

9. In Kuwait, for example, there is an active Constitutional Court, which has issued numerous judgments—see Ballantyne (1986, Appendix XIII).

10. Case No. 20 of Judicial year 1. For an extract (translated into English) from this judgment and a Note thereon by this author, see *Arab Law Quarterly* (1985, (I), p. 100 et seq.).

11. The great debate as to what constitutes riba in this context has filled, and will continue to fill, many volumes (Al Sanhuri 1953, Vol. III, p. 176 et seq.; Ballantyne 1986, p. 121 et seq.; Mallat 1988). For modern authority on the principles involved

see Nabil Saleh (1992), Al Sanhuri (1953, Vol. III, p. 196 et seq.), Mallat (1988, p. 69), and Ballantyne (1986, p. 121 et seq.).

12. Emphasis added throughout (see Ballantyne 1986, p. 50 et seq.).

13. Kuwaiti Civil Code Law 67/80 Article 1(2) as ammended by Law 15/96. As to "custom" see above, and generally see Ballantyne (1986, p. 52 et seq.).

14. See Ballantyne (1987, p. 354 et seq.).

15. The great importance of ensuring that a loan is "commercial" within the definition in the applicable code, as distinct from "civil," will be apparent.

16. For the text of these codes see Ballantyne (1986, Appendices XVII and XVIII). For the extraterritorial jurisdiction, see Ballantyne (1986, chap. II).

17. Law 7/1987. See Ballantyne (1986, p. 352 et seq.) where the index to the Code is translated in full.

18. For up to date authority on this, see Shirazi (1990), Ray (1995), Mirakhor et al. (1988).

REFERENCES

Abu Zahra, M. 1938. *Property and Contract in Muslim Law* N. Place: N. Pub.

Al Majallat Al-Ahkam Ash-Shari'a. 1981. Definition, Articles 162-165. Jeddah: Al Tuhama Press.

Al Muhaimeed, A. 1993. "The Constitutional System of Saudi Arabia: A Conspectus." *Arab Law Quarterly*, VIII: 30.

Al Sanhuri, A. 1953. *Musadir Al Haqq Fil Figh Al Islami*. Beirut: Manshurat Mohammad Addaiyat.

Anderson, J.N.D. and N. Y. Coulson. 1958. "The Moslem Ruler and Contractual Obligations." *New York University Law Review* 33.

Anderson, N. 1980. Foreword. In *Legal Development in Arabia*, by W. Ballantyne. London: Graham and Trotman.

Arabian American Oil Company [ARAMCO] v. *Kingdom of Saudi Arabia* 1958:27. *International Law Reports* 117.

Alawi. 1953. "Principles of the Islamic Law on Contracts." *George Washington Law Review [Symposium on Muslim Law]*.

Badawi, Th. 1971. *Al Qanun adDusturi*. Cairo.

Bahraini Commercial Code. 1987. Decree Law 7/1987.

Ballantyne, W. 1980. *Legal Development in Arabia*. London: Graham and Trotman.

________. 1985a. *Arab Law Quarterly* I:158.

________. 1985b. "The New Civil Code of the United Arab Emirates: A Further Reassertion of the Shari'a." *Arab Law Quarterly* I: 245.

________. 1986. *Commercial Law in the Arab Middle East: The Gulf States*. London: Kluwer Law.

________. 1987. "Index to the Commercial Code of Bahrain." *Arab Law Quarterly* II: 245.

________. 1988a. "Coulson Memorial Lecture." *Arab Law Quarterly* III: 317.

________. 1988b. "Islamic Law and Financial Transactions in Contemporary Perspective." In *Islamic Law and Finance*, edited by C. Mallat. London: Graham and Trotman.

———. 1994. "The Shari'a: Bridges or Conflict?" A Foreword to *The Newsletter of The Arab Regional Forum of the IBA* I(1).

———. 1995. "Syndicated Loan Agreements." In *Commercial Law in the Middle East*, edited by C. Mallat. London: Graham and Trotman.

Barthelemy, J. and P. Duez. 1933. *Traité de Droit Constitutionnel*. Paris: Dalloz.

Gutteridge, H.C. 1949. *Comparative Law*. Cambridge: Cambridge University Press.

Kuwait Civil Code. 1980.

LAOUST. 1948. *Le Traité de Droit Public D'Ibn Taimiyya*. (Translation of *Siyasa Shar'iya*) Beirut.

Mahmasani, S. 1955. "Transactions in the Shari'a." In *Law in the Middle East*, edited by M. Khaddouri and H. Liebesny. Washington, DC: The Middle East Institute.

Mallat, C., ed. 1988. *Islamic Law and Finance*. London: Graham and Trotman.

Mirakhor, A. et al. 1988. "Financial Perspectives." In *Islamic Law and Finance*, edited by C. Mallat. London: Graham and Trotman.

Petroleum Development (Trucial Coast) Limited v. *The Shaikh of Abu Dhabi*. 1951. International Law Reports, Case Number 37. *International Common Law Quarterly*.

Qatar Civil and Commercial Code. 1971. Law 16/1971.

Ray, N.D. 1995. *Arab Islamic Banking and the Renewal of Islamic Law*. London: Graham and Trotman.

Rector of Al Azhar University v *The President of the Republic, The President of the Council of Ministers, The President of the Legislative Committee of the People's Assembly, and Atef Fouad Goudah*. Supreme Constitutional Court of Egypt; Case No. 20 of Judicial Year 1. [For an extract, translated into English, from this judgment and a Note by W. Ballantyne, see *Arab Law Quarterly*, 1985, (I): 100 et seq.]

Ruler of Qatar v. *International Marine Oil Company Limited 1953*. *International Law Reports* 20: 534.

Saleh, N.A. 1992. *Unlawful Gain and Legitimate Profit in Islamic Law*. Cambridge: Cambridge University Press.

Schacht, S. 1959. "Islamic Law in Contemporary States." *American Journal of Comparative Law*.

Shirazi, H., ed. 1990. *Islamic Banking*. London: Butterworths.

UAE Code of Civil Transactions. 1985. Law 5/1985.

UAE Commercial Code. 1993. Law 18/1993.

TAXATION AND PUBLIC FINANCE

Volker Nienhaus

INTRODUCTION

In an Islamic state, legislation and policy must not fall into conflict with general rules and specific regulations of the Quran and the Sunna. General rules have to be applied to specific cases, and specific regulations of the past may no longer be applicable under changed socioeconomic circumstances today. Islamic economics is the perception and application of economic theories with an explicit consideration of the value system of Islam. Jurists (theologians) have been, for centuries, the experts for the interpretation and application of Islamic values and norms. But today, the monopoly of Islamic jurists to apply Islamic values and to give guidance in all aspects of economic, social, and political life has been challenged by Muslim social scientists and economists. They are familiar with contemporary theories and can claim a higher competence for the explanation of complex economic and social relations. Since understanding is the precondition for judging, the recommendations and assessments of Islamic economists should be taken seriously if they are explicitly based on the Islamic value system that, for centuries, found its manifestation in the Islamic law. Islamic economists have come up with proposals for the design of a system of taxation and public finance in a modern Islamic state that differ

International Review of Comparative Public Policy, Volume 9, pages 249-275.

ISBN: 0-7623-0268-2

from the traditional views of Islamic jurists. This paper will comment on the quality of the new contributions and point to some implications that have not yet attracted due attention. Although the basic principles of taxation and public finance in Islam will be presented in the following sections, the paper is not meant to be encyclopedic but interpretive. Further, it is mainly based on contributions of authors belonging to Sunni Islam. A basic difference between Sunni and Shi'i Muslims is that the latter do not recognize the right of an Islamic state to organize the compulsory levy of *zakat*, which is the core element of the Sunni system of taxation. Therefore, the Shi'i concept of public finance, and its contemporary implementation in the Islamic Republic of Iran, differ in several respects from the Sunni model that is discussed in this paper.

METHODOLOGY FOR AN "ISLAMIC SYSTEM"

For economics in general, and taxation and public finance in particular, the Quran contains no technically clear instructions but only some principles and rules that need an interpretation before they can be applied in practice. What the Prophet began in this respect was continued by the Four Rightly Guided Caliphs. The policies and regulations they implemented during the "Golden Age" of Islam became the prime reference for all later jurists.

Three decades after the Prophet began the rule of an Islamic state, approximately fifteen different types of compulsory levies and sources of public funds were known. It is neither possible nor necessary to deal with all in this paper. First, a considerable number of revenue categories were related to the conquest of new territories that is no longer relevant today. Second, the most important conceptual and practical problems of taxation and public finance in a contemporary Islamic state will be covered by an analysis of four major taxes, namely zakat (poor tax on wealth and agricultural produce), *jizya* (poll tax on non-Muslims), *kharaj* (land tax), and *ushur* (import tax).

The Early Islamic State as the Period of Reference

The socioeconomic conditions today differ fundamentally from those of the early Islamic period. In contemporary society, interdependent anonymous markets are crucial for the coordination of individual production and consumption decisions, technological progress enhances productivity and causes far-reaching structural changes, the importance of agriculture is decreasing and that of industry and services is increasing, and the majority of people work on a contractual basis for a fixed income. In the early Islamic period, economic transactions were largely based on personal relations, dynamic changes were rare and the economy was basically static, agriculture and trade were dominant, and self-employment or slave work were the rule and hired labor the exception.

But in spite of such fundamental changes, Islamic economists, to be Islamic, must rely on, or at least reckon with, the past developments and the methodology of Islamic law (Behdad 1993; Kuran 1992; Nomani and Rahnema 1994; Zarqa 1992). Whenever clear instructions for a specific issue exist in the Quran, which is the primary source of knowledge and guidance, they must be implemented. If only related rules and principles can be found, they must be considered for application in analogy. When nothing particular can be found, general Quranic principles related to goals and functions of an Islamic state must be interpreted. For implementing regulations, the practice of the Prophet becomes crucial because he was the authorized interpreter of the Quran. In addition, the practices of the Rightly Guided Caliphs must be considered carefully because of their outstanding reputation and closeness to the Prophet. Islamic jurists of later periods tried to deduce principles from the Quran and from the practice of the early Islamic period. However, taxation and public finance were developed on a case by case basis that led to a considerable diversity of practices in the early period. For example, only two of the four most important taxes have an explicit Quranic base and were applied systematically by the Prophet (zakat and jizya). Of the other two, the Prophet had applied only one tax (kharaj), but merely in a single case and not systematically. In particular, the second caliph, Umar, changed the tax system in many respects (Qureshi 1978; Ra'ana 1970; M.A.S. Siddiqi 1983). He introduced or systematically applied ushur and kharaj and amended zakat by including additional items to the tax base. Later caliphs continued with this system, but they also reversed some decisions of their predecessors.

The "Divine" Character of Zakat

Islamic economists often create the impression that the Quran and the Prophet established—at least for zakat as the most important tax—a detailed and conclusive system with a definite tax bases and tax rates that cannot be changed by any individual or state authority (Salama 1995b, p. 389; Afzal-ur-Rahman 1976, p. 197; El-Din 1986, p. 327; Metwally 1983, p. 303). Although this opinion is widely held, it is nevertheless faulty. Contrary to what is maintained, the Quran does not describe the zakat system fully. It gives only some very general rules and does not specify any source or rate of zakat. The Quran is explicit only with respect to the groups of beneficiaries. The sources and rates of zakat were specified by the Prophet. To declare them immutable by any individual or state authority gives them the same divine epistemological status as the Quran. Seemingly, this position was not shared by the companions of the Prophet and by the early caliphs who changed sources, as well as rates, only a few years after the death of the Prophet (for details see below). Most of these changes were sanctioned later by Muslim jurists who themselves proposed further modifications of the zakat system.

It seems that the early jurists had a perception of the methodological status of the zakat regulations of the Prophet which differs substantially from that of many contemporary writers. Because the Prophet received revelations from God, his zakat regulations could have the same epistemological status as the Quran. They could be the word of God, which is eternal, and cannot be modified by human beings. But there is no tradition that the Prophet himself held this view, and the conduct of the companions also does not indicate that they treated the worldly zakat regulations as being divine revelations. If the regulations are not divine or revealed, then they can only be of human origin. They are the result of the Prophet's efforts to exercise ijtihad, that is, to interpret and apply Quranic principle (M.A.S. Siddiqi 1983, pp. 54-62). The ijtihad of the Prophet is different from that of any other human being in one important respect. In several cases, God has corrected the Prophet through a special type of revelation when he made a mistake in his ijtihad, and the Prophet changed his position afterward. Faulty ijtihad is irreconcilable with the religious and political role of the Prophet. This implies that any non-corrected ijtihad of the Próphet has an implicit divine sanction and approval. It is the best possible interpretation and implementation of Quranic principles and can claim validity and obedience—but not for all times to come. Only the word of God is eternal truth, but the word of man (even if he is the Prophet) can claim truth at best for a certain place in time and space. The human mind is too limited and fallible to discover eternal truth, and the Prophet was a human being with a limited and fallible mind. But his divine guidance provided a guarantee for the believers that the efforts of the Prophet lead to the best possible solutions at the given time and space where the Prophet exercised the ijtihad. Succeeding Muslim rulers did not have this privileged access to truth.

Because of the outstanding epistemological qualities of the ijtihad of the Prophet, his zakat regulations must be carefully studied and evaluated. If circumstances change, amendments and adjustments may become not only permissible but even necessary in order to serve best the eternal objects of the Quran. However, even if today's socioeconomic situation were totally different from that in the days of the Prophet, a completely free reconstruction of zakat only "inspired by the Quran," but without any explicit reference to historical practices, cannot be reconciled with the elaborate methodological concepts of the institutionalized Islam.

Case Law Instead of General Principles

The zakat regulations of the Prophet were meant for an unspecified time but not for eternity. They are examples for the application of the principle of "realism" in legislation that "was meant to deal with actual events only. Presupposition, speculation and hypothetical issues were excluded from the philosophy of legislation" (M.A.S. Siddiqi 1983, p. 67). Unfortunately, the explication of principles was also excluded because this was not necessary. The detailed regulations covered all actual cases; an explication of principles would have been necessary only when the legislation was meant to deal with presently unknown (future) cases.

The early caliphs have amended the zakat regulations of the Prophet (see in general H.S.M. Zaman 1991; Duri 1985; Salama 1995a). They not only included items in the zakat base that were not included by the Prophet, but they also excluded an item on which zakat was levied under the Prophet and in one case, they changed the rate (see below). The reasons for these modifications are not explicitly reported.

It was not before the last quarter of the first Islamic century that Muslim jurists became more systematic. They began to develop legal theories and made efforts to identify principles for the implementation and modification of zakat regulations. The problem with these legal theories is that none of them can explain all modifications with only one single principle; all theories have to combine several principles. This reduces the precision and informative value of each theory, and it allows several theories to coexist.

The traditional theories developed by jurists of past centuries discuss at great length cases that are, at best, of minor importance today. The "modernization" of the legal theories and the integration of knowledge and insights of modern social sciences and economics has just begun, and apparently is an arduous undertaking. Many economists seem to be more interested in propagating what they, as trained economists, believe zakat should be than in debates for advancing legal theories. But on the other hand, many Muslim jurists are also reluctant to enter into a dialogue with social scientists and economists. The result of such a dialogue could be a confirmation of the devaluation of their conventional wisdom for the adaptation of Islamic institutions to the modern world. Islamic economics must become an interdisciplinary science, but monodisciplinary educated scientists have tremendous difficulties in its development.

TAXATION

Zakat

Zakat is the most prominent and most elaborate element of an Islamic system of compulsory levies. Zakat is mentioned several times in the Quran (e.g., II:177, IX:60 [Ali 1983/1934]), and the beneficiaries are listed. However, nothing detailed can be found in the Quran regarding the base for the levy, the rate(s), the methods of payment, and the exemptions. This was not a serious problem as long as zakat was a voluntary payment, but particulars had to be fixed when it was transformed into a compulsory levy collected by the state. This was done by the Prophet in Medina.

The Prophet enumerated items on which zakat had to be paid (zakat base), namely (1) animal properties (camels, sheep, cows), (2) gold, silver, and coins, (3) agricultural produce (wheat, barley, dates, and grapes), and (4) buried treasures. It is doubtful whether zakat was also levied on merchandise and honey. The Prophet specified for each item a minimum quantity below which no zakat was due (*nisab*), and the rates for quantities above that minimum. From today's perspective, the most important rates are 2.5 percent for gold, silver, and coins, 5 percent for agricultural produce from irrigated land, and 10 percent from rainfed land. The Prophet also determined the timing for the zakat payment, namely: for animal properties and gold, silver, and coins—once a year; for agricultural produce—at the time of harvest; for buried treasures—at the time of extraction.

It is important to note that the Prophet did not give a general definition of the items subject to zakat. Such a definition could have been used after his death when new items emerged or known items changed their character. The early caliphs continued with his "pragmatic legislation." When they were confronted with a specific new case (e.g., whether zakat should be paid on horses or not) they decided this case only. They did not establish principles reaching beyond the particular case. It were the jurists who later, approximately since the end of the first century, evaluated the practice and the case law of the Prophet and the early caliphs by applying rational arguments (ijtihad) to formulate more general principles. The problem is that the legal principles are not sufficiently general and abstract and that different principles conform with the practices of the Prophet (and the early caliphs). These different principles (must) coincide for past cases, but they often come to different conclusions for new cases. Further, the principles of the early jurists do not cover all issues dealt with by contemporary economists. For example, no clear principle was established to decide whether assets or income were or should become the tax base for zakat, and what are the appropriate rates for the zakat levy. This is of particular significance today when a large variety of assets (in particular financial assets and ownership rights) and forms of income (in particular contractual income) exist that were unknown in the early Islamic period.

Tax Base

There is a considerable diversity of views—if not confusion—regarding the nature and scope of the tax base of zakat. It is neither clear whether it is primarily a kind of wealth or income tax, nor is it clear exactly what types of wealth or income are to be included in the tax base.

Wealth or income tax. From many texts of Islamic economists, one gets the impression that zakat is basically a kind of wealth tax with the addition of ushr as a levy on agricultural produce. A reservation is that not all categories of wealth are subject to zakat but only those that are actually or potentially productive (Afzal-ur-Rahman 1976; Chaudhry 1992, p. 25; M.A. Abdullah 1995, pp. 18-19; Shaik 1980, pp. 19-20).

The items on which the Prophet charged zakat have been enumerated above. There is no doubt that gold, silver, coins, and buried treasures are forms of wealth that economists would classify as "stock variables." This is also the case for animal properties, but it is doubtful with respect to agricultural produce. Agricultural produce is more income (in economic terminology: a "flow variable") than wealth, and it is often argued that the difference in the zakat rate for rainfed and irrigated land reflects (in a rudimentary way) the difference in the production costs. From this some jurists and economists conclude that zakat was levied not only on wealth but also on income—at least on income from agriculture. This view is sometimes supported by reference to the practice established under the early caliphs to subtract zakat at source from (annually paid) state pensions. However, for both examples, a more "wealth-oriented" interpretation can be given.

If the final base for zakat is not the (flow variable) agricultural produce, but the (stock variable) land that yields the agricultural produce, then the value of the wealth category "land" is approximated by the agricultural produce. In the case of rainfed land, the whole produce is attributed to the land which is fertile by itself. In contrast, irrigated land becomes fertile only when combined with considerable human efforts. Thus, in a rough approximation, only half of the produce can be assigned to the land, while the other half is due to the landowner who installs and maintains irrigation facilities. The value of the asset "irrigated land" is only half the value of rainfed land. Under this perspective, the charge of 10 percent, or 5 percent of the produce of land, is not an income but a wealth tax.

A similar argument applies to the state pensions. The pension payment is based on an individual's claim against the state which, economically, is a valuable asset (because it involves annual payments). As such, the pension payment is part of the individual wealth. Therefore, not the annual income but the possession of the right or claim (i.e., the underlying wealth category) constitutes the zakat duty.

For the modernization of the traditional zakat system, a clarification of the character of the tax is of high importance and has far-reaching consequences for the fiscal potential and its allocative and distributive effects. It is surprising how divergent opinions still are. This is probably largely due to the vagueness of the crucial Quranic verses and the ambiguity of the practice of the Prophet and the early caliphs.

Productive and unproductive wealth. There is a broad consensus on the inclusion of productive and exclusion of unproductive forms of wealth. However, in a developed market economy this distinction does not make much sense since any type of wealth can be transformed into any other type (e.g., from "unproductive" jewelry into "productive" agricultural land). This does not even necessitate a sale of wealth items: "unproductive" assets can be used as collateral for the financing of productive assets. Suppose persons A and B both own jewelry worth 100,000 Dinar. As long as they keep their wealth in this form, it is "unproductive" and exempt from zakat. If person A sells his jewelry and buys agricultural land, his new wealth will become subject to zakat. Person B keeps his jewelry and obtains a bank credit backed by the jewelry as collateral to purchase agricultural land. He will not pay zakat because the debt can be subtracted from the value of the purchased land so that his productive net wealth is zero. The result is very unsatisfactory: two people with the same wealth do not pay the same zakat.

One Full Year of Ownership

Abu Bakr gave a very specific interpretation of the practice of the Prophet to collect zakat on properties only once a year, namely, that no zakat was due before a full (lunar) year of ownership has passed. Some Islamic economists emphasize this principle again today (e.g. Chaudhry 1992, p. 25; Abdullah 1995, p. 19; El-Badawi and Al-Sultan 1992, pp. 70-71). It is surprising that its far-reaching implications are not discussed.

According to the principle of "one full year of ownership," assets will not be subject to zakat in the first and in the last year of ownership (if bought after the first of the first and sold before the last day of the last year). Zakat will be charged only on assets owned from the first to the last day of a zakat year. If this principle is applied for each asset separately, it becomes very easy to avoid compulsory zakat payments: the respective assets are sold before a full year of ownership is completed and repurchased after a short while. If the problem of zakat evasion is taken seriously, institutional arrangements should not be designed in such a way that they greatly facilitate evasive behavior.

There are two ways out of the problem. First, to give up the somewhat strange principle of "one full year of ownership." In this case, zakat would be charged on total net wealth at the end of each zakat year irrespective of how long a particular asset is owned. Second, to keep the principle of "one full year of ownership" but to apply it not to each asset individually, but to the net wealth in total (i.e., to the aggregate net value of all assets). A sale of a particular piece of property changes the structure of wealth but not its aggregate value. Real assets would only be substituted (temporarily) by financial assets both of which are components of total (net) wealth. Practically, the principle of "one full year of ownership" will then mean that the basis for the zakat calculation is the minimum value to which total wealth drops within a year. Amazingly few reflections are found in the Islamic economics literature on the economic consequences and problems of this old legal principle applied in complex market environments. Here, Islamic economics is prescriptive instead of analytical.

Immutable Rates?

The rates of zakat were fixed by the Prophet. This alone is a sufficient justification for most authors to maintain these rates for all times. They discuss rates only when new items of wealth or income have to be included in the zakat base. The question then is: which of the established rates should be applied in analogy? But such a restrictive stance, with regard to modifications of rates, is not very convincing. Although the rates set by the Prophet have a divine sanction, they are not of divine origin. They were the best possible rates for achieving the aims and objectives of zakat at the time of the Prophet, but it does not mean that they are the best rates for all times and under all conceivable circumstances.

The Prophet did not establish principles but implemented solutions for practical problems. With respect to zakat, the Quran enumerates the categories of beneficiaries for whom the zakat proceeds have to be spent. It is consistent with the approach of pragmatic legislation to assume that the Prophet determined the basis and the rates of zakat in such a way that the proceeds were sufficient to allow a financial support for the beneficiaries which was held adequate at that time. This was the main problem that had to be solved by his pragmatic legislation. Because the ruler could not spend zakat proceeds at his discretion, and because a compulsory levy is a burden for the people—some see it even as a kind of violation of private property rights—it seems plausible that the Prophet fixed the zakat rates close to the minimum necessary to raise sufficient funds for reasonable support of the specified beneficiaries. Compared to tax rates of 50 percent and more for agricultural produce, which were not uncommon at that time, a zakat rate of 5 or 10 percent is very moderate.

Empirical data on the level of support per capita for each category of beneficiaries, relative to the income of the "average man" of the early Islamic period, are not available. A very rough mark is offered by nisab, that is, the minimum of wealth or agricultural produce exempted from zakat. If people with less wealth (or income) cannot maintain a reasonable standard of living from their own resources, zakat should fill the gap between the actual wealth (or income) and the minimum. Today, absolute amounts set by the Prophet may no longer be sufficient to secure a reasonable standard of living or the minimum necessary for mere survival. The same absolute amount (e.g., US$100) allows very different standards of living in different countries. Prices and costs of living differ in time and place, but also the socially accepted minimum or reasonable standard of living. Therefore, all figures of the early Islamic period should be translated into relative terms—e.g., compared to the standard of living of a small trader, craftsman, or farmer—before implementations in modern societies are considered. Since it is implausible to assume that the rates set by the Prophet are appropriate for all times and under all conceivable socioeconomic circumstances, zakat rates should be reconsidered today. Methodologically, the modification of zakat rates is not different from a modification of the zakat base; both are based on the ijtihad of the Prophet. The inclusion of an item in the zakat base, with a specified rate, is the same as changing the zakat rate from zero to a positive value. Strangely, considerable deviations from the Prophetic tradition are widely accepted only for modifications of the zakat base but not for the rates.

A conclusive determination of zakat rates, under fundamentally changed socioeconomic conditions, requires further research. For example, by his fixing of individual rates, the Prophet defined a specific structure of zakat rates. The rate for irrigated land was double the rate for gold, silver, and coins, and the rate for rainfed land was double the rate for irrigated land. It is not obvious why he determined such a structure, and it needs more reflection whether only the level, or also the structure, of zakat rates could be adjusted to modern circumstances, and what should be guiding rules for restructuring.

Only a few Islamic economists seriously proposed a revision and adaptation of zakat rates (for example Mannan 1986, pp. 258-260), but there is not much debate on this in the Islamic economics literature. The vast majority just repeat legal opinions of former centuries without deeper methodological and economic reflection. But not all of these opinions are compatible with each other. This generates the confusing picture of a consensus on generalities and conflicts in nearly all practical details.

Contemporary Practice: Confusion and Deficiencies

The confusion is not reduced but compounded if the present zakat practice of Muslim countries is considered. A comparative study of Al-Omar (1995) on general, administrative, and organizational aspects of zakat in 13 Muslim countries (most of which, however, do not claim to be "Islamic states"), discloses a very wide variety of practical arrangements. For example, the payment of zakat is voluntary in seven countries (Bahrain, Bangladesh, Egypt, Iran, Iraq, Jordan [compulsory 1944-1953], and Kuwait), and compulsory in six (Libya, Malaysia, Pakistan, Saudi Arabia, Sudan, and A.R. Yemen). The zakat base for the compulsory levy differs substantially among the respective countries. All six countries that have compulsory zakat, charge it on agricultural produce. But in Malaysia, only rice is subject to zakat. Livestock is subject to zakat in five of these countries (Libya, Malaysia, Saudi Arabia, Sudan, and Yemen); only one country (Libya) includes camels and cows used for agricultural work (ploughing, irrigation, etc.). Zakat on gold, silver, cash and other forms of invisible wealth is collected in four countries, but they assess these assets differently. Three countries (Libya, Saudi Arabia, and Sudan) collect zakat for buried treasure. Only one country (Sudan) imposes zakat "on wealth that yields income, such as rented property, factories, farms, etc." (Al-Omar 1995, p. 39). In two countries (Sudan and Saudi Arabia), "regulations provide for the collection of zakah [zakat] in respect of factories, hotels, art producing companies, taxi owners and offices of real estate agents" (Al-Omar 1995, p. 39). Income is subject to zakat in two countries: "The Saudi and Malaysian regulations are…the same as to the imposition of the zakah [zakat] in respect of free vocational jobs and employees' salaries" (p. 39). For further details on contemporary practices, see the country studies in El-Ashker and Haq (1995).

Zakat and Redistribution

Suppose there is a consensus in each Muslim country on the exact definition of the groups of zakat beneficiaries and on the reasonable level of support. Then the funds for the beneficiaries could be raised through an adapted zakat system with a "modernized" zakat base and modified zakat rates. The base and rates could be optimized (for each country) in such a way that the desirable level of support can be granted to the beneficiaries. Alternatively, a traditional zakat system, with an old, relatively narrow zakat base and rigid rates, could be implemented. In this case, zakat revenues will probably not be sufficient to satisfy all reasonable claims of the beneficiaries. Then, either claims can be satisfied only partially, or the traditional zakat system has to be supplemented by "secular" taxes or compulsory social security contributions. Both cases do not throw a favorable light on the genuine Islamic institution of zakat. It can guarantee, by itself, merely an insufficient support for the beneficiaries, and a satisfactory support level is achieved only by the superimposition of secular elements.

The risk of insufficient revenues is minimized (but not excluded) if a broad zakat base (e.g., productive wealth plus all types of income) is applied. A broad zakat base could also prevent injustices that may occur if zakat is levied on a limited number of items only. According to Muslim jurists, the category of agricultural produce does not include items such as tea or rubber. The application of a narrow and rigid interpretation then leads to the strange situation where small farmers, who produce wheat or barley, have to pay zakat while rich landlords, who own large plantations and produce tea or rubber, are exempted from zakat. In this case, zakat does not allow the poor a share in the wealth of the rich, but facilitates a redistribution from low to the lowest income groups, or from the poor to the poorest sections of the society. It is obvious that such an implementation of zakat neither complies with ideological claims nor contributes to a popularization of a genuine Islamic concept.

It seems that a thorough revision of traditional Islamic concepts and institutions is needed whereby the opinions of traditional jurists should be considered seriously, but not taken as the final word. Economic and political consequences of alternative implementations must be studied in a more comprehensive, systematic, and, perhaps, radical manner than in "mainstream Islamic economics" today.

Other Taxes of the Early Islamic Period

The tax system of the early Islamic state was not confined to zakat. For the state budget, other compulsory levies were even more important than zakat. Their revenues were considerably higher, and the use of the tax receipts was not restricted to specific beneficiaries.

Kharaj

Kharaj was a land tax levied on conquered land. It had no explicit Quranic foundation and the Prophet did not apply kharaj systematically (Chaudhry 1992, p. 70). Kharaj was introduced as a systematic land-tax by the second caliph Umar after the conquest of Iraq. Instead of distributing the conquered land among his officers, the former non-Muslim owners could keep their land against the payment of a tax. Initially, kharaj was determined as a fixed amount of money per acre, and was differentiated according to the type of produce of the land. By relating the fixed amounts to the average total produce, one can calculate implicit tax rates that increase when yields decrease. In such cases, they could reach excessive dimensions. Therefore, seemingly, the system of fixed amounts was moderated by the introduction of an upper limit at half of the produce (Mannan 1986, p. 248). Later, the system was changed to proportionate rates. The farmers had to pay not a fixed amount, but a specified percentage of the harvest. This proportionate rate was determined by the ruler in view of the financial needs of the state (Salama 1995a, p. 44). The tax rate was differentiated with respect to the type of produce, the method of irrigation, and the proximity to the market in the case of orchards. The rates ranged from 20 to 50 percent or sometimes even more.

In terms of receipts, kharaj was the most important tax, and its revenues could be used by the ruler at his discretion. Kharaj was considerably higher than zakat on agricultural produce. Since initially Muslims were obliged to pay zakat only, many landowners in conquered territories converted to Islam in order to reduce their tax burden. The loss of noticeable amounts of disposable tax revenues, as a consequence of these conversions, led to an important modification of the tax. It was no longer tied to the status of the owner of conquered land; all conquered land became kharaj land, and Muslim owners of kharaj land had to pay this tax, too.

Without a substantial modernization, kharaj, being a land tax, suffers from a deterioration of the macroeconomic significance of its tax base. In the process of development, the agricultural sector shrinks while industry and services expand. Income from the industrial and service sector gains rapidly in importance. Further, if, in industrializing economies, only agricultural production or income is subject to a tax, this is a discrimination against the farmers that can hardly be justified by any Islamic rule or principle. On the contrary, justice demands that same cases should be treated equally. Economically, people with the same income should pay the same tax.

While the vast majority of Islamic economists simply recapitulate the historical kharaj without reflection on the problems of its application today, Kahf has proposed that in cases where kharaj land is no longer used for agriculture but for other commercial purposes (e.g., the construction of a factory, a warehouse, or a shopping center), the non-agricultural users of the land should become subject to kharaj. The implementation of such a proposal requires the determination of an adequate tax base (e.g., profit or turnover) and of respective tax rates. This cannot be done without (very) free analogies interspersed with speculative elements.

Although kharaj is an early Islamic tax, it has neither a Quranic origin nor a firm foundation in the Sunna. Today, it is outdated and can be modernized only with the help of excessive analogies. The primary goal of kharaj, to provide the (Islamic) state with revenues, can be achieved much better, for example, by implementing a secular income (and profit) tax. A modern income tax is much better suited to prevent injustices than an artificially "upgraded" land tax. The clear consequence would be to give up kharaj altogether. Such an obvious conclusion, however, is not drawn or discussed by Islamic economists. The reason may be that they rarely analyze kharaj under contemporary conditions, but refer to early Islamic practices and repeat old legal doctrines without much reflection.

Jizya

Jizya was a poll-tax for non-Muslims living in an Islamic state. It was based on the Quran (IX:29) and applied by the Prophet. He charged one Dinar on every non-Muslim male adult; women and children and old men were exempted from jizya. Umar differentiated the rate according to the income of the taxpayer. He charged one Dinar from the poor, two Dinars from the middle class, and four Dinars from rich non-Muslims. The reasons for this tax structure are not known (Oran and Rashid 1989, p. 88; see also Chaudhry 1992, p. 59).

The discriminating character of the poll-tax for non-Muslims has been criticized in the West, but Muslim jurists and economists point out that jizya was introduced as a modest compensation for privileges granted, and services provided, by the Islamic state to the non-Muslims. For example, they were exempted from military service, they were not subject to zakat, and they were entitled to the same public goods and social services of the Islamic state (including support from zakat funds) as the Muslims. Further, it is argued that jizya was not a heavy financial burden, but more a symbol of the recognition of the supremacy of the Islamic state.

Contrary to what is maintained by some Islamic economists (Kahf 1995b, p. 121; Oran and Rashid 1989, p. 88), jizya can hardly be considered an equivalent of zakat for the non-Muslims in the early Islamic period. It was charged on a totally different base and with different rates, namely as a fixed amount per capita instead of a percentage of wealth and agricultural produce. Thus, the relative tax burden of jizya was inversely related to wealth, while zakat increased roughly proportionate to wealth. But even if the historical jizya was not an economic equivalent of zakat for non-Muslims, the contemporary proponents of jizya argue in favor of such an equivalent because otherwise non-Muslims would have a tax advantage over Muslims. This goal, however, could be achieved simply by abolishing the old jizya rules and applying the techniques of zakat to non-Muslims, too. Then all residents of a country (irrespective of their religion) would become subject to the same nondiscriminatory compulsory levy. This levy may be called zakat for Muslims (for whom it has an additional religious dimension) and jizya for non-Muslims, but only the name would be left from the original poll tax.

Ushur

The last of the early Islamic taxes is ushur. This foreign trade tax had neither a Quranic basis nor was it applied by the Prophet. Ushur was introduced by Umar in response to similar taxes imposed by other countries on Muslim traders. Non-Muslim traders from abroad had to pay a tax of 10 percent on the value of their merchandise when they entered the Islamic state (i.e., for imports). According to some authors, 10 percent was also collected from non-resident non-Muslim traders for exports. Later, ushur was charged on imports (and exports) of non-Muslim residents of the Islamic state at a rate of 5 percent, and of Muslim residents at a rate of 2.5 percent. "(I)t turned out to be a particularly important source of revenue for the Islamic State after its introduction by the Caliph Umar" (Oran and Rashid 1989, p. 89).

Seemingly, there was considerable difference between the treatment of non-Muslim and Muslim traders: the former had to pay ushur for each cross-border transaction, while the Muslims had to pay ushur only once a year. The payment once a year, and the tax rate of 2.5 percent, indicate that for Muslim traders ushur was not really meant as an import duty, but as a special form of the collection of zakat on merchandise. In the early Islamic period, the levy of ushur from Muslim traders might have been justified by technical reasons regarding the tax collection, but today, no serious argument is given why zakat on merchandise could not be fully integrated into the general zakat system.

The higher ushur rates for non-Muslim traders clearly discriminated against this group without a convincing explanation. This is an obstacle to a "modernization" of ushur. Today, nowhere in the world are import duties based on the religion of the trader; any attempt in this direction is in conflict with existing international trade agreements. Further, if the trader is not a natural but a legal person, the trader does not have a religion. If the religion of the owners of a legal person (e.g. the shareholders of a joint stock company) are taken instead, the religion may be mixed if Muslims and non-Muslims jointly own a company. Finally, in the case of companies whose shares are traded on anonymous stock exchanges, it is nearly impossible to identify the religions of the owners. For all of these reasons, the traditional ushur should not be revived in a modern Islamic state.

Additional Secular Taxes?

There is an ongoing debate among Muslim jurists and economists whether an Islamic state has the right to impose taxes in addition to zakat (Kahf 1995b, p. 117). In this formulation, the question seems to be exaggerated because the Prophet himself and the Rightly Guided Caliphs—in particular Umar—have imposed additional taxes, namely: jizya, kharaj and ushur. Thus, the real question is not whether, but under what conditions, additional taxes can be imposed. To raise this question implies that the legislative body of an Islamic state is not free to impose new taxes as it deems necessary.

The Islamic legislation and government is limited by rules and principles laid down in the Quran, applied by the Prophet and explicated and developed further by the early caliphs and the Muslim jurists. Authors who are very reluctant about the imposition of new taxes, emphasize the importance of the traditional Islamic wisdom. They warn to ignore the body of knowledge accumulated over the centuries and to rush to conclusions based on nothing but "Islamically inspired" individual reasoning of modern economists. For them, a Quranic verse, a Prophetic hadith, or a legal opinion of an eminent jurist is more than just a decoration or "fashionable" dress (or disguise?) for long-held personal sentiments based on conventional economics.

It is pointed out by Kahf (1995b, p. 119) that jurists have allowed additional taxes only "in cases of defence needs, assurance of a subsistent living of the poor and the indispensable expenses that safeguard the collective interests of Muslims." "Specifically, purposes of welfare, development, and/or economic equilibrium have no references in texts of Quran or Sunnah whether explicit or implicit." This is a very strong position against a large number (if not the majority) of Islamic economists who base their proposals for (additional secular) taxes on exactly such purposes and aims of an Islamic state which they deduce from the Quran and Sunna. Because Kahf is very strict concerning the justification of additional taxes, his criteria could constitute a kind of lowest common denominator for additional secular taxes.

In his view, the imposition of additional taxes could be justified if the revenues from zakat and alternative sources of finance are not sufficient for the following purposes (Kahf 1995b, pp. 119-120): provision of internal and external security; financing the political, legal, and administrative apparatus of the state; relief in cases of emergencies; maintenance of a minimum standard of living for all; development of a production capacity adequate to provide the subsistence living if the private sector is not providing such production. For Kahf, the alternative sources of finance are: kharaj, jizya, revenues of public enterprises utilizing the public ownership of natural resources, interest-free borrowings from domestic and international sources, revenues from equity financing of income generating development projects, fees for government services, and voluntary contributions.

Kahf's concept of public finance limits the state to core functions, but he does not advocate a "minimal state" in the Western libertarian sense. He excludes subsidies for producers, but most other activities of modern governments are covered (although they are not explicitly listed as purposes of an Islamic state): A large variety of "social goods" provided by modern states can be financed from different sources, including even zakat, as long as the goods fall into the category of "minimum standard of living." This comprises, inter alia, minimum education, transportation, and health care; the state must also look after housing for the poor. Other activities should not be financed from zakat, but they nevertheless could be accepted tasks of the state. Infrastructure and development projects are the most important cases here. It follows that a restriction of the power to impose additional taxes besides zakat makes an Islamic state neither a minimal nor a poor state.

The Islamic state can legitimately perform most of the tasks of contemporary secular states, and the financial resources at its disposal may be plentiful if zakat is levied on all modern forms of wealth (including, in particular, financial assets and fixed capital of companies) and income (including contractual income), if jizya is transformed into an equivalent of zakat for non-Muslims, if kharaj is charged not only for agricultural but for any use of land at relatively high rates, and if infrastructure and development projects are financed by borrowing or equity participation.

Other Islamic economists do not ignore the non-tax revenues emphasized by Kahf, but it seems that they are more skeptical regarding the revenue capacity of the non-tax sources. They usually start from a discussion of necessary or legitimate expenditures of an Islamic state and concede the power to introduce additional taxes if needed to cover the necessary or legitimate expenses. This approach is exemplified below.

PUBLIC EXPENDITURES

There are two types of public expenditures in an Islamic state. First, the Quran enumerates the groups of people for whom proceeds from the compulsory levy zakat have to be spent. Second, the Quran and Sunna require certain tasks (such as defence, civil administration, the propagation of Islam) to be fulfilled by the Islamic state. In this case, the legitimacy of public expenditures follow from the legitimacy of the tasks themselves.

Beneficiaries of Zakat

The Quran (IX:60) states that zakat is "for the poor and the needy, and those employed to administer the (funds); for those whose hearts have been (recently) reconciled (to Truth); for those in bondage and in debt; in the cause of Allah; and for the wayfarer." Discussions are mainly on the exact historical and contemporary definition and demarcation of these groups, on their relative share in the zakat revenues (equal or with priorities), and on the level and form of support that should be given (in cash or in kind, only direct and personal or also indirect for larger groups).

Definitions and Demarcations

Without going into details, the following definitions can be given (Muhammad 1993, pp. 80-85): The "poor" are those who do not possess any properties and do not have an income from employment. The "needy" are those who have some possession or income, but not enough for their sustenance. "Those employed to administer zakat" were employees of public zakat institutions, but today this category would also include employees of private zakat organizations, in particular in countries where no public zakat systems have been established.

"Those whose hearts have been reconciled" means converts to Islam. The Prophet paid generous amounts to tribal leaders who converted to Islam, but this type of spending was reduced under the first and terminated under the second caliph. "Those in bondage" were the slaves, and zakat was used to purchase their freedom. In our times, this category has lost its literal relevance, but some authors suggest—in a rather "free" interpretation—that today zakat should be used to free the Muslim world from "mental slavery" (e.g., intellectual and cultural domination by the West). Such a view is not strongly rejected but also not enthusiastically supported by the majority of Islamic economists.

"Those in debt" are people who are unable to repay a debt which they incurred either to help others (e.g., as a guarantor) or because they had no other possibility to secure their own survival. Debts incurred to finance a "luxurious lifestyle" should not be included in this category, while some authors suggest that commercial debts caused by a bankruptcy should be included. "In the cause of Allah" is a very comprehensive and vague category. Initially, under this category, compensations were given to people who (voluntarily) fought for Islam in a war and were not paid by the state. Literally, the meaning of "in the cause of Allah" is much wider and not confined to military struggle. It covers all efforts to seek the pleasure of Allah and to establish truth and justice. Muslim jurists have included the propagation of Islam as well as salaries for teachers and scholarships for students or even health facilities. "(T)he zakat share here is allocated in the general welfare, just like any modern state which uses taxes for the same purpose" (Oran and Rashid 1989, p. 93).

The "wayfarer" means a person stranded on a journey that he or she undertook for a respectable cause; in particular, to perform *hajj* or to earn a living or to seek knowledge. A support from zakat funds should enable travelers either to return home or to reach their destination. Some contemporary authors have widened this category vastly and include transport and communication facilities in general. This, however, has not found a broad consent.

The debate on a "wide" or "narrow" interpretation of the beneficiary categories has lost its acuteness when it became generally accepted that an Islamic state can levy secular taxes besides zakat. If this were not permissible, most Islamic economists would favor an extensive interpretation of the categories of beneficiaries so that most, if not all, public expenditures of a secular state would be covered by them. A problem with such an extensive interpretation of zakat beneficiaries is that zakat revenues would in most cases be totally insufficient to cover all proposed expenses if the traditional base and rates are maintained. Since the rates are not questioned by most Muslim jurists and economists, proponents of a broad definition of beneficiaries favor a wide zakat base that includes all forms of wealth and, more important, income. But, with an average wealth tax rate of roughly 2.5 percent and an average income tax rate around 10 percent, the allocative and distributive potentials of the budgets of Islamic states would, in general, be far more limited than the present potentials of governments in Muslim, as well as Western, countries.

Allocation and Priorities

Neither the Quran nor the Prophet gave an explicit ranking of the eight categories of zakat beneficiaries. But this does not mean that they all should receive equal amounts. It seems that the Prophet did not treat all categories equally; he spent relatively large amounts of zakat for converts and warfare. The companions also did not treat all categories as equally important. For example, Umar stopped the payment of zakat for converts. The concept of pragmatic legislation suggests that the allocation of zakat was in such a way that it contributed best to the solution of actual problems. When problems changed, the allocation of funds changed, too. In the light of widespread poverty in most of the densely populated Muslim countries today, it would be reasonable to give priority to the poor and the needy.

Level of Support

With respect to the level of support for the poor and the needy, it is argued that the maximum benefit is the Zakat-free minimum (nisab) of twenty Dinars, which may be too little to survive in some countries today; Oran and Rashid (1989) calculated that the nisab would amount to US$1,400 in 1984. While some assert that zakat should secure the mere survival of the poor and needy, others are far more generous and maintain that "the Shari'a does not only advocate the provision of the bare necessities of life to the poor and the needy, but goes much further and advocates that they should enjoy what is now termed as a reasonable 'standard of living'" (Muhammad 1993, p. 81).

The view that a nisab of twenty Dinars determines the level of support (because otherwise a zakat recipient would turn into a zakat payer) raises several questions. There is not one nisab for total wealth; for each category of wealth a separate nisab is defined. Twenty Dinars is the nisab for wealth in gold, silver, and coins only. It could be argued why all nisab values should not be added, or why the nisab for another category of wealth is not taken. What is more important, nisab is a minimum wealth. Zakat payments to the poor and needy are not wealth but income transfers which are needed to cover the regular cost of living. Even if Zakat payments were a multiple of nisab but are needed for subsistence, no wealth will be left at the end of the year that could turn a zakat beneficiary into a zakat payer.

A limitation of zakat payments to (the equivalent of) twenty Dinars could lead to a widening gap between the poorer and the wealthier population if the average income and the average cost of (a reasonable standard of) living increase while the value of the transfer income is fixed. The question of an adequate level of support today cannot directly be answered by reference to historical nisab. However, the economic significance of nisab in the early Islamic period should be considered carefully to see what standard of living zakat allowed in this period.

Public Expenditures and Goals of an Islamic State

The transfers to zakat beneficiaries are public expenditures in countries where the state has made the payment of zakat mandatory. Zakat expenditures are financed from compulsory levies, and the care for the poor and the needy is a recognized public duty. Besides this, the state has to finance other expenditures. Islamic economists have dealt at length with the question, what are necessary or legitimate or permissible tasks for which an Islamic state can rightly incur public expenditures and charge taxes. A typical Islamic theory of public expenditures starts with need fulfillment, reduction of inequality, security, and development as the main goals of an Islamic state. Subsidiary goals are employment, stability, and scientific progress (M.N. Siddiqi 1995, p. 149).

Mandatory Functions and Expenditures

Islamic economists can provide convincing evidence from the Quran, the Sunna, the early caliphs, and the Muslim jurists that need fulfillment, or the guarantee of a minimum standard of living, and internal and external security are main goals of an Islamic state. For "distribution" and "development," however, the foundations in the Quran and Sunna are considerably weaker. Goals that are anchored in the Quran and Sunna lead to permanent functions of an Islamic state and corresponding categories of expenditures are also permanent. They can be specified as expenditures for defense, law and order, justice, civil administration, need fulfillment (support of the poor), and the propagation of Islam.

Some authors (e.g., M.N. Siddiqi 1995) add regulations and guidance for the fulfillment of socially obligatory duties if the private sector fails. Regulations and guidance are "a wide category including market regulation, supervision of the common man's behavior in public places and enforcing Islamic practices in general" (M.N. Siddiqi 1995, p. 157). The fulfillment of socially obligatory duties by the state, if the private sector fails to honor this duty, is the essence of *fard kifaya*. "A fard kifaya is a duty...[that] is not addressed to individual persons but to the community. The Law-Giver wants the duty done, irrespective of who does it" (M.N. Siddiqi 1995, p. 155). The scope for fard kifaya is very wide. It ranges from prayers for the deceased to the establishment of essential industries. Only some socially obligatory duties were specified in the Quran and the Sunna while most others were later added by the Muslim jurists. This raises doubts whether all socially obligatory duties should be considered in the category of mandatory and permanent expenses. Functions mandated by the Quran and Sunna express the explicit will of Allah and lead to permanent categories of expenditures with highest priority ("mandatory expenditures").

Necessary Functions and Circumstantial Expenditures

A second category of public expenditures comprises those expenditures which experts deem necessary for realizing the objectives of Sharia under the present socioeconomic circumstances. This category comprises, in particular, expenditures for the protection of the environment, for public goods in addition to those included in mandatory expenditures, for scientific research, capital formation and infrastructure, for economic development, for subsidies for private activities, and for stabilization policies (M.N. Siddiqi 1995, p. 158). It is maintained that each of these categories of expenditures is related to a duty of the Islamic state derived from objectives of the Shar'ia (e.g., preservation of life, need fulfillment, and strengthening the Muslim community). In contrast to the mandatory functions that express the will of Allah, the necessary functions express the will of experts and lead to temporary categories of expenditure, which can vary in importance with changing circumstances ("circumstantial expenditures").

Assigned Functions and Expenditures

To extend the scope of legitimate expenditures of an Islamic state even further, Muslim economists introduce a third category of "assigned functions" that should be based on the will of the people (M.N. Siddiqi 1995, p. 152). The corresponding "assigned expenditures" are revocable if the people so decide.

In the light of the economic theory of politics, it is an idealistic belief that the people assign tasks to the state: The logic of collective action explains why one should not expect that "the people," as a very large group, will organize themselves and assign well-defined activities to the state. Those who "assign" tasks to the state are much smaller groups looking for exemptions from general rules, for privileges and support. The most efficiently organized groups in Western democracies are producers (capital owners and workers) in ailing industries who approach politicians for subsidies, protection against external competitors, and preferences in government procurement programs. Politicians (have to) take up such demand if the lobby groups can mobilize considerable support or opposition at the next election. The costs of a privileged treatment of producers are dispersed widely and borne mainly by consumers and taxpayers.

Usually the privileged treatment of well-organized lobby groups is justified by politicians with (vague) references to the general welfare, to employment, or to a mandate given by the voters. In fact, "the voters" never gave a specific mandate, and "the people" never assigned a specific task to the government. Any claimed mandate can, at best, be the mandate of a majority of voters or people. But depending on the voting system, political parties can win elections and form majorities in legislative assemblies with considerably less than 50 percent of the vote. Considering that not all people participate in the polls, the agreement ratio for the government and the dominant party (or coalition) can be much less than 50 percent of the population.

Further, even the absolute majority of votes is not a guarantee that specific policies are really legitimized by an "assignment" by the majority of the people. Political parties are not very definite in their election programs, and it is an exception and not the rule that the voters precisely know what specific policies will be implemented after the election. At best in these rare cases, the majority vote could be interpreted as the people's "assignment of functions" to the state. But even then reservations must be made: The voters are not asked for their opinion on individual items of a party program but only for the program in total. Thus, a party may win the consent of a majority of voters not because of, but in spite of, specific policies or "assigned functions" in its policy package.

Thus, if politicians claim that they implement a particular policy because they have been entrusted this task by the voters, this may well be a pretended argument to disguise policies which basically are to the advantage of some interest groups.

Another set of problems is caused by the fact that circumstantial and assigned functions can come into conflict with each other. For example, experts could argue that the state should not protect jobs in ailing industries that have lost their markets to foreign competitors. Instead, the state should use funds to accelerate the development of industries with good growth prospects and job potentials. But the people may give more importance to the conservation of existing jobs rather than to the creation of new ones. Therefore, they assign the function "conservation of traditional jobs" to the state while economists are opposed to this. In such a case, a political decision is necessary; the classification of public expenditures in mandatory, circumstantial, and assigned does not solve the problem.

Practical Significance of a Categorization of Public Expenditures

The value of a classification and categorization of public expenditures for the intellectual development of a coherent Islamic concept of public finance is not questioned here, but doubts arise concerning its practical significance. It would be significant if the Islamic concept calls for expenditures that a secular state would not commence by itself, or prevent expenditures that a secular state would have made. It seems, however, that practical differences in this sense are marginal.

Relevance of classifications. It is obvious that a secular state will not spend money on the propagation of Islam. But besides this, every category of expenditures listed above can be found in most contemporary secular states. Further, the classifications of the Islamic theory do not prescribe the instruments with which the goals of public expenditures shall be achieved. For example, need fulfillment could be achieved, in the short run, by direct public transfers to the poor or, in the long run, by their gainful employment in expanding industries. The latter requires private and public investment. However, resources spent for capital formation and development (classified as "circumstantial expenditures") cannot be used for direct transfers to the poor (classified as "mandatory expenditures"). Thus, a government that spends large amounts on development projects and subsidies for industries, but little on direct transfers, may argue that it is best serving the long-term need fulfillment although it allocates much less resources to mandatory than to circumstantial expenditures. In addition, if the government spends enough to secure the survival of all, although it could spend much more if it reduces defense or investment expenditures, it can point out that public funds have been allocated to all mandated categories of expenditures and that there is no rule determining the proportion in which the funds have to be allocated.

In sum, the practical significance of a categorization of public expenditures is very limited. Either a government already allocates public funds to mandatory and circumstantial functions—then the categorization is redundant—or, a government has neglected mandatory functions and is not really willing to change its pattern of expenditures—then it can find many excuses or justifications for its practice and the categorization becomes ineffective.

Limitations for public expenditures? It seems that Islamic economists are more concerned with the inclusion of categories of expenditure in, than with their exclusion from, the categories of permissible expenditures. Muslim economists claim that an Islamic economic system is superior to any secular system in terms of allocation, distribution, and stability. Everything that is not explicitly forbidden is allowed in Islam, so all except a few types of public expenditures (e.g., interest payments or subsidies for prohibited productions such as wine or pork) must be permissible. The superiority was deduced from the Islamic value orientation and from the replacement of interest based financing by profit and loss sharing techniques. However, the realities of Islamic states such as Pakistan and Iran, and Islamic banks in many Muslim countries, have revealed a multitude of reasons why the ideal of the Islamic economists could not be realized in practice. A convincing theory on how to achieve a (gradual or evolutionary and not enforced or revolutionary) transformation from the status quo to the Islamic ideal is still missing.

There is a remarkable contrast between the efforts of Muslim economists to show the modernity of an Islamic economic system and modern developments in the Western theory of public finance. First, many Muslim economists, with a preference for a Keynesian type of stabilization policy, tried to expound why a counter-cyclical fiscal policy would not contradict the concept of public finance in Islam. In contrast to this, Western mainstream economics turned away from macroeconomic fine tuning and favored a noninterventionist type of monetary and fiscal policy. Second, Muslim economists are, in fact, rather permissive with respect to public expenditures, while Western economist have become increasingly worried about escalating successes of rent-seeking groups, and the factually unlimited power of the state to raise taxes and to spend public money according to political priorities and preferences. These priorities and preferences are more and more determined by the interests of politicians who want to be (re-)elected and less and less by ideals or principles of a market economy.

Contrary to most Muslim economists, who still consider state interventions to be the solution for social and economic problems, many Western economists consider interventions and regulations to be the origin of the problems. State interventions, government regulations, and successful rent-seeking negatively affect allocation, growth, stability, and distribution. An increasing number of Western economists give priority not to the inclusion but to the exclusion or limitation of categories of expenditure from the state budget. Their concern is to limit, rather than further extend, the state's power to tax and spend. Besides the analysis of dysfunctionalities and injustices in rent-seeking societies, proposals for reforms of the fiscal constitution have been made. Here, Western economics is not less normative or prescriptive than Islamic economics, but it has a more profound analytical basis. Unfortunately, there is not much intellectual exchange and stimulation between Western and Islamic economists in this field.

CONCLUSIONS

Islamic economics is defined by an explicit value orientation and a particular methodology for legitimizing legal systems and policies. Not the technicalities but the underlying principles and the method of legitimization constitute the Islamic substance of institutional arrangements. Against this background, the evaluation of the literature on taxation and public finance exposes a surprisingly large number of methodological inaccuracies, free interpretations, and even errors. It seems that the serious academic discussion among Islamic economists and between them and Islamic jurists has not yet started.

For this discussion, Islamic economists should become familiar with more recent developments in "Western" economics. This could help the transformation of Islamic economics from a largely normative and prescriptive discipline into an empirical social science with the potential to explain social phenomena. Neither recitations of old legal texts nor classifications are economic theories; this does not mean that they are irrelevant for the development of a consistent Islamic worldview. Nevertheless, a more convincing case for an Islamic system of taxation and public finance could be made if the real behavior of people and politicians in the contemporary Muslim world, and possible transformation strategies toward an Islamic ideal, would become a major issue for research in Islamic economics. The economic theory of politics and constitutional economics applied to the particular conditions of developing Muslim countries, and combined with the value system of Islam, would pose a much stronger intellectual challenge to Western economics than classificatory or prescriptive contributions or designs of utopian societies of ideal Muslims.

REFERENCES

Abdullah, M.A. 1995. "Introduction." Pp. 17-33 in *Readings in Public Finance in Islam*, edited by M.A. Gulaid and M.A. Abdullah. Jeddah: Islamic Development Bank, Islamic Research and Training Institute.

Afzal-ur-Rahman. 1976. *Economic Doctrines of Islam*, Vol. III. Lahore: Islamic Publications.

Ahmad, K., ed. 1980. *Studies in Islamic Economics*. Jeddah and Leicester: International Center for Research in Islamic Economics and The Islamic Foundation.

Ahmed, Z., M. Iqbal, and M.F. Khan, eds. 1983. *Fiscal Policy and Resource Allocation in Islam*. Islamabad: Institute of Policy Studies.

Al-Omar, F.A. 1995. "General, Administrative and Organizational Aspects." Pp. 21-64 in *Institutional Framework of Zakah: Dimensions and Implications* (Proceedings of the Third *Zakah* Conference, Kula Lumpur 1990), edited by A.A. El-Ashker and M.S. Haq. Jeddah: Islamic Development Bank, Islamic Research and Training Institute.

Ali, A.Y. 1983/1934. *The Holy Qur'an–Text, Translation and Commentary*. Brentwood, CA: Amana Corp.

Behdad, S. 1993. "Appendix: Fundamentals of Islamic Jurisprudence." Pp. 188-190 in *Islamic Economic Alternatives–Critical Perspectives and New Directions*, edited by K.S. Jomo. Kuala Lumpur: Ikraq.

Chapra, M.U. 1992. *Islam and the Economic Challenge*. Leicester: The Islamic Foundation.

Chaudhry, M.S. 1992. *Taxation in Islam and Modern Taxes*. Lahore: Impact Publications International.

Duri, A.A. 1985. "Taxation in Early Islam." *Journal of Islamic Banking and Finance* 2(April-June): 61-70.

El-Ashker, A.A. and M.S. Haq, eds. 1995. *Institutional Framework of Zakah: Dimensions and Implications* (Proceedings of the Third Zakah Conference, Kula Lumpur 1990). Jeddah: Islamic Development Bank, Islamic Research and Training Institute.

El-Badawi, M.A. and S.M. Al-Sultan. 1992. "Net Working Capital versus Net Owner's Equity Approaches to Computing Zakatable Amount: A Conceptual Comparison and Application." *The American Journal of Islamic Social Sciences* 9: 69-85.

El-Din, S.T. 1995. "Allocative and Stabilizing Functions of Zakat in an Islamic Economy." Pp. 327-355 in *Readings in Public Finance in Islam*, edited by M.A. Gulaid and M.A. Abdullah. Jeddah: Islamic Development Bank, Islamic Research and Training Institute.

Faridi, F.R. 1980. "Zakat and Fiscal Policy." Pp. 119-130 in *Studies in Islamic Economics*, edited by K. Ahmad. Jeddah and Leicester: International Center for Research in Islamic Economics and The Islamic Foundation.

———. 1995. "A Theory of Fiscal Policy in an Islamic State." Pp. 129-148 in *Readings in Public Finance in Islam*, edited by M.A. Gulaid and M.A. Abdullah. Jeddah: Islamic Development Bank, Islamic Research and Training Institute.

Gulaid, M.A. and M.A. Abdullah, eds. 1995. *Readings in Public Finance in Islam*. Jeddah: Islamic Development Bank, Islamic Research and Training Institute.

Jomo, K.S., ed. 1993. *Islamic Economic Alternatives–Critical Perspectives and New Directions*. Kuala Lumpur: Ikraq.

Kahf, M. 1995a. "The Islamic State and the Welfare State: Similarities and Differences." Pp. 57-70 in *Readings in Public Finance in Islam*, edited by M.A. Gulaid and M.A. Abdullah. Jeddah: Islamic Development Bank, Islamic Research and Training Institute.

———. 1995b. "Taxation Policy in an Islamic Economy." Pp. 105-127 in *Readings in Public Finance in Islam*, edited by M.A. Gulaid and M.A. Abdullah. Jeddah: Islamic Development Bank, Islamic Research and Training Institute.

———. 1995c. "Applied Institutional Models for Zakah Collection and Distribution in Islamic Countries and Communities." Pp. 197-228 in *Institutional Framework of Zakah: Dimensions and Implications* (Proceedings of the Third *Zakah* Conference, Kuala Lumpur, 1990), edited by A.A. El-Ashker and M.S. Haq. Jeddah: Islamic Development Bank, Islamic Research and Training Institute.

Kuran, T. 1992. "The Economic System in Contemporary Islamic Thought." Pp. 9-47 in *Islamic Economic Alternatives–Critical Perspectives and New Directions*, edited by K.S. Jomo. Kuala Lumpur: Ikraq.

Mannan, M.A. 1986. *Islamic Economics: Theory and Practice*, rev. ed. Sevenoaks: Hodder and Stoughton.

Metwally, M.M. 1995. "Fiscal Policy in an Islamic Economy." Pp. 297-326 in *Readings in Public Finance in Islam*, edited by M.A. Gulaid and M.A. Abdullah. Jeddah: Islamic Development Bank, Islamic Research and Training Institute.

Muhammad, A.A. 1993. *Zakat and Rural Development in Malaysia*. Kuala Lumpur: Berita Publishing.

Nomani, F. and A. Rahnema. 1994. *Islamic Economic Systems*. London: Zed Books.

Oran, A. and S. Rashid. 1989: "Fiscal Policy in Early Islam." *Public Finance* 44(1): 75-101.

Qureshi, A.I. 1978. *Fiscal System of Islam*. Lahore: Institute of Islamic Culture.

Ra'ana, I.M. 1970. *Economic System under 'Umar the Great*. Lahore: Sh. Muhammad Ashraf.

Salama, A.A. 1995a. "Fiscal Policy in an Islamic State." Pp. 37-55 in *Readings in Public Finance in Islam*, edited by M.A. Gulaid and M.A. Abdullah. Jeddah: Islamic Development Bank, Islamic Research and Training Institute.

———. 1995b. "Empirical Economic Effects of Obligatory and Non-Obligatory Payment of Zakah to the State." Pp. 151-161 in *Institutional Framework of Zakah: Dimensions and Implications* (Proceedings of the Third Zakah Conference, Kula Lumpur 1990), edited by A.A. El-Ashker and M.S. Haq. Jeddah: Islamic Development Bank, Islamic Research and Training Institute.

Shaik, A.A. 1980. "Concept of Zakah: A Survey of Qur'anic Texts and their Explanations in Shariah and Contemporary Economics." Pp. 2-65 in *Some*

Aspects of the Economics of Zakah, edited by M.R. Zaman. Plainfield, NJ: The Association of Muslim Social Scientists.

Siddiqi, M.N. 1980. "Muslim Economic Thinking—A Survey of Contemporary Literature." Pp. 191-351 *Studies in Islamic Economics*, in edited by K. Ahmad. Jeddah and Leicester: International Center for Research in Islamic Economics and The Islamic Foundation.

________. 1995. "Public Expenditure in an Islamic State." Pp. 149-179 in *Readings in Public Finance in Islam*, edited by M.A. Gulaid and M.A. Abdullah. Jeddah: Islamic Development Bank, Islamic Research and Training Institute.

Siddiqi, M.A.S. 1983. *Early Development of Zakat Law and Ijtihad*. Karachi: Islamic Research Academy.

Zaman, H.S.M. 1991. *Economic Functions of an Islamic State–The Early Experience*. Leicester: The Islamic Foundation.

Zarqa, M.A. 1992. "Methodology of Islamic Economics." Pp. 49-58 in *Lectures on Islamic Economics* (Papers and Proceedings of a Seminar on "Teaching Islamic Economics for University Teachers," Islamabad 1987), edited by A. Ahmad and K.R. Awan. Jeddah: Islamic Development Bank, Islamic Research and Training Institute.

THE PROBLEM OF INTEREST AND BANKING IN ISLAMIC PUBLIC POLICY

Farhad Nomani

INTRODUCTION

The elimination of *riba* (literally, an excess or increase) and *gharar* (literally, hazard) from financial transactions, and the establishment of interest-free Islamic banking based on legally recognized Islamic contracts, are two important and controversial issues of contemporary Islamic public policy. There is a consensus among Islamists that riba and gharar are prohibited by the Shari'a, and that banking should be in conformity with it. However, there has always been a debate over the meaning of these concepts and the acceptable method of Islamic banking. In addition, governments in Muslim countries of the twentieth century have had different positions on the problem of riba and gharar all along, and some have pursued different types of Islamic banking experiments since the 1970s.

The objective of this paper is to show the sources of the difficulty in drawing a unanimous Islamic public policy for the elimination of riba and gharar and the establishment of an Islamic banking system. The first section of this paper summarizes the position of classical Shari'a sources on riba, gharar, and Islamic

International Review of Comparative Public Policy, Volume 9, pages 277-310.

ISBN: 0-7623-0268-2

partnership contracts in the Quran, the Tradition of the Prophet (Sunna) and the juristic views of major schools of Islamic law or jurisprudence (Hanafi, Shafi'i, Maliki, Hanbali, and the Ja'fari Shi'i). The second and third sections review the major debates among Muslim scholars and examine the positions that the governments of Egypt, Iran, and Pakistan have taken for the elimination of riba, and partial or "thorough" Islamization of the banking system. This paper demonstrates that difference of opinion on public policy concerning the problem of interest-free Islamic banking exists not only between policymakers and Islamic scholars in different countries, or belonging to different schools of thought, but also among those in the same country or school of thought.

CLASSICAL PROHIBITIONS AND CONTRACTS IN THE SHARI'A

Riba

The technical meaning of riba is a controversial issue, especially since the development of modern banking. For some Muslims riba signifies only excessive interest rates, for others it refers to any interest or other prohibited gains.

Riba was a common practice in the Arabian Peninsula at the advent of Islam. It mainly took the form of barter exchange. It was either in the form of an extra amount in the quantity of an article or a delay in its delivery. Thus, the exchange of articles of the same, and sometimes different, kinds gave rise to an unequal exchange. Also, the failure to fulfill a sale contract at the time of maturity led to the continuous multiplication of the obligation in kind. It happens that the Quran explicitly prohibits the latter practice, which in this paper will be called the Quranic riba, or simply riba (see XXX:39, IV:161, III:130-132, II:275-280).[1]

The Traditions of the Prophet are also explicit about the injunction against riba. Yet the reports of the sayings of the Prophet (*hadith*) are not unanimous about what constitutes riba and what forms of contracts come under the Quranic prohibition. One of the most widely accepted sayings of Muhammad about riba, reported in most compilations of the hadith, asserts that riba exists when six articles of the same kind are either bartered unequally or not delivered immediately. These articles were identified as gold, silver, dates, salt, barley and wheat (Qadri 1986, p. 332; Motahhari 1991, pp. 220-234).

According to the Quran and the Tradition of the Prophet, the ultimate punishment for the practice of riba among Muslims would be the loss of God's blessing in this world and the torture of hell in the next. However, both sources leave entirely open the question of whether Muslims may or may not practice usury with people of another religion.

All schools of Islamic law, condemn riba and prohibit transactions that are susceptible to riba. Classical jurists were interested in the concept of riba within the framework of the Islamic law of contract. They categorized various type of contracts as sales and asserted that the Quran and the Sunna had permitted *bay'* (selling or trading) and prohibited riba. Thus, based on the opposition of bay' and riba, these jurists discussed the legality and illegality of actual and hypothetical commercial practices. In this endeavor, early jurists classified riba into two categories: *riba al-fazl* (riba of increase or riba by way of excess) and *riba al-nasia* (riba of delayed payment or riba by way of deferment). Riba al-fazl is the unlawful excess of the countervalues in a "hand-to-hand" transaction and all jurists prohibit it because it does not realize the balance of the exchanged countervalues. However, early jurists differ from each other in detail concerning the conditions of sale and the articles the exchange of which were susceptible to riba.

Riba al-nasia arises when the exchange of countervalues is delayed, since a deferred exchange could produce a gain in one of the countervalues. Nevertheless, such an exchange was unlawful whether or not it gave rise to a gain (Saleh 1992, pp. 24-33; Algabid 1990, pp. 43-45; Motahhari 1991, pp. 204-205). The notion of the interdiction of deferred delivery is closely related to the concept of gharar (hazard), and for this reason the idea in riba al-nasia is to effectuate the immediate transfer of ownership at the time the contract is drawn up.

Thus, the early dominant trend among Hanafi, Maliki, Shafi'i, Hanbali, and Ja'fari Shi'i jurisprudence considered riba as a prohibition affecting sales. This view ignored the discussion of loans based on riba in which an advantage or an increase was stipulated for the lender, mainly because it could not find an authoritative justification for such a prohibition except in one narration of the sayings of Muhammad. Only later, some jurists explicitly categorized riba differently and recognized riba in both sales and debts. In fact, Muslim scholars of modern times have followed the latter classification (Homoud 1985, pp. 54-55). Early jurists recognized a loan mainly as a gratuitous transaction which excludes riba. Consequently, these jurists accepted the validity of giving a premium or an "increase" to a lender if such a premium was not a requisite of a loan.

Moreover, some jurists, mostly Hanafis, Shafi'is, and Shi'is, allowed legal tricks that would circumvent the prohibition against riba, provided such tricks and conditions were possible, reasonable and lawful (Homoud 1985, pp. 100-110; Saleh 1992, pp. 44-61). There are several devices for the evasion of the prohibition of riba. One of these tricks is the use of a pledge (*rahn*), such as the use of real estate to secure the repayment of a debt. Such a device enabled the lender to derive some advantage from the pledge during the period of the loan. Riba in debts could be legalized by different kinds of double sale contracts. One of these double sales enables a prospective debtor to sell to a creditor something for cash, and buy the article back from the lender for a greater price payable at a future date. To avoid the prohibition of riba by the sale of one money for another money on credit, the legal trick would be that one buys an article from a seller and pays in one money; then the seller shall purchase back the article on credit (Qadri 1986, pp. 335-337; Schacht 1964, pp. 79, 138; Motahhari 1991, pp. 265-269; Doi 1984, pp. 334-345). Such multiple transactions, each of which valid by itself, together can realize the objective of the contractors. In fact, many practical contracts that facilitate Islamic banking in modern economies, such as resale with a stated profit (*murabaha*) or demanding a pledge to secure the repayment of a loan, as will be seen, are all legal devices. Some of the methods of legal reasoning with the help of which Islamic jurists, and in the modern world, Islamist states, have resorted to legal tricks are: *maslaha*, istihsan and *istihbab* in order to take into account the interest of the public and the preference of the jurist (See the article by Rahnema 1997, this volume; Coulson 1990, p. 100; Kamali 1991, pp. 246-309).

Gharar

Islamic law considers all sale or hire contracts that involve speculative risk (gharar) invalid. To avoid gharar in a sale contract, the following must be known: the object and its quantities; the price or the countervalue; the time of delivery and sale condition of payment (Doi 1984, p. 399). The requirement that the object and its quantity in a contract must be known is very sensitive for articles which can be measured or weighed in order to avoid the prohibition of riba (Schacht 1964, p. 147).

Thus, classical jurists approve of future sale contracts only when such contracts are "free" from riba or gharar. For example, sale of most foodstuff, but not of gold or silver (as money), for future delivery with prepayment in cash (*salam* or *salaf*), delayed payment for goods delivered at once (*nasia*), and hire and lease (*ijara*) of services of human beings, lands or buildings are authorized. In general, such sales are valid if they are sales of a known benefit in return for its known equivalent (Qadri 1986, pp. 325-326; Saleh 1992, pp. 62-105). In other words, most Islamic schools of law do not permit contracts with an unspecified price that will be determined in the future by the market or by another person. However, some jurists have allowed a "small" element of risk (gharar) and uncertainty if a greater part of the contracted goods is delivered. Contracts that are not free from the element of riba and gharar, but are allowed by many jurists, are agricultural sharecropping (*muzara'a*), sharecropping in the case of orchards and gardening (*musaqat*), fixed land rent, and fixed wages in return for unspecified work or unrelated to profit and loss of the enterprise. The contract is valid in the first two cases simply because the Prophet explicitly allowed them and himself had entered into such a contract. The last case is, however, a matter of controversy in Islamic law (Afzal-ur Rahman 1982, pp. 143-153, 166-167). In contrast to the views of the classical jurists of the Hanbali, Shafi'i, and Hanafi schools, some jurists, like Ibn Taimiyah, approve of transactions based on gharar and maisir (gambling or speculation) if their prohibitions lead to hardship for people (Islahi 1988, pp. 166-168).

Muzaraba and Musharaka

Laws that are based on ethical controls of transactions, including Islamic law, do not recognize the liberty of contracts in general, and contracts of association in particular. As regards our concern for the modern operational framework of Islamic banking, the relevant and most recognized forms of associations are *musharaka* (or *shirka*, partnership), especially the limited type, and *muzaraba* (or *qiraz*, dormant or silent partnership). These are transactions through partnership rather than through exchange of goods (see Schacht 1964, pp. 155-157; Siddiqi 1987, pp. 11-18; R.S. Khan 1987, pp. 48-49; Islahi 1988, pp. 155-156).

The Quran does not refer to muzaraba, although the contract was used before the advent of Islam. However, the Tradition is explicit about the approval of the Prophet and his companions in such a contract (Udovitch 1986, p. 132). For this reason, all Islamic schools of law permit the use of muzaraba and some other forms of partnership.

Inasmuch as muzaraba combines the advantages of partnership and loan contracts, it can be used as a device for avoiding the prohibition of riba. In fact, jurists identify muzaraba only as a special form of partnership. In general, partnership involves participation of at least two persons with defined amount of capital for jointly running a business and for sharing profit and loss in predetermined proportions. Muzaraba is not a partnership proper since it consists of a fiduciary relationship and a procuration. It is similar to a partnership because the profit is to be shared between partners, although only the dormant partner or the owner of the capital, and not the agent (or active partner), bears the loss; the agent loses only his time and effort after the deduction of all expenses. Like a loan contract, the investor in a muzaraba agreement is not directly associated with the agent who is responsible for the management of the capital, but is entitled to a certain proportion of profit of the enterprise. In practice, this predetermined share of profit could correspond to interest. It is for this reason that classical jurists have been careful defining the nature and terms of the legality of the contract. In general, the jurists who allowed the use of analogy and ignored the importance of the motive in a contract, were more lax and flexible about the requirements of muzaraba and musharaka contracts. Of course, whether one is bound by the verdict of this or that school or jurist is a matter of continuous dispute and conflict between Muslim economists. This conflict, however, is acute in the enactment of a law and the legislation of rules and regulations of Islamic financial contracts, as will be shown.

In any case, as far as modern Islamic banking is concerned, the most relevant disagreements between the classical jurists are over the following points: the two-tier muzaraba; the kind of business permitted in muzaraba; the form of capital involved (i.e., cash or in kind); the distribution of profit and financial liabilities of partners; the duration of muzaraba and musharaka contracts (Siddiqi 1987, pp. 58-64, 67, 86-90).

In the case of two-tier muzaraba (or muzaraba under muzaraba), which is the only form of muzaraba that would suit modern Islamic banking, the active partner (i.e., the bank) receives capital from a dormant partner (i.e., the bank's client), and, in turn, engages in another muzaraba contract with another active partner (i.e., an entrepreneur). Related to this case is the use of muzaraba capital for a musharaka contract with another active partner.

The use of these two contracts in industry, services, and agriculture is vital for modern Islamic banking. However, the Maliki, Shafi'i, and classical Shi'i jurists restrict the use of these contracts only to short-run trade. The reasoning is that only in commerce the result of the activity is unpredictable or unknown, and hence riba is absent (Siddiqi 1987, pp. 99-101; Abu Saud 1981, p. 70).

In short, it is clear that the application of the rulings of the classical Islamic schools of law to modern socioeconomic conditions, including modern civil and commercial laws, in general, and financial markets, including Islamic banking, in particular, is not easy. This is due to the historical specificity of the Law and the fact that the definition and scope of contracts and injunctions are not always clear and/or unanimous. In developed market systems, barter sales, in which currency is not one of the countervalues (riba al-fazl), and barter sales, by way of deferment (riba al-nasia), are rare and hardly relevant. Financial markets and their products such as stocks, bonds, and insurance commodities in which the object of transaction and the price or both are not determined and fixed in advance, could be in the jurists' list of various cases of gharar. Besides, the confinement of muzaraba to short-run trade and the application of classical requisites for similar partnerships are not reasonable and practical at present. Nevertheless, the Shari'a, interpreted differently in different regions and even within the same region, has always been part of the social consciousness of Muslim communities and, in the modern times, Muslim countries. Therefore, historically the contradiction between theory and practice has given rise to Islamic legal devices, Western-inspired rules and devices, legal innovations, and new interpretations; all of which have incited religious and popular criticism and disapproval in the twentieth century.

MODERN INTERPRETATIONS OF RIBA

The Limited and Broader Definition of Riba in Muslim Countries

The debate on the interpretation of riba among Islamist intellectuals and ulema was intensified in the early twentieth century at the advent of modern banking in Muslim countries by Western banks. These banks, which mainly served the governments and the trade with the European countries, were eventually replaced by state or privately owned national banks and savings institutions. All of these banks paid interest on savings accounts and charged interest on their loans. This practice led to dissatisfaction among ulema and Islamist intellectuals. However, it is interesting to note that anti-riba slogans have never mobilized Muslim masses, despite fierce rhetoric of radical ulema and Islamist intellectuals in condemnation of riba. From the beginning, two major views have been shaped concerning the legal meaning of riba. One has argued for a limited definition, and the other has insisted on a broader meaning of riba. The former view defines riba as usury, and, therefore, asks only for the prohibition of excessive interest. Proponents of this view argue that riba leads to exploitation of the poor by means of usurious consumption loans. Despite differences in detail, the views of advocates of a limited definition of riba have influenced and shaped public-civil laws, codes, and policies of almost all Muslim countries in the twentieth century, especially until the 1980s. Thus, the Iranian civil code that was adopted in 1927-32, and the civil codes of the majority of Arab countries, adopted in 1930-60, all recognize riba implicitly or explicitly as usury or excessive interest. All the other Muslim countries that became independent during the first two decades after World War II, have also opted for a tacit recognition of "reasonable" interest as lawful.

The civil codes of all Arab countries, especially those adopted with the help of the Egyptian lawyer Abd al-Razzaq Sanhuri, except Lebanon, forbid compound interest on the ground that this rate is close to the Quranic riba. Many of these countries, including Egypt, determine maximum contractual rates that are normally lower for civil, compared to commercial, matters. Banks are governed by separate laws allowing them to charge higher rates. Besides, commercial laws permit interest not only in loan contracts, but also of pledges in addition to the expenses relating to both contracts, the cost of enforcement and compensation for losses resulting from defects in the item pledged. However, in some of the Persian Gulf states, such as Kuwait and Bahrain, interest is authorized in commercial codes but forbidden in civil codes. Thus, such countries officially restrict interest payment to commercial debts. In Libya, where the state banks pay interest on the savings accounts, interest on a loan is allowed only between legal entities. Algerian law is silent on the issue of interest, but interest is charged and paid and the central bank sets the rate of the banks (Financial Institutions 1990, pp. 1-250; El Ahdab 1992, pp. 55-

56; Davies 1984, pp. 199, 220-221; Hill 1989, pp. 159-163; Ballantyne 1990, pp. 153-157). The Saudi Arabian "Commercial Regulations" avoids the use of the word "interest" and replaces it with "profit" and "commission" for all commercial transaction including loans. The banking (or monetary agency) rules of Saudi Arabia explicitly bars the central bank from charging interest. However, alternative solutions exist, for example, by paying "commissions" on savings accounts (Kay 1979, pp. 34-36, 74-80). The banking laws of Iran and Pakistan prohibit riba, defined as all forms of interest, but many legal tricks soften the prohibition (Nomani and Rahnema 1994, pp. 129-136, 177-182). In Southeast Asian Muslim countries, the law is secular and there is no substantive law against interest. However, some of these countries, for example Malaysia, have introduced legislation for Islamic banking that are regulatory and operate along interest-based banking system (Connors 1986, pp. 57-66).

Since the 1970s, the supporters of the limited definition of riba have been on the defensive and their views were eclipsed by the rise of the Islamic resurgence movement. This movement is associated with the extended view of riba, espoused either by pure traditionalist or modern radical Muslims.

The traditionalist and modern advocates of a broader definition of riba assert that any interest is riba, and, therefore, forbidden by the Shari'a. The traditionalists, including the majority of Sunni and Shi'i ulema, accept legal tricks from a juristic point of view, while, modern Islamists deny the legitimacy of such devices on ethical grounds. Still, however, a minority of ulema and Islamist intellectuals define riba as any monopolistic profit, speculation, and hoarding, and reject the use of legal tricks. There is also a very small group of Islamist intellectuals who reject capitalism because of its exploitative characteristic that is explained by the concept of riba.

Since the 1970s, the supporters of the extended definition of riba, especially the modern faction, have had an effective mobilizing influence in many Muslim countries, effectuating minor or drastic changes in the codes related to riba and banking rules and institutions. The relationship of the majority of ulema with this trend has been influenced by the strength and the radicalism of the Islamist movement. In general, the enthusiasm of the majority of ulema for a complete Islamization of the commercial laws and the banking sector, in glaring contrast to their insistence on Islamic personal and criminal rules, has been lukewarm. It seems that unlike Islamist intellectuals, the majority of ulema are satisfied with a formalistic introduction of Islamic banking nomenclature, and, therefore, are not enthusiastic about radical demands for a strict interpretation of transactions that are capable of riba.

The Debate on Riba in Egypt

The problem of riba since the latter part of the 1970s has been part of a broader problem of the application of the Shari'a in Egypt. With the amendment in 1980 of the 1971 Constitution, the Shari'a is the major source of legislation in Egypt. Therefore, it is mandatory that all laws, including the banking laws of Egypt, be in conformity with Islamic principles and rules. To the present, no major banking legislation has been enacted, partly because the public, the government, and ulema are divided on the issue. On riba, however, the issue has a long history in Egypt. It began in the first years of the twentieth century. In those years, many of the owners of the savings accounts of the newly established Postal Savings Fund refused to receive interest from the Fund. The new grand mufti of Egypt (chief jurisprudent), Muhammad Abduh, one of the most influential leaders of Shari'a reform movement and modernist approach to Islamic law, was put in a delicate position when he was asked to present his legal opinion on the matter (Homoud 1985, p. 122). He was confronted with a choice between his deep loyalty to the authority of Islam and his desire to modernize Islamic teaching (Kerr 1966; Haddad 1994, pp. 30-63; Cragg 1995, pp. 11-12). Abduh identified the interest paid by the Postal Fund as riba, and argued against the use of money, not as a means of exchange, but as a means to unfair increase of wealth by riba. He also asserted that the Postal Fund was not borrowing money from the public (depositors) out of necessity, and, therefore, the prohibition of the payment of interest could not be lifted. However, in order to facilitate economic activities, Abduh proposed the replacement of the savings accounts by muzaraba contracts in order to avoid riba. Even though the role of muzaraba in banking practices was not further elaborated by Abduh and the Fund's operation continued as before, his reasoning, and the idea of the use of muzaraba, have inspired all kinds of Muslims who favored a broader definition of riba. At the same time, Abduh's emphasis on the division of Islamic prohibitions into absolute and "adjustable" or less rigid injunctions that could be lifted under certain conditions due to "necessity," was utilized by later modernists who advocated a limited definition of riba. Besides, it is important to note that the views of Muslim scholars concerning Abduh's position on riba are not unanimous. Some, like Rashid Reza who narrates the view of Abduh, and Saleh, insist that Abduh considered only the pre-Islamic riba, and, therefore, compound interest as the absolutely forbidden riba (Saleh 1992, pp. 35-36). Others, such as Homoud and Mallat, find this view as an uncritical acceptance of Rashid Reza's interpretation of Abduh's position and assert that the Grand Mufti identified the interest received by the Postal Fund depositors as riba (Homoud 1985, pp. 115-119, 122-126; Mallat 1986, pp. 72-74).

The second important debate over the meaning of riba took place around the draft of the new civil code of Egypt in the 1930s and 1940s. The civil code, which was drafted by a commission led by Sanhuri, authorized a predetermined and fixed return on money. Precisely, the debate focused on articles of the code related to the payment of interest in case of delay of payment and the ceiling on interest rate beyond which the rate was considered as usury or riba.

These articles aroused a politically mitigated, yet analytically forceful, opposition among the traditional jurists. After all, in the introductory articles of the code, Shari'a was considered as a major source of the law, and its interpretation was the responsibility of the muftis (Ziadeh 1968, pp. 141-147; Hill 1989, pp. 163-179). Thus, Abu Zahra, a leading jurist of Egypt in that period, questioned the validity of the articles by the following argument. He stated that the reliance of the advocates of a limited definition of riba on the principles of necessity and public purpose or interest (maslaha) as a justification for the payment of interest is wrong. He argued that these two principles cannot be used in the case of riba because "necessity" applies only at the level of individuals, and in the case of loans with interest, it is relevant for a single borrower and not lenders of money. In addition, in his view, since the Islamic principle of necessity is applicable to individual cases and not to the society in general, the principle of the public interest cannot be invoked in defense of interest rates for the whole society (Kamali 1991, pp. 273-280). According to Abu Zahra, it is true that the act of borrowing for riba is not absolutely forbidden as a "first-degree sin." However, lending for riba is an absolute prohibition and is forbidden per se. In his opinion, this is because the Quranic riba is riba al-nasia, which is like a modern loan. Therefore, all kinds of interest are riba and forbidden. To avoid the prohibition of riba, muzaraba can be used in the banking system (Homoud 1985, pp. 62, 121-122; Mallat 1986, pp. 70-77).

In reaction to this argument, Ibrahim Zaki al-Badawi, who was working on the draft of the code in relation to Islamic law, presented the following rebuttal. He asserted that riba al-nasia, based on a continuous multiplication of the principal, is the only absolutely prohibited Quranic riba. However, he insisted that riba al-fazl for six articles has been forbidden by the Prophet in order to prevent the circumvention of the Quranic prohibition of riba. Echoing the views of Ibn Taimiyah and Ibn Qayyim on the degrees of prohibition in Islam, Badawi states that the prohibition of riba al-fazl is lighter. This riba is allowed in case of need (*haja*), while the Quranic riba can be permitted only in cases of absolute necessity. Besides, a loan based on interest for production purposes that eventually creates benefit, is valid. According to Badawi, this is because unlike the "excess" or increase in the Quranic riba which periodically grows at the time of maturity, the principal of the loan increases with interest at the moment of contracting based on mutual consent (and not compulsion). Most financial transactions and instruments (e.g., bonds, debts, stocks, and insurance) are beneficial and facilitate economic activities of Muslims, and, therefore, should be considered as necessity in Islamic public policy and codification. In Islam nothing is forbidden unless explicitly prohibited in the Shari'a. Thus, Badawi concludes that by allowing simple interest rates and forbidding compound interest rates, the proposed civil code of Egypt is in conformity with the Shari'a (Mallat 1986, pp. 79-81).

Badawi's argument was elaborated by Sanhuri in the 1950s. Sanhuri's contribution to the debate, in addition to what he approves of in Badawi's argument, is the following. He argues that riba has been forbidden in Islam for public policy considerations (i.e., preventing hoarding of necessities, speculation on currency, and unfair sale transactions). However, he asserts that there is a difference between the riba al-fazl and riba al-nasia of the Tradition on the one hand, and the Quranic riba; the latter is forbidden per se, while the first two are only forbidden to close the roads that might lead to pre-Islamic riba. As a result of such degrees of prohibition, the ban on the Quranic riba can be lifted only in case of pressing necessity, while the ribas of the Tradition can be permitted in case of need (haja) for public policy purposes. Therefore, Sanhuri believes that jurists like Ibn Qayyim erred by not differentiating between the Quranic riba and riba al-nasia. The Quranic riba was based on compound interest and assigned equal importance to the original capital and the interest received. Besides, modern loans are not the same as the aforementioned prohibited ribas, and could not lead to riba. Modern loans with interest, not being the aim of the ban on riba, cannot be forbidden per se, and, therefore, the principle of "need" can be invoked in their favor, enabling the law to lift the ban. In a market system, the need for loans is natural and common. Therefore, as long as loans based on simple interest are not changed into loans based on compound interest, they would be acceptable and are in conformity with the Shari'a (Saleh 1992, pp. 36-37; Homoud 1985, pp. 58-63, 120-122; El Ahdab 1992, pp. 48-49, 51-53; Mallat 1986, pp. 83-85).

Ironically, a rebuttal to Sanhuri's analysis was presented by Badawi, who, in 1964, revised and denounced his former position on simple interest. In his self-criticism, Badawi took side with the traditional ulema and claimed that any kind of interest is riba. His criticism of Sanhuri's argument is basically methodological. Badawi implicitly opposes not only Sanhuri's eclectic method, but also that of Abduh's. According to Badawi, Abduh and Sanhuri's reformist methodology calls for returning to the primary sources, especially the Quran. This methodology emphasizes the time-specificity of the rulings and prohibitions of Islam, and divides them into absolute and more flexible rules, and, therefore, unadjustable and adjustable. Badawi insists that Abduh and Sanhuri introduce the principle of necessity and public interest in order to justify the lifting of some of the prohibitions and orthodox rulings. This approach is eclectic, in Badawi's view, because it arbitrarily disregards the rulings of the classical jurists. The method is based on rationalism and innovates legal solutions or reinterprets the old ones by using a comparative evaluation and selection of rules of the doctrines of four Sunni schools of law. In this effort, in Badawi's opinion, Abduh and Sanhuri are particularly inspired by Ibn Taimiyah, Ibn Qayyim, Wahhab, and one of the companions of the Prophet, Ibn Abbas. What Badawi insinuates is that Abduh and Sanhuri rely on the principle of necessity and public interest of Ibn Taimiyah and Ibn

Qayyim and their independent and interpretive spirit (Islahi 1988, pp. 57-74; Mallat 1986. pp. 81-82); that they are reiterating the call of Wahhab for the return to the Quran and the Tradition of the Prophet (Islahi 1988, pp. 57-74); and, that they are influenced by Ibn Abbas' emphasis on the importance of knowing the circumstances in which Quranic prohibitions were revealed in order to avoid misunderstanding and his erroneous interpretation of the Quranic riba (Kamali 1991, pp. 40, 43).

Since 1970, several attempts to change the Egyptian Civil Code in relation to interest charges have not been successful. One such attempt was made in 1985. Based on the fact that the amended constitution of 1980 identifies the Shari'a as "the" principal source of legislation, the Rector of the al-Azhar brought a suit before the High Constitutional Court, demanding the prohibition of the payment of interest on delay of loan repayments. The Court dismissed the case, arguing that the Shari'a rulings cannot be applicable to the already existing legislation retroactively without creating confusion and instability for the commercial and judicial process (Edge 1986, p. 45).

The next round of the debate on riba was ignited by the decision of the government to issue three types of government "investment certificates" to be sold to the public by banks in 1988. The first was a ten-year investment certificate with a higher value at the date of maturity than the purchase price. The second type of certificate had a predetermined return, while the holders of the third type of certificates could receive a prize (bonus).

In December 1988, the Grand Mufti of Egypt identified the third one as licit, but insisted that he and other ulema find the first and second types capable of riba due to their explicit predetermined positive rate of return. Almost a year later, the same mufti presented a different opinion on the same issue, maybe after some discussions with government economists and ulema. In the new fatwa, it is asserted that all parties to a banking loan contract may benefit from the transaction (something that Sanhuri had insisted on in the 1950s). This fatwa implied that simple interest is in conformity with the Shari'a. In the same fatwa, interest-bearing savings accounts and government "investment" or "development" bonds (certificates) were considered lawful and not usurious (Galloux 1993, pp. 56-57; *The Economist*, September 16, 1989, p. 42). Immediately, the mufti was criticized by the advocates of a broader definition of riba who were not satisfied with the government's treatment of the financial problem of Islamic investment companies in 1988, claiming that the mufti is legalizing what God has prohibited. However, the Central Bank of Egypt, encouraged by the fatwa, allowed banks to raise interest rates to encourage savings in the local currency and attract funds to the banking system. The central bank's concern was due to the fact that private Islamic investment companies were paying a much higher rate of profit to depositors than banks before the change of their status into shareholding companies. The central bank's argument for this change was that Islamic companies were giving savers only a receipt in exchange for deposits and not share certificates. Apparently, this practice was not in conformity with the Shari'a, as interpreted by legal advisors of the central bank (Galloux 1993, pp. 57-61; Najjar 1992, pp. 65-68).

The Debate on Riba in Pakistan

Before the 1980s, the problem of riba in Pakistan was overshadowed by a more crucial constitutional debate over the role of the Shari'a as the source of public and private laws. When in 1956 the first constitution was approved, the state was to be the Islamic Republic of Pakistan and no law was to be promulgated contrary to the Quran and the Tradition of the Prophet. Naturally, there were also clauses in the constitution for the removal of riba and the establishment of Islamic tax, zakat. However, Pakistan was not transformed into an "Islamic state" in the strict sense of the term, and, in fact, in 1962, there was even an unsuccessful attempt to drop the word "Islamic" from the official name of the republic and eliminate the clauses requiring laws to conform with the Shari'a. Subsequently, bowing to public pressure, the Advisory Council of Islamic ideology was formally given the function of bringing existing laws into conformity with the Shari'a (Mahmood 1973, pp. 25-31). The Council's most important concern in the second half of the 1960s and 1970s was the problem of the ban on riba. The question was asked by the government in 1964, and the response of the Council in 1960, 1966, and 1969, despite the lack of participation of the ulema in the decisions, was clear. The Council asserted that riba is forbidden in Islam and Pakistan should have an interest-free banking system (Malik 1989, pp. 253-254). In the meantime, there was an attempt to eliminate riba from the banking system, but the problem was ignored due to political upheavals and the constitutional question of the early 1970s.

In the post-independence period, the debate on riba in Pakistan was shaped by two opposing views on Islam that may be regarded as traditionalist and modernist. The most influential representative of the traditionalist view was Abdul Ala Maududi, and, since the latter part of the 1950s, the modernist point of view has been represented by Fazlur Rahman.

Maududi's traditionalist position on riba as all kinds of interest, presents not only his interpretation of the problem in Islam, but also his reaction to modern economic theories of interest in the West. In his view riba is a predetermined fixed rate on the use of money that is claimed by any lender. A guaranteed and fixed positive return on money is unjust in view of the uncertainty of the entrepreneurial profit. Maududi argues that the modern loan based on interest is just like the riba al-nasia or the Quranic riba. In his view, risk taken by the lender of money cannot justify interest since risk can lead to positive or negative return. Abstinence and waiting do not justify interest unless money is used for productive activities that can lead to a profit or a loss. The lender cannot be rewarded for time, because time cannot be owned. Interest is not like rent, because the concept of rent applies to durable goods and it is a return for the depreciation of these goods. However, in his opinion, a reward for the owner of money is justified only when the owner is ready to enter in a muzaraba contract and share the risk of the enterprise with the agents or partners (Siddiqi 1981, pp. 237-238, 253-258; Pal 1994, pp. 65-67).

This view was left theoretically uncontested in Pakistan until 1963, when Fazlur Rahman expressed views similar to those of the Egyptian modernist Muslim scholars. As a liberal reformer, Rahman's analysis of riba is only an integral part of his general view on Islam. He is against literal interpretation of the Quran which, according to him, prevented the adaptation of Islam to modern circumstances. Like Abduh, he believes that Muslim scholars must study the historical circumstances of the Quranic verses, the specificity and the requirements of the modern era in order to be able to apply the principles behind the revelations (Sonn 1995, p. 408). Accordingly, he questions the reliability and the importance of the hadith, emphasizing the understanding of the spirit of the Quran (Commins 1995, pp. 119-122). In his lengthy argument on riba in 1963 and his study of the Quran, Rahman asserts that the Quranic riba is the continuous multiplication of the sum of capital "against a fixed extension of the term of payment of debt" following a default (Rahman 1964, p. 5). He finds the division of riba into riba al-fazl and riba al-nasia in the hadith unreliable, and he insists that only the Quranic riba is prohibited per se. In his view the prohibition of the Quranic riba is essential for the public welfare; however, the medieval jurists wrongly drew the conclusion "that all forms of interest are banned" (Rahman 1980, p. 41). He asserts that the rate of return on a modern loan is determined by the market. In modern economics interest is just like any other price, determined by demand and supply for loans. Yet, Rahman suggests, an Islamic economic policy must be based on the establishment of the spirit of cooperation and solidarity demanded by the Quran. Therefore, the Islamic government should regulate the market forces against excesses because the elimination of interest in modern economies is theoretically possible only if the supply of money is infinite at any level of demand (Rahman 1980, pp. 34-41; 1964, pp. 1-43; Pal 1994, pp. 65-72).

In 1973, Pakistan had a third constitution in which ulema and Islamic political parties, including Maududi's Jama'at-i Islami, limited the legislative power of the parliament. The new constitution granted the Council of Islamic Ideology (CII) the function of providing legal Islamic opinions (fatwas) when necessary. At the same time the CII was asked to prepare a report for the elimination of Islamic prohibitions, including recommendations for bringing all laws in conformity with the Shari'a (Nomani and Rahnema 1994, pp. 134-136). However, Islamization of the economy was initiated seriously only under the presidency of General Zia in 1977. Accordingly, the CII, which by now had changed the composition of its membership to reflect the views of the new government, was made responsible for a report on an Islamic restructuring of the economy (Malik 1989, pp. 257-259). The CII report of 1980 demanded the restructuring of the banking sector and taxation. This report, which was prepared with the help of economists and experts of the Central Bank and the Ministry of Finance, was a blueprint for the reorganization of banking practices and procedures on the basis of Islamic legal concepts (Gieraths 1990, pp. 173-175). It also relied on an earlier attempt by Sunni religious scholars and economists from Muslim countries who, in 1977, started the preparation of a Handbook of Islamic Banking (HIB) in Arabic. The HIB was published with the technical advice and "financial assistance recommended or channeled through" the International Association of Islamic Banks (Kazarian 1993, pp. 5-6, 246; Malik 1989, pp. 257-259). In addition, the CII report benefited from the discussion of religious scholars who worked out the operational framework and banking instruments of two Egyptian Islamic banks in 1979 and 1980 (see Algabid 1990, pp. 88-112). Nevertheless, the importance and the scope of the issue of Islamization inevitably gave rise to disagreements among Sunnis, Shi'is, and even "secularists" of the CII having traditional, moderate, sectarian, isolationist, and integrationist Islamic views. The Shi'i ulema were particularly concerned with the dominance of the Sunnis in the CII, eventually leaving the Council over the problem of zakat and the Sunni-Shi'i interpretation of the tax (Malik 1989, pp. 264-265). In fact, the ideological and Islamic disagreements among the members of the CII partially explains the inability of the Council to become a political platform of the traditionalists in the past fifteen years.

In any case, as regards riba and Islamic banking, the CII report reflected the opinion of Muslim economists who, following Maududi and other Islamist intellectuals, advocated a broader definition of riba (Ahmed et al. 1983, pp. 103-257). In fact, in a seminar on the subject of monetary and fiscal economics of Islam in Pakistan in 1981, the participants overwhelmingly supported the views expressed in the report. They emphasized that all kinds of interest charged on consumption, as well as investment loans, are riba, and, therefore, are normally and legally invalid. The general line of the reasoning and the rationality of this position was very similar to the opinion of Maududi on riba expressed in the 1930-70 period (Ahmed et al. 1983, pp. 2-14, 103-111).

Since the Islamization of the banking system in Pakistan, the debate on riba has focused on the "impurity" of the interest-free banking practices and its formalistic aspect. In fact, the CII report itself anticipated the debate by explicitly casting doubt over the potential widespread use of legal tricks in Islamic banking despite their acceptability in most Islamic schools of law (Ahmed et al. 1983, pp. 118-119). Even though the majority of ulema have been content with the existing setup, many Muslim intellectuals, economists, judges, and Islamic political parties are not at ease with the institution and publicly express their dissatisfaction (R.S. Khan 1987, pp. 136-152; Siddiqi's comment in Ahmed et al. 1983, pp. 223-231). One view would go even further, arguing for a broader definition of riba, demanding the elimination of all kinds of injustice and exploitation prevalent in Pakistan or other Muslim countries. According to this view, the CII report only took position against riba al-nasia, ignoring the prohibition of riba al-fazl (see Umar Chapra's comments in Ahmed et al. 1983, pp. 212-214). In the meantime, advocates of the limited view of riba did not present a comprehensive opinion on the CII report, despite their occasional, timid and theoretical criticism of the elimination of interest in a defensive manner (Pal 1994, pp. 73-76).

Since 1990, the traditionalist view and supporters of a broad definition of riba have continued their pressure for a "complete" end to interest-based banking by demanding the application of the Shari'a to different spheres of life. In fact, Khurshid Ahmad, a writer and political activist of Jama'at-i Islami, insisted that "99 percent" of Islamic banking in Pakistan is based on interest (Kuran 1993, p. 315). In the summer of 1990, Islamic parties sponsored a bill in the Senate, where they had majority, demanding the reimposition of the amputation of hands as a criminal punishment, the elimination of interest rates, and the subservience of the Assembly to the Shari'a court. In 1991 their candidate for premiership modified the bill, emphasizing generalities such as Islamization of education and guaranteeing the validity of all existing contracts until the development of an alternative Islamic system. The bill also envisioned the abolition of riba in three years, with a possibility of the extension of the period by the Assembly. In the meantime, an advisory Commission for the Islamization of the Economy (CIE) was to define usury and pass judgment over all the existing banking instruments with regard to riba (S. Ali 1991a, pp. 20-21).

Independently, the supreme religious court, the federal Shari'a bench in the High Court working on 119 cases that challenged various economic legislative bills and statuary provisions as un-Islamic, conducted an extensive survey and a study on different economic topics such as the Islamic nature of the banking system. In their testimony to the Court, the government's lawyers argued for a moderate view on interest, insisting that the Quranic riba is different from modern interest, especially interest on productive loans. They insisted that the Tradition of the Prophet is not conclusive over interest, and, therefore, riba is under *mutashabihat* (words or phrases in the Quran the meaning of which are not known, the so-called gray areas that have led to different interpretations by scholars). They demanded that the issue be resolved by the commission for the Islamization of the system. Nevertheless, in November 1991, the court declared 22 legislative enactments and statutory provisions as "repugnant to Islam." Thus, all kinds of interest, several legal tricks in banking, noncooperative insurance, and negotiable instruments were identified as riba, even in transaction with non-Muslims. The verdict approvingly referred to a conclusion reached by the Islamic Academy of Jurisprudence in Saudi Arabia that was constituted under the

Organization of Islamic States in 1985 (S. Ali 1991b, p. 27). The director of the School of Economics at Islamabad's International Institute for Islamic Economics and the court advisor, M. H. Choudhury, asserted that there could not be any other judgment and that the Quran "does not leave any room for interpretation" (*World Press Review* 1992, p. 43). He declared that Islamic banking in Pakistan has been realized only on paper and that all kinds of legal devices, such as markup, housing loans made by the state based on a rent, and interest on government bonds are all un-Islamic. The only concession that Choudhury made, however, was that the ruling of the court against the payment of interest on foreign loans is not realistic in the short run (*World Press Review* 1992, p. 43). On the other side of the debate, the Finance Minister, S. Aziz, reiterated that "although riba is totally prohibited in Islam, unfortunately there is no universally acceptable definition of riba in the Muslim world according to which existing financial practices can be tested on the basis of Islamic law" (*The Economist* Jan. 18, 1992, p. 34). The government did not appeal the judgment in order to avoid a direct clash with its Islamist supporters and only asked the court to reverse the verdict, calling it "an archaic interpretation of the Islamic injunction" (R.A. Khan 1993, pp. 134-135). Finally, the government encouraged three banks to appeal to the Supreme Court in order to reverse the Shari'a Court's ruling on interest and postpone the crisis (Ali 1992b, pp. 18-19). In the same year, the Commission for the Islamization of the Economy came out with a report, reiterating the CII 1980 report's recommendation, including a profit-and-loss sharing scheme for government borrowing. The CIE also ruled that the existing insurance system was contrary to the Shari'a, and called for a system based on cooperative insurance. In reaction to the CIE report, one of the chief economists of the Planning Commission asserted that such recommendations were neither Islamic nor suitable for Pakistan in view of the fact that more than 85 percent of depositors had very little in their accounts, and they needed security and a positive return rather than the risk of a loss. It was argued that if riba is an exploitative income, it should apply more to a banking system based on profit-and-loss sharing scheme than to a system based on a fixed return (R.A. Khan 1993, p. 135).

The debate on riba is still unresolved in Pakistan and awaits the verdict of the Supreme Court. In the meantime, banks have adopted many Islamic contracts, as well as legal tricks, for their operation. However, the specter of a rigid interpretation of riba still hangs over them. Again, as in Egypt, the problem is not only legal or juristic; it is mixed with politics, complicating the problem of public policy even more than before.

The Pragmatic View on Riba in Iran

The problem of riba has never been an important issue in Iran of the twentieth century, before, as well as after the revolution. This does not imply that Iranian Shi'i ulema approve of riba or Islamist groups, and, finally, Islamist movements in Iran have not demanded the abolition of riba; it only means that unlike Egypt and Pakistan, political and economic debates over Islamization were never focused on the issue of riba. For this reason, one finds very little discussion, not to speak of a debate, on riba from a legal, juristic, constitutional, and, after the implementation of Islamic banking, from a technical or economic point of view. It seems that Iranian Islamists and the Islamic Republic of Iran itself, have been content with mere formalistic changes and have left it at that. This situation might be envied by the Egyptian and Pakistani modernist Muslims, but, curiously enough, has also been referred to as an ideal case by many Islamist intellectuals who are not well informed about what has happened in Iran, partly due to the uncritical idealization, historical and sometimes ideologically motivated research on the subject.

Different political tendencies of Shi'i Islam have always rejected riba as an exploitative phenomenon. One tendency which was in reaction to the onslaught of modernization after World War II, that was uprooting traditional Islamic values, notably, Navab-Safavi's political movement, was very explicit about the abolition of riba from the modern banking sector. According to Navab-Safavi, in a truly Islamic state, Muslims would automatically, and without any coercion, accept obligatory financial prohibitions. Nevertheless, this group was for the forceful imposition of Islamic values and injunctions before the advent of an ideal Islamic state. For this reason, there was not much persuasive analysis about the prohibition of riba even within an Islamic intellectual framework.

All the other important personalities of Islamic tendencies before the revolution, namely Morteza Motahhari, Ali Shariati, and Mehdi Bazargan concentrated their effort in one way or another on an Islamic response to the spread of communism in Muslim Iran among intellectuals. Shariati's tendency was an attempt at accommodating, coopting and eventually dissolving Marxism-Leninism in Islam. Motahhari's and Bazargan's tendencies were nonaccommodating, and sought to prove that Islam is different from communism and is superior to it. It is only within this perspective that Shariati is interested in the concept of riba. For him, and all the other leftist Muslims, riba in Islam represents the exploitation of one class by another (Rahnema and Nomani 1995, pp. 69-80). This conceptualization could not be left unanswered by religious scholars like Motahhari, who had to debate not only against secularists but also against leftist tendencies within the Islamic movement. Thus, Motahhari explicitly professes that the exploitation explanation of riba is not an Islamic but a socialist view. According to him, riba and exploitation are two different concepts, and the absolute prohibition of riba does not imply that Islam is against capitalism: "riba is unjust, but all injustices are not riba" (Motahhari 1991, p. 89).

In general, Motahhari's arguments about the problem of riba are very similar to the opinions of the majority of the Shi'i ulema. These arguments can be summarized in the following manner. Money is only a means of exchange and its transaction must not lead to an increase in the form of interest. However, if money capital in commercial or productive activities leads to an increase called profit, this profit is lawful because in such cases profit is a return to labor and the "excess" in commercial and productive activities is not a mere exchange of money. If bills of exchange and discounting are lawful in Shi'i Islam, it is because in such acts the lender "does not receive something extra, but less" (Motahhari 1991, p. 40). According to Motahhari that which is forbidden in Islam is to demand something more than what has been given. Besides, sale on credit is not riba and is valid in Shi'i Islam because the substance of the act is the freedom of the seller to increase the price in case of a delay in payment. Nevertheless, all kinds of riba, riba al-fazl, and riba al-nasia, are prohibited in Islam. Modern loans are equally forbidden as riba because the lender asks for a predetermined and fixed positive return. Thus, loans for consumption and for productive purposes, based on interest, are riba. For this reason the Egyptian Muslim scholars' differentiation between consumption and productive loans, "good" or "bad" riba, interest and riba is not valid. Shi'i jurists do not consider one type of riba to be less unjust than the other, and, consequently, acceptable.

Motahhari regards all legal tricks as invalid, even though many religious Shi'i scholars find them legal. Therefore, for him the only legal way to increase a sum of money is by means of a muzaraba contract in which a "lender" shares in profit or loss of the enterprise of a "borrower." He asserts that the payment of interest on savings accounts and government bonds is riba, while the payment of bonuses on deposits are lawful if such bonuses are not predetermined. A guaranteed rate of return on the stock of an investment bank is legal since the bank is committing itself to a payment without asking the owner of the stock to agree to any condition in return; such a commitment is like what takes place in insurance contracts, which are in general legal. Finally, like many Shi'i ulema, he permits the payment of interest to and receiving interest from non-Muslims (Motahhari 1991, pp. 1-155).

Motahhari was a follower of Khomeini, and, therefore, the latter's legal opinion on the subject of riba was more or less the same. For example, according to Khomeini paying an extra sum that is not to be determined at the time a savings account is opened or a promissory note is made, or giving a prize to depositors are all lawful (Rahnema and Nomani 1990, pp. 149-150). However, the views of some other jurists in Islam, such as Milani and Shariatmadari, were even more pragmatic in relation to the needs of the modern financial system. Such jurists had come up with many legal tricks that would facilitate the effective payment of interest on loans, much to the dissatisfaction of clerics like Motahhari. Such acceptable devices were management or service fees, discounting, markups, and hire-purchases, that were approved of even by Khomeini. However, the pragmatic ulema came up with legal tricks that turned loans into a sale in a very formalistic manner. For example, according to Milani, instead of stating that I lend you \$A and receive in a year \$A + B, one could say I sell this \$A for \$A + B (Motahhari 1991, p. 141). Since in Shi'i law the exchange of unequal quantities of things that are countable (e.g., money) are permitted, the act would be lawful.

The Iranian liberal modernist view on riba in Islam was presented by Bazargan. Bazargan as a non-cleric Islamic scholar advocated a tolerant religion compatible with liberalism and political democracy in a market system. He was familiar with the arguments of liberal modernists of Egypt and Pakistan and his position on riba follows their arguments. According to Bazargan, borrowing for consumption purposes by the needy should be free of interest. However, productive loans can receive simple interest because the interest on a loan is normally less than the profit received from the enterprise. He reiterates that only usurious interests have been forbidden by the Quran, and for this reason sales on credit with a rate of return, which are like loans based on interest, are permitted by jurists. Nevertheless, Bazargan finds devices such as muzaraba and cooperative banks useful for modern banking (See Motahhari).

In 1979, the Shi'i clergy, under the leadership of Khomeini, came to power in Iran and the name of the country was changed to Islamic Republic of Iran. During the uprising against the Shah in 1978, the religious movement did not present any Islamic economic program except general simplistic slogans about the end of "plundering and usurpation of wealth," autarky and self-sufficiency; there were no slogans against banking based on riba or any reference to interest. The movement was focused against the Shah, and the clergy avoided concrete and controversial economic programs that has divided them since 1979 (see Rahnema and Nomani 1990).

According to the Constitution of the Islamic Republic the official religion of Iran is the Twelver Shi'i and all civil, penal, financial and economic laws and regulations are to be based on the Islamic criteria of the Ja'fari school of law. Thus, jurists of Council of Guardians were to judge the conformity of the laws and regulations with the Shari'a according to the Ja'fari school of law. The followers of Sunni schools of law were only given the right to conduct the personal affairs (marriage, divorce, inheritance, and wills) according to the rulings of their own schools.

Riba is prohibited and mentioned in two articles (43 and 49), along with gambling, "monopoly, hoarding" and other "evil practices." Before the vote on the relevant Article 43 on riba in the Assembly of Experts, only two comments were made. One was suggesting that the prohibition of riba is legally so evident that one could drop it from the Constitution. The second comment equated riba with exploitation and charging high prices. The article prohibiting riba was passed without any debate (61 representatives voted for, 1 against, and 2 abstained) (*Soorat-e Mashrooh-e...*, 1985, pp. 1483-1485).

Unlike the criminal law, the civil code and the general theory of contract of Iran, which were adopted in 1927-1932, were Islamic in substance. Thus, the Islamic Republic did not have to change much in the civil and commercial laws (*Majmo'eh-ye Nazariyyat-e*... 1992, pp. 292-293; Amin 1986, pp. 36-101). All private banks and insurance companies were nationalized in 1979 (mainly because their owners had left Iran and all were nearly insolvent). The Law of Riba-free Banking was approved by the Parliament in August 1983. This law was similar to what the CII had presented in Pakistan in 1981, except that it was more pragmatic, allowing many legal tricks and few definitional differences.

The Islamization of banking took place in a gradual manner, again, without any meaningful debate over the process or the articles of the Law. First, in 1981 interest on all asset-side transactions were replaced by a maximum "service charge" and a minimum "profit" rate. On the side of liability of banks, the interest on deposits was replaced by a "guaranteed minimum profit." The Council Guardians approved the law immediately, except for one point related to gradual elimination of interest within a three-year period, which was accordingly dropped by the Parliament (*Majmo'eh-ye Nazariyyat-e*... 1992, pp. 379-380). In 1988, the Council of Guardians finally approved banking based on riba with non-Muslims, even though Iranian banks and authorities had been paying and receiving riba in their dealings outside of Iran without much fuss. Thus, according to the Council "collecting interest...from foreign governments, organizations, companies and persons...is permitted by Shari'a" (Amin 1989, p. 136).

The pragmatic approach of the Shi'i Islamic state in Iran is also reflected in the acceptance of a charge or paying of a "penalty" or "damages" for delay payment of debt. Thus, in a ruling in 1983, the Council approved of the inclusion of a "penalty" clause for unpaid debt at the rate of 12 percent per year by the consent of the borrower. Nevertheless, the same Council on different occasions in 1985, declared all forms of interest forbidden. The decision of the Council on the penalty clause instead of interest, has theoretically enabled the Islamic courts to accept a contractually arranged 12 percent rate of interest as a legal trick (Amin 1989, pp. 128-130). This pragmatism has encouraged even the official journals of the Islamic Republic to ignore the nomenclature of the Islamic banking and to talk about "interest rate" instead of the "rate of profit." Besides, there is the "un-Islamic" guarantee of a "positive"' rate of profit on the savings and term investment deposits by Islamic banks, or payment of interest by municipal and government bonds that have rarely offended the Islamic clergy and public (see, Abrandabadi, 1994, for a rare critical response to this issue). Such common practices based on explicit and implicit legal tricks and interest rates, led a reputed clergy, Va'ezzadeh-Khorasani, in his capacity as the chair of a research group on Islamic economics, to propose an explicit payment of interest. His proposal, which has not given rise to any debate for or against, is based on the following reasoning. He argues that the Quranic verses on riba are all related to private loans and exchanges, and contemporary riba is not like the riba that was practiced before the twentieth century. He asserts that in Iran all banks are state owned and the state is "totally" Islamic. Therefore, an Islamic state could pay and charge interest; the Islamic state bank can receive interest from loans granted to private or government sectors. "After all," he insists, "Shi'is do not consider riba to be exploitative when the practice is within the family," and state loans can be perceived as the transfer of money from the public to the representative of Muslims (i.e., the Islamic state) (Va'ezzadeh-Khorasani 1991, p. 118). He believes that such a change in the existing institution of the banking system eliminates the "unethical legal" tricks like muzaraba or all kinds of banking partnerships and devices that are "formalistic," "cosmetic," and "farcical" in comparison to "real" riba-free contracts (Va'ezzadeh-Khorasani 1991, pp. 118, 119, 120). However, he neither discussed nor specified "real" riba-free contracts.

MODERN INTERPRETATIONS OF ISLAMIC FINANCIAL CONTRACTS AND INSTITUTIONS

The advocates of a limited definition of riba did not oppose modern banking when this institution was introduced in Muslim countries; they were aware of its usefulness in a modern market system. Accordingly, in response to skeptics, they argued that the system was in conformity with the Quranic prohibition against usury, and that the institutions of check and bill of exchange were the financial innovations of medieval Islam. This view insisted that the Shari'a, which is a mercy from God, is to facilitate human welfare and not to create obstacles for it, and that ulema could use the principles of maslaha and istihsan to legitimize modern usury-free banking. They argued that modern banks receive money not as borrowers but as trustees. In turn, banks facilitate investment and receive part of the profit earned in investment as interest, and share this interest with original depositors (Qadri 1986, pp. 337-339). However, neither the traditional ulema nor the advocates of a broader definition of riba have been convinced by such arguments. Thus, it was natural that they ask for the Islamization of the banking sector in Muslim countries. In this endeavor, Islamist intellectuals and to a lesser degree, the traditional ulema have benefited from the innovative ideas of modernist Muslims like Abduh. This is because at the advent of modern banking in the Middle East, modernists were concerned with the rationalization and development of the Shari'a, and, therefore, one of their innovations was the idea of applying medieval contracts of profit sharing and partnerships to modern banking.

Innovation of Financial Instruments

If interest is not to be earned on savings and loans, profit should replace it as an incentive device, and hence, the usefulness of Islamic contracts of muzaraba and musharaka. However, the applicability of such classical contracts, with their strict medieval terms and conditions, to modern requirements of banking would be very limited. In order to lift this serious constraint, the advocates of a broader definition of riba have always pleaded with the ulema for new legal judgments on such financial instruments and contracts (Siddiqi 1987, pp. 99-104). Much to their disappointment, the traditional ulema have not developed the Shari'a in relation to Islamic banking to a great extent mainly because of their continuous disagreements on many financial issues. To date, in a country like Pakistan, the agency contract or capital trust financing is called profit-and-loss sharing and not muzaraba, in order to avoid a conflict with the traditional Islamic law (Attia 1986, p. 117). For this reason, Muslim economists, scholars, and politicians, and in many cases with the help and backing of governments, have bypassed Islamic jurisprudence to come up with their own Islamic legal innovations. The major practical efforts of this type, such as the CII report, the Riba-free Banking Act of the Islamic Republic of Iran, and the Handbook of Islamic Banking, all lacked the support and participation of the mainstream ulema in their formulation. At the same time, these innovations, especially in relation to banking, have not fully satisfied the idealistic views of many Islamist intellectuals because of the inevitable element of realism related to legal tricks in such schemes.

As regards Islamic financial contracts, the two most important medieval contracts that have been adopted and renovated by the "consensus" of Islamist economists and Islamic institutions, such as the International Association of Islamic Banks, are muzaraba and musharaka. Islamists believe that with the help of these two contracts, banks (mainly private) are able to operate as financial intermediaries and change the main relationship between the conventional bank and the demanders of funds from that of creditor-debtor to partners. However, the classical jurists' insistence on such things as the use of muzaraba for trade, the independence of the agent or active partner in the management of the business, and strict rules for consecutive partnerships, had to be changed to reflect modern banking needs.

Ideally, in the case of muzaraba (agency or capital trust financing in modern usage), the public enters into an unrestricted contract with a bank concerning the kind of activity to be pursued by the bank, duration and location; the bank in turn gives this capital to entrepreneurs in a restricted muzaraba contract or an unrestricted one in the form of general investment fund, without requiring any guarantee or collateral. The business activity of the first working partner (i.e., the bank) is the selection of the second working partner or an investment project. The profit of the business activity of the second working partner is distributed among all partners at the end of the contract according to the conditions settled among partners in advance. The loss is born by the owners of capital in proportion to the amount of their capital, and the liability of the entrepreneur is restricted to their labor and effort. This general scheme provides its associated banking accounts: demand deposits, savings, and investment accounts. In principle, demand deposits and savings accounts are guaranteed without any rate of return; while investment accounts must not be guaranteed, but can share in the risk of a profit or a loss of an enterprise. Ideally, the latter contract should be the leading banking instrument for different types of Islamic productive activities, but it has been shunned by banks and their clients in practice (Roy 1991; Kazarian 1993; Nienhaus 1986). In Iran, however, muzaraba is restricted to short-term internal trade financing (Heshmati-Moulai 1991, p. 77).

A second modified contract in Islamic banking is "equity participation" or musharaka. In this case the bank and the public jointly contribute capital, with or without joint management of business ventures. Banks could also buy equity shares in an already established enterprise. The profit and loss of such a contract are shared by all partners according to a predetermined share. Again, in the practice of Islamic banking, a very small part of assets are related to such a mode of financing despite its desirability for ideal Islamic behavior (Nienhaus 1986). Nevertheless, partnership contracts in Islamic banking are innovative in the sense that they are neither fully an equity nor a debtor-creditor relationship. In the case of muzaraba one takes part in profit, without being partner in the equity, even though the legal implications of such an arrangement are not very clear from an Islamic point of view. Musharaka is a relationship between shareholders of equity and lenders or depositors in conventional systems. For practical purposes the matchup between the duration of projects in which funds of depositors are used and the financial period of the deposit, should be ignored for a juristically correct sharing of profit and loss. Besides, practical considerations must take into account the problem of the need of depositors to withdraw funds, disregarding the liquidation of investments for which their money is used. The end result of such problems is the requirement of continuous innovation on these issues.

Controversial Financial Instruments and Institutions

Ironically, some of the Islamic financial instruments that are very popular among Islamic banks and their clients, and are recognized as legal by many jurists, are controversial for purist Islamist intellectuals. First, there are those instruments that are definitely forbidden by Sunni legal schools, even though some of them are permitted by most Shi'i jurists. Discontinuing commercial papers is prohibited by Sunni Islam, but accepted by Islamic banking in Iran.

For Sunnis, commercial papers represent a debt or a transfer of debt with a premium, while for Shi'i pragmatists, a discount is accepting a decrease and, therefore, is not capable of riba. In the same manner, factoring (as a sale by the client of trade debts to the bank or the factoring company) against a discount is forbidden by Sunnis and permitted by Shi'i jurists. However, Islamic banks and their religious advisors have to come up with solutions to these problems, which cannot be anything but legal tricks.

Other controversial transactions that are used for financing working capital, commercial, housing and services and are known to conventional banking, are markup (murabaha), leasing (ijara), lease-purchase, and letter of guarantee. The letter of guarantee would not be legal if the bank receives a service fee in proportion to the amount of the letter of guarantee. However, some jurists insist that such a service by banks should not be remunerated, unless the letter is covered by the client. An alternative is to classify the letter of guarantee as a lucrative contract, which is contrary to the classification of traditional jurisprudence, and allow the bank a percentage of the profit of the business for which the letter is issued (Attia 1986, pp. 103-105). In Iran, banks can receive commission fees for such a service according to Non-riba Banking Act in a contract called *jo'aleh.*

Markup, leasing, and lease-purchase contracts are all based on predetermined fixed rate of return, and in most cases, such contracts require a collateral. These contracts are legally valid, even though they are legal tricks, because unlike riba, the rate of return is tied to each transaction, a given object, and not to the time period or financial asset. However, the purists assert that the predetermined fixed rate and the collateral tend to eliminate the risk that in principle banks and entrepreneurs must share in terms of profits or losses.

Post-delivery or financial future transactions (salaf or salam) are problematic in Islam basically because of the prohibition of any deferred payment in the case of monetary exchange and certain foodstuffs. Some Sunni jurists prohibit any forward or future transactions in currencies. Some other jurists allow such transactions if the rate of exchange is determined in advance and both currencies are paid simultaneously later. This is for avoiding interest that might be realized in case of deferred payment if one currency is paid on the spot and the other is paid later (Attia 1986, p. 106).

In cases of the pre-purchase of products at a certain cash price, all jurists allow such transactions if they are not for speculative purposes, and are not based on expectation of price changes. However, the Iranian Islamic banking regulation is not very strict on this point. In Iran, the price of pre-purchase of goods by banks is fixed in view of the factors affecting price determination, including forecast of their selling price at the date of delivery as well as the profit expected by banks (Amin 1989, p. 159).

To avoid the controversy of salam and currency risk in matters of swaps, Sunni Islamic banks have developed the following device. One bank agrees with another on a temporary exchange of deposits in different currencies, both refunded in the same currency after the agreed period of time. In the meantime, "each bank is utilizing the deposit it received at its own risk" (Attia 1986, p. 107).

The question of new financial instruments in modern markets, such as currency and commodity options, is not still clear in the modern Islamic jurisprudence. This is due to the problem of not exchanging the two currencies at the time of agreement and the inability to specify the rate for currency exchange in the future. In addition, option contracts provide a form of insurance, and, therefore, purchasers must pay a price for them called a premium, which in itself is a doubtful contract in Sunni Islam. In general, however, financial futures and options for debt instruments that are all used by conventional banks to manage their interest-rate risk, must be used by Islamic banks with caution.

All existing Islamic countries have tacitly (Saudi Arabia) or explicitly (Iran and Pakistan) permitted the operation of stock markets on the general principle of salam sale (see Khalatbari 1995, pp. 177-206; Zeinelabdin 1991, pp. 51-743; Nomani and Rahnema 1994, pp. 155-156). The major problem in relation to the stock market is the advisability of having a stock market free from speculation, that is, a stock price that reflects only the underlying economic conditions of the enterprise. In an Islamic stock market, shares could be transacted by the public and the price of a stock would fluctuate around some established level. However, the purist view insists that Islam prohibits the selling of things that one does not possess and transactions that have the element of chance (Afzal-ur Rahman 1982, pp. 155-22). Accordingly, a transferable contract (a share) can lead to unearned income or riba due to capital gains. This view would not allow such a transaction and, therefore, is an obstacle toward efficient development of a secondary capital market for the issuing muzaraba and musharaka certificates for equity financing and the central bank's open market operations (Ahmed el al. 1983, p. 252).

There are certain other financial practices that are frowned upon by some Sunni jurists as legal tricks but are, in most cases, permitted by Shi'i ulema. One of these practices is bonuses paid on current and savings accounts, which often become a fixed and expected practice. The other is the tacit guarantee of a minimum positive return on investment and savings account, even though the rate is determined at the end of the financial period. Both devices are used by all Islamic countries like Iran and Pakistan (Nomani and Rahnema 1994). Another debatable practice, which is recognized as valid in civil laws of Muslim countries, is the penalty for the delay of payment in favor of creditors (Amin 1989, pp. 128-130; Davies 1984, pp. 212-213).

A real problem in many developing countries is that the nominal rate of interest paid to depositors in the formal banking sector, unlike the informal financial markets in bazaars, has been in many cases less than the inflation rate. Such a negative return on savings deposits is against many "fair" economic principles that are upheld by all Islamist intellectuals and jurists. However, in none of the Islamic countries are depositors compensated for this unfair practice, and the ulema have opposed indexation of such deposits against inflation for technical reasons (Mannan 1986, p. 192).

On the issue of legality of an Islamic banking system based on a fractional reserve system, the ulema have been silent, even though in many Arab countries and Pakistan, Islamist intellectuals have addressed the issue. Such a system enables banks to create money by a multiple of an original sum of money of a demand deposit. To remedy this, Islamist economists have suggested a 100 percent reserve system and the nationalization of the banking system (see Ahmed et al. 1983, pp. 47-48).

An Islamic government can borrow from the public and banks by issuing securities and bonds based on a profit sharing (or benefit sharing) principle, even though the Sunni ulema have not been at ease with this idea. Recently, however, all Islamist governments have been issuing "development" bonds, granting lenders a positive nominal rate of return for specified and unspecified government projects. In Saudi Arabia such a practice was criticized by the ulema who later capitulated. In Pakistan, such loans are still earning interest, but the Shari'a Court has ordered the elimination of interest in such transactions.

To summarize, it is obvious that there is a trade-off between the strict and inflexible interpretation of the Shari'a and the requirements of a smooth functioning modern economy. Besides, there has never been a common and unique understanding of the Shari'a and the requirements for its development. In one country the interpretation would not permit legal devices, and in another it is interpreted in an arbitrary and inconsistent manner to suit the needs of the Islamic government. Thus, the government of the Islamic Republic of Iran has continuously financed its budget deficit by borrowing from the Central Bank. This borrowing has been interest free and the loan has never been paid back, even though the same central bank receives interest from state banks and interbank loans are based on interest (see Coville 1995, pp. 215-228).

The Insurance Dilemma

Insurance, as a legal contract providing a kind of financial security, has been one of the most controversial issues in Sunni Islam since the mid-nineteenth century. Insurance has been popular among businesses, and almost every Muslim country has some forms of insurance, except life insurance, in their civil and commercial codes. Yet, traditionalist ulema and purist Islamist intellectuals doubt the validity of insurance.

Some Sunni jurists have relied on *ijtihad* and have concluded the validity of the contract by an analogical deduction. One judgment, notably that of Abduh's, has found insurance loosely akin to muzaraba. In this view, policyholders contribute their capital in installments and the insurer acts as their active partner, sharing with them part of the profit of the enterprise. As such, both parties benefit from the contract, which excludes riba as injurious to one party. Besides, "necessity" and "public purpose" principles can be used to legitimize the contract (Qadri 1986, pp. 339-340).

The opponents of insurance find the similarity of insurance and muzaraba unwarranted. They assert that insurance does not conform to any contract validated by the Shari'a and is an unnecessary innovation. The general argument of this view is based on the following legal points. Insurance is similar to riba, gambling (*maysir*), wagering contracts, gharar (risk), and uncertainty and, therefore, is forbidden. Insurance, especially life insurance, has the element of riba, because the capital of the policyholders increases by interest. Such a contract is close to gharar and gambling because the subject-matter is uncertain and the insurer (the "winner") cannot recover his "bet" or the money invested despite the declaration of loss and that the benefits to be received by the insured are not clearly defined. In addition, Sunni Islam rejects insurance due to a philosophical-moralistic point of view, that is, risk coverage is against the will of God and the ultimate destiny of Man's life and possessions determined by God (Afzal-ur Rahman 1982, pp. 11-176). Nevertheless, such arguments have been ignored by civil and commercial laws in the Muslim world, even in a traditionalist country like Saudi Arabia, since 1987. Thus, Sanhuri, the architect of many such codes in Arab countries who refuses to justify insurance contract by analogy, insists on the novelty of insurance in the Shari'a emphasizing its indispensability for the welfare of the public. Many other Sunni jurists have also approved of many types of general insurance, except life insurance, by stressing the investment of the capital received by the insurer in business. In fact, with the resurgence of Islamic movements in the 1970s, the emphasis has shifted to an Islamic insurance contract, called "solidarity" or "mutual insurance" (*ta'awuni*). It is asserted that the legal framework of such a "solidarity" insurance is a muzaraba company that should not engage in speculative activities and interest-based transactions (Tawfiq Attar 1976, pp. 41-45). As such, the action of an insurance company will be based on "public interest." The equivalent of a "premium," or the contribution of the insured, is invested in a musharaka or muzaraba according to which any loss will be reimbursed from the accruing profits or from the reserve funds. It remains to be seen how such an arrangement will work in practice (Klingmüller 1990, pp. 154-160).

Contemporary Shi'i jurists have rarely raised doubt about the validity of insurance contracts, and Shi'i intellectuals have ignored the problem altogether (Motahhari 1991, p. 275). In an analytical presentation of the problem of insurance, a Shi'i jurist, Muhammad Vahidi, legitimizes the contract by the following argument that is similar to Sanhuri's reasoning. He argues that insurance is not gambling and is not capable of riba and gharar. Insurance is an independent contract by itself and consists of offer and acceptance, and, thus, is lawful. The legal Islamic principle of "fulfill your obligations" applies equally to classical Islamic contracts and lawful modern contracts. Insurance is a conditional contract of gift in which the insurer agrees to pay the premium to the insurer free of charge but specifies that the insurer shall compensate him in case of the occurrence of specified event (Amin 1985, p. 80). Khomeini's view on insurance is similar, and he, too, identifies insurance as a new independent contract. He emphasizes the similarity of insurance to the Islamic contract of compromise or settlement (*solh*) and gift. However, he insists that the subject matter of insurance (person, and all other kinds of property), the identity of the parties to the contract, the premium payments, and the length of insurance must be known in an insurance contract (Biazar-e Shirazi 1983, pp. 126-130). Life insurance is not particularly discussed by the contemporary Shi'i ulema. However, before the Islamic revolution, Motahhari was skeptical of the validity of life insurance. He states that if contemporary Shi'i jurists have not explicitly forbidden life insurance it is because they were not familiar with all kinds of insurance. The problem in this case, according to Motahhari, is that in such a contract, unlike all valid contracts, the benefits to both sides of the transaction are not always known and riba becomes the main objective of insurance. Motahhari, however, insists that other types of insurance contracts are free form riba, gambling and gharar, and, for these reasons are lawful according to Shi'i ulema.

According to Motahhari, insurance is free from riba because both parties benefit from the exchange. One party receives security (*ta'min*) and tranquillity of mind and the other capital. An insurance contract is not capable of gharar because the subject matter of the transaction is not the money that would be received, but the person or the movable or immovable property insured, which is known in order to avoid gharar. This contract is not gambling if the risk that is a fact of life is not sought, in which case insurance reduces risk and improves the welfare of the insured (Motahhari 1991, pp. 275-310).

Some Problems and Prospects

Islamic banking is growing in many Muslim countries, although it is still underdeveloped and fragmented by international standards. When Islamic banks came into existence in the 1970s, it was a phenomenon based on the combination of religion, politics, and oil money. Such a characteristic inevitably leads to controversies, varying interpretations and religious compromises. This is demonstrated in the case of nationalized banking in Iran and Pakistan, and Islamic commercial banks in some Muslim countries, as well as some international and national banks that offer certain Islamic financial services and Shari'a and even some Shari'a advice.[2]

The future expansion of Islamic banking and finance partially depends on the establishment of financial capital markets, where long-term bank assets are invested in tradable securities and where banks can secure daily liquidity in money markets and interbank markets. However, none of these markets are faultless markets according to a strict interpretation of the Shari'a. And since there is no common interpretation of the law, the innovation of such new Islamic financial services will be continuously debated and contested. Yet, lack of innovation will be a serious economic constraint for the development of Islamic capital markets. Some Islamic banks are slowly expanding into long-term investments based on Islamic contracts, but still a very important part of Islamic business is in debatable short-term finances, such as markup and leasing. To some extent this is because markup, which presents the lowest risk of all methods of Islamic finance available, might be cheaper than other finances, and traders are more at ease with it. Since most of the bank's liabilities in these contracts are short term, banks are hesitant to tie up funds in medium- or long-term assets. Besides, secondary markets for Islamic instruments are not yet developed and Islamic banks do not have access to the central bank lending facilities in countries where the whole banking industry is not Islamized.

Presently, some Islamic countries, such as Iran, Pakistan, and Sudan, claim to have Islamized the entire national banking, and others, like Saudi Arabia, insist that their banking has "always" been Islamic. However, the majority of Muslim countries have only permitted Islamic banks to co-exist with conventional banks since the 1980s. Nevertheless, all these countries have been wary of the expansion of unregulated small private Islamic muzaraba funds. The high risk exposure and returns of the latter, compared to regularized or state banks, has led to widespread bankruptcies and embezzlement. At least, this is what many full-fledged Islamic governments claim, and for this reason countries like Iran have recently forbidden their formation.

The most important financial activity of Islamic banks and conventional banks that offer Islamic services has been concentrated on short-term and less risky operations. Muzaraba and musharaka contracts for productive investment are rare. Markup and leasing prevail and the costs for providing funds are fixed in advance and limited to a fixed period reflecting interest and traditional financial practices. The level of transparency of these banks, especially in countries like Iran where all banks are state banks, is almost nonexistent. However, depositors are guaranteed their funds and are assured of receiving some nominal positive returns that in many of these countries is less than the rate of inflation. Despite the unconditional enthusiasm of some Muslim economists for these Islamic experiences in Iran and Pakistan, it is known by now that whatever these experiences are, they are not recognized as Islamic. In Iran, in view of "excesses" and "flexible" interpretation of the prohibition against riba by state banks in recent years, Khamene'i, the leader of the country, has asked for "reviving the interest-free loan system" (*MEED*

January 7, 1994). In 1993, in Pakistan, with the consent of the government, some commercial banks have appealed the verdict of the Shari'a Court that demanded the elimination of riba from all financial instruments.

At the same time, hundreds of millions of medium- and low-income individuals, who are victims of inflation and negative real returns on their small deposits, should wonder whether it has not been the wrong historical moment for renouncing interest. Nevertheless, one must not forget that the problem of interest in Muslim countries is not only an economic question capable of social scientific scrutiny, but it is also a religious dictum that can be interpreted differently, with varying politics and economic interest. Therefore, the weight and importance of the former in relevant public policies rises with the relative decline of the legitimacy of the existing Islamic and semi-Islamic political and economic powers, despite the relative independence of the religious factor in Muslim countries. Yet, one must guard against passing judgment on the level of Islamization of the banking system in this or that Muslim country, because such a judgment wrongly presupposes a unique concept of the Shari'a.

NOTES

1. All references to the Quran are from A.Y. Ali (1989).
2. See Roy (1991), Neinhaus (1986), Kazarian (1993), Kuran (1993), O'Sullivan (1994), Wilson (1993), and Khan and Mirakhor (1990).

REFERENCES

Abdeen, A.M. and D.N. Shook. 1984. *The Saudi Financial System*. New York: John Wiley & Sons.

Abrandabadi, H. 1994. "Molahezati Dar Bareh-e Oragh-e Musharakat." *Resalat*, September 23.

Abu Saud, M. 1981. "Money, Interest and *Qirad*." In *Studies in Islamic Economics*, edited by K. Ahmad. London: The Islamic Foundation.

Adams, C.C. 1968. *Islam and Modernism in Egypt*. New York: Russel and Russel.

Afzal-ur Rahman. 1982. *Economic Doctrines of Islam,* Vol. 3. Lahore: Islamic Publications Limited.

Ahmed, Z., M. Iqbal, and M.F. Khan, eds. 1983. *Money and Banking in Islam*. Islamabad: Institute of Policy Studies.

Algabid, H. 1990. *Les banques islamiques*. Paris: Economica.

Ali, A.Y. 1989. *The Holy Qur'an*. Maryland: Amana.

Ali, S. 1991a. "Shariah Strains." *Far Eastern Economic Review*, May 2.

________. 1991b. "Question of Interest." *Far Eastern Economic Review*, December 12.

________. 1992. "State and Religion." *Far Eastern Economic Review*, February 13.

Amin, S.H. 1985. *Islamic Law in the Contemporary World*. Glasgow: Royston Ltd.

________. 1986. *Commercial Law of Iran*. Tehran: Vahid Publications.

———. 1989. *Islamic Law and Its Implications for Modern World.* Glasgow: Royston Ltd.

Attia, G. 1986. "Financial Instruments Used by Islamic Banks." In *Islamic Banking and Finance.* London: Butterworths.

Ballantyne, W. 1990. "A Reassertion of the Shari'ah: The Jurisprudence of the Gulf States." In *Islamic Law and Jurisprudence,* edited by N. Heer. Seattle: University of Washington Press.

Biazar-e Shirazi, A.K. 1983. *Resaleh-e Novin,* Vol. 2. Teheran: Nashr-e Farhang-e Eslami.

Commins, D. 1995. "Modernism." In *Oxford Encyclopedia of the Modern Islamic World,* Vol. 3. Oxford: Oxford University Press.

Connors, J. 1986. "Towards a System of Islamic Finance in Malaysia." In *Islamic Law and Finance,* edited by C. Mallat. London: Graham & Trotman.

The Constitution of the Islamic Republic of Iran. n.d. Teheran: Islamic Propagation Organization.

Coulson, N.J. 1990. *A History of Islamic Law.* Edinburgh: Edinburgh University Press.

Coville, T. 1994. "La Banque Centrale d'Iran et la politique de liberalisation." In *The Economy of Islamic Iran: Between State and the Market,* edited by T. Coville. Teheran: Institute Français de Recherche en Iran.

Cragg. K. 1995. "Abduh, Muhammad." In *The Oxford Encyclopedia of the Modern Islamic World,* Vol. 1. Oxford: Oxford University Press.

Davies, M.H. 1984. *Business Law in Egypt.* Deventer: Kluwer Law and Taxation Publishers.

Doi, A.R.I. 1984. *Shari'a: The Islamic Law.* London: Ta Ha Publishers.

The Economist. 1989. September 16, p. 42.

———. 1992. January 18, p. 34.

Edge, I. 1986. "Shari'a and Commerce in Contemporary Egypt." In *Islamic Law and Finance,* edited by C. Mallat. London: Graham & Trotman.

El Ahdab, A.H. 1992. "Interest in Islamic Law." *The ICC International Court of Arbitration Bulletin* 3(May).

Financial Institutions in OIC Countries of the Middle East. 1990. *Journal of Economic Cooperation Among Islamic Countries* II(January-April).

Gallaux, M. 1993. "Réforme bancaire et finance islamique." *Monde arabe Maghreb Machrek* 141.

———. 1994. "Economie islamique/occidentale." *Les cahiers de l'Orient* 34.

Gieraths, C. 1990. "Pakistan: Main Participants and Final Financial Products of Islamization Pracess." In *Islamic Financial Markets,* edited by R. Wilson. London: Routledge.

Haddad, Y. 1994. "Muhammad Abduh: Pioneer of Islamic Reform." In *Pioneers of Islamic Revival,* edited by A. Rahnema. London: Zed Book Ltd.

Heshmati-Moulai, H. 1991. "Bahsi Piramoun-e So'allat-e Matrahe dar Mored-e Nezam-e Bankdari-ye Eslam." In *Majmo'eh-ye Magalat-e Farsi: Siyasatha-ye Mali.* Mashhad: Astan-e Gods-e Razavi.

Hill, E. 1989. "Islamic Law as a Source for the Development of a Comparative Jurisprudence, The Modern Science of Codification: Theory and Practice in the Life and Work of Abd al-Razzaq Ahmad Al-Sanhuri." In *Islamic Law: Social and Historic Contexts,* edited by A. al-Azmeh. London: Routledge.

Homoud, S.H. 1985. *Islamic Banking*. London: Arabian Information Ltd.

Islahi, A.A. 1988. *Economic Concepts of Ibn Taimiyah*. London: The Islamic Foundation.

Kamali, M.H. 1991. *Principles of Islamic Jurisprudence*. Cambridge: Islamic Texts Society.

Kay, E. 1979. *Legal Aspects of Business in Saudi Arabia*. London: Graham & Trotman.

Kazarian, E.G. 1993. *Islamic Versus Traditional Banking: Financial Innovation in Egypt*. Boulder, CO: Westview Press.

Kerr, M.H. 1966. *Islamic Reform: Political and Legal Theories of Muhammad Abduh and Rashid Rida*. Berkeley: University of California Press.

Khalatbari, F. 1994. "The Tehran Stock Exchange and Privatization." In *The Economy of Islamic Iran: Between State and Market*, edited by T. Coville. Teheran: Institut Français de Recherche en Iran.

Khan, M.S. and A. Mirakhor. 1990. "Islamic Banking: Experiences in the Islamic Republic of Iran and Pakistan." *Economic Development and Change* 38(January).

Khan, R.A. 1993. "Pakistan in 1992." *Asian Survey* 33(February).

Khan, R.S. 1987. *Profit and Loss Sharing*. Karachi: Oxford University Press.

Klingmüller, E. 1990. "Islam et assurances." In *Les capitaux de l'islam*, edited by G. Beaugé. Paris: Presses du CNRS.

Kuran, T. 1993. "The Economic Impact of Islamic Fundamentalism." In *Fundamentalism and the State*, edited by M. M. Marty and Appleby. Chicago: University of Chicago Press.

Mahmood, M. 1973. *The Constitution of the Islamic Republic of Pakistan: 1973*. Lahore: Pakistan Law Times Publication.

Majmo'eh-ye Nazariyat-e Shora-ye Negahban. 1992. Vol. I., edited by H. Mehrpour. Teheran: Keyhan.

Malik, S.J. 1989. "Legitimizing Islamization—The Case of the Council of Islamic Ideology in Pakistan." *Orient* 30(2).

Mallat, C. 1986. "The Debate on Riba and Interest in Twentieth Century Jurisprudence." In *Islamic Law and Finance*, edited by C. Mallat. London: Graham & Trotman.

Mannan, M.A. 1986. *Islamic Economics: Theory and Practice*. Cambridge: Hodder and Stoughton.

MEED. 1994. January 7.

Motahhari, M. 1991. *Mas'aleh-e Riba*. Teheran: Sadra.

Najjar, F.M. 1992. "The Application of Sharia Laws in Egypt." *Middle East Policy* 1(3).

Neinhaus, V. 1986. "Islamic Economics, Finance and Banking—Theory and Practice." In *Islamic Banking and Finance*. London: Butterworths.

Nomani, F. and A. Rahnema. 1994. *Islamic Economic Systems*. London: Zed Books Ltd.

O'Sullivan, E. 1994. "A Market in Search of an Industry." *MEED* August 26.

Pal, I. 1994. "Pakistan and the Question of Riba." *Middle Eastern Studies* 30(January).

Qadri, A.A. 1986. *Islamic Jurisprudence in the Modern World*. New Delhi: Taj Company.

Rahman, F. 1964. "Riba and Interest." *Islamic Studies* 3(4).

________. 1980. *Major Themes of the Qur'an*. Minneapolis, MN: Bibliotheca Islamica.

Rahnema, A. and F. Nomani. 1990. *The Secular Miracle: Religion, Politics and Economic Policy in Iran*. London: Zed Books Ltd.

________. 1995. "Competing Shi'i Subsystems in Contemporary Iran." In *Iran After the Revolution: Crisis of an Islamic State*, edited by S. Rahnema and S. Behdad. London: I.B. Tauris.

Roy, D.A. 1991. "Islamic Banking." *Middle East Studies* 27.

Saleh, N.A. 1986. "Financial Transactions and the Islamic Theory of Obligations and Contracts." In *Islamic Law and Finance*, edited by C. Mallat. London: Graham & Trotman.

________. 1992. *Unlawful Gain and Legitimate Profit in Islamic Law*. London: Graham & Trotman.

Schacht, J. 1964. *An Introduction to Islamic Law*. Oxford: Clarendon Press.

________. 1987. Riba. In *First Encyclopaedia of Islam, 1913-1936*. Leiden: E.J. Brill.

Siddiqi, M.N. 1981. "Muslim Economic Thinking: A Survey of Contemporary Literature." In *Studies in Islamic Economics*, edited by K. Ahmad. London: The Islamic Foundation.

________. 1987. *Partnership and Profit-Sharing in Islamic Law*. London: The Islamic Foundation.

Sonn, T. 1995. "Rahman, Fazlur." In *The Oxford Encyclopedia of the Modern Islamic World*, Vol. 3. Oxford: Oxford University press.

Soorat-e Mashrooh-e Mozakerat-e Majles-ye Baresi-ye Nahai-e Ghanoon-e Assasi. 1985. Vol. 4. Tehran: Majles-e Shora-ye Eslami.

Tawfiq Attar, A.N. 1976. "The Rulings of the Islamic Shari'a Concerning Insurance." In *Islamic Law and Change in Arab Society*. Ceman Reports.

Udovitch , A.L. 1986. "Kirad." In *Encyclopédie de l'islam*, Vol. 5. Leiden: E.J. Brill.

Va'ezzadeh-Khorasani, M. 1991. "Tarhha-ye Jadid Baray-e Hall-e Moshkel-e Bankha." In *Majmo'eh-ye Magalat-e Farsi: Siyasatha-ye Mali*. Mashhad: Astan-e Gods-e Razavi.

Wilson, P.W. 1991. *A Question of Interest: The Paralysis of Saudi Banking*. Boulder, CO: Westview Press.

Wilson, R. 1993. "Islamic Banking and Finance." In *The Middle East and North Africa*. London: Europa Publications Ltd.

World Press Review. 1992. April, p. 43.

Zeinelabdin, A. 1991. "Stock Markets in Selected OIC Countries." *Journal of Economic Cooperation among Islamic Countries* 12(July-October).

Ziadeh, F. 1968. *Lawyers, the Rule of Law, and Liberalism in Modern Egypt*. Stanford, CA: Stanford University Press.

www.ingramcontent.com/pod-product-compliance
Lightning Source LLC
LaVergne TN
LVHW011914200926
848484LV00002B/617
9780762302680